T0396620

Radiological Safety and Quality

Lawrence Lau • Kwan-Hoong Ng
Editors

Radiological Safety and Quality

Paradigms in Leadership and Innovation

 Springer

Editors
Lawrence Lau
International Radiology Quality Network
Reston, VA, USA

Kwan-Hoong Ng
Department of Biomedical Imaging
University of Malaya
Kuala Lumpur, Malaysia

Dr. Shrimpton is an employee of the Crown.
The copyright on his chapter belongs to Her Majesty of the UK.

ISBN 978-94-007-7255-7   ISBN 978-94-007-7256-4 (eBook)
DOI 10.1007/978-94-007-7256-4
Springer Dordrecht Heidelberg New York London

Library of Congress Control Number: 2013953543

Printed on acid-free paper

Springer is part of Springer Science+Business Media (www.springer.com)

Foreword

The messages are listed in alphabetical order according to the name of the organizations and agencies.

"Image Wisely" is a social marketing campaign to raise awareness about adult radiation protection in medical imaging. It was recommended by a joint task force of the Radiological Society of North America (RSNA) and the American College of Radiology (ACR). Building upon the success of Image Gently for radiation protection in children, the task force named its new initiative "Image Wisely" and recommended widening the circle to include medical physicists, as represented by the American Association of Physicists in Medicine (AAPM), and imaging technologists, as represented by the American Society of Radiologic Technologists (ASRT).

Image Wisely was launched at the annual meeting of the RSNA in November 2010, and initially sought to raise awareness about adult radiation protection issues in computed tomography. In 2011–2012, the second phase was completed, targeting nuclear medicine. Planning for the third phase is under way, targeting radiography and fluoroscopy. Now, more than 17,000 individuals have pledged to adhere to the principles of Image Wisely:

1. To put patient's safety, health, and welfare first by optimizing imaging examinations to use only the radiation necessary to produce diagnostic quality images
2. To convey the principles of the Image Wisely Program to the imaging team in order to ensure that the facility optimizes its use of radiation when imaging patients
3. To communicate optimal patient imaging strategies to referring physicians, and to be available for consultation
4. To routinely review imaging protocols to ensure that the least radiation necessary to acquire a diagnostic quality image is used for each examination

As co-chairs of the Image Wisely campaign, we are highly supportive of the multi-disciplinary and inter-sectorial approaches to leadership and innovation for advances in radiological quality and safety. The 26 chapters in this publication describe a wide range of such activities, and the editors and contributors are to be congratulated.

James A. Brink, MD
Richard L. Morin, PhD
Co-Chairmen
Image Wisely

This book, *Radiological Safety and Quality – Paradigms in Leadership and Innovation*, is the product of a wonderful collaboration of a spectrum of experts and leaders from the radiological safety field. Contributors come from academia and various international organizations dealing with radiation safety. This book focuses on the use of radiation in medicine. Emphasis was put on crosscutting activities between professionals and experts in various clinical contexts. The authors share tremendous practical experiences that reflect the best practices in the field. The sharing of knowledge, experience, and best evidence-based research facilitates the development of innovative solutions. This book displays the deep commitment of global leaders to improve radiological safety and quality in practice.

The content of this book does not necessarily reflect the official position of the International Atomic Energy Agency (IAEA), but its goal to promote safe and effective use of radiation in medicine aligns with the mission of our Division. Drs. Lau and Ng are to be congratulated for their initiative to gather an international team of renowned experts to produce this book that will soon become a reference document for those who wish to learn more about radiation safety and quality from the perspective of clinical practitioners. Clinical practitioners will become better informed of the challenges, emerging trends, and possible solutions, which in turn will assist the planning of future actions to improve the care of their patients.

Rethy K. Chhem, MD, PhD (Edu), PhD (History)
Director, Division of Human Health
International Atomic Energy Agency

In 2013, there are many organizations working directly and indirectly in the field of radiological protection. These organizations, both national and international, focus on different aspects of radiological protection including fundamental science, recommendations and guidance, standards, and regulation and everyday practices that affect workers, patients, and the public around the world. The common aim is to promote radiation safety, with an appropriate level of protection, without limiting the benefits associated with the use of radiation.

The role of the International Commission on Radiological Protection (ICRP) is to provide recommendations and guidance based on scientific knowledge, evolving social values, and practical experience. ICRP has developed certain objectives to improve radiation safety and these include raising awareness of radiological protection in medicine through the education of health professionals and improving dissemination of information, considering the health effects of radiation, dealing with natural sources of radiation and integrating protection of the environment into the general system of radiological protection. As the number of stakeholders in radiological protection continues to rise, these objectives can only be achieved through wider communication and collaboration between the different organizations. This is essential to avoid duplication and overlap in specific areas of work.

The challenges faced by radiological protection organizations currently and in the future are not easy to overcome, but with dedication, collaboration and ongoing effort, it should be possible to continue to make improvements in radiation safety for the benefit of all.

Claire Cousins, FRCR, FRCP
Chairwoman
International Commission on Radiological Protection

Medical physicists play a key role in, amongst a wide range of professional services, medical imaging, and radiation safety in healthcare. They are trained to face new challenges in radiation medicine, especially when new technologies and innovative imaging procedures are introduced. The International Organization for Medical Physics (IOMP), in collaboration with other relevant organizations, institutions, and individuals, is committed to foster and enhance radiological safety in healthcare on a global basis.

This book addresses the important issues of safety and quality in diagnostic radiology. It contains 26 chapters covering different aspects of these subjects, including the important role of the leadership in formulating strategies and policies in achieving quality service. These are written by renowned experts in the field, many of them medical physicists. The book consists of a collection of valuable educational and guidance materials for medical physicists, doctors, and radiographers. In particular, medical physicists should find the book relevant and valuable for their work in improving the quality and safety of their services. This book also provides the motivation and stimulation medical physicists need in making their own innovation in the development of better quality radiology service to improve patient care. This is in line with the objectives of IOMP. Credits are given to all the authors for their contributions in this project and sharing their knowledge and experience on radiation safety and quality improvement in radiology. The editors, Dr. Lawrence Lau and Professor Kwan Hoong Ng are congratulated for their great efforts and success in putting up an outstanding publication with exceptional academic and practical value.

Kin Yin Cheung, PhD
President
International Organization for Medical Physics

The International Radiation Protection Association (IRPA) promotes excellence in the practice of radiation protection through national and regional associate societies by providing benchmarks of good practice and enhancing professional competence and networking. The IRPA has declared its vision to be recognized by its members, stakeholders, and the public as the international voice of the radiation protection profession in the enhancement of radiation protection culture and practice world-wide. The value and strength of IRPA are the enormous resources of practical knowledge and experience in radiation protection and related specialist fields of its almost 18,000 professionals in 61 countries.

Increasing globalization in medicine, industry, research, and other areas brings great benefits, but is not without challenges. For example, it is essential that proper safety culture should accompany these advances. With this in mind, the IRPA initiated a process to enhance radiation protection culture among radiation protection professionals in different organizations and companies worldwide, and to provide a forum for discussion in an open-ended process.

Radiation protection professionals have the responsibility to make their voices heard, ensure that the message is understood, and assume a leadership role in improving radiation safety in practice, e.g. in medical applications of ionizing radiation. Recognizing the great benefits achieved through the use of ionizing radiation in medicine, it is essential to foster an increased sense of responsibility of "non-radiation" professionals and/or direct managers for the radiation protection of patients and staff.

The system of radiation protection depends not only on scientific stringency, but also accepted ethical values, which encompass both political and societal considerations. The credibility of the system of protection, a key issue from IRPA's perspective, is based on a global understanding and acceptance of radiation protection principles such as justification, optimization, and individual dose limitation, and the recognition of the importance of a shared objective to implement these principles.

Activities and goals as mentioned above are central to collective international efforts. The IRPA delivers the platform to share knowledge and expertise necessary to achieve these common goals.

Renate Czarwinski, MSc
President
International Radiation Protection Association

The identification, development, implementation and ongoing improvement of actions to strengthen radiological quality and radiation safety involve many stakeholders from different disciplines and sectors. Professional experts research the evidence and prepare the recommendations and guidance tools. Professional organizations and international agencies advocate their adoption by the regulatory authorities and use by the practitioners. Any disconnection or deficiency in this process could jeopardize their eventual use in practice to improve patient care and safety. The aim is to bridge the gaps and strengthen the weak links.

While the challenges are similar in many countries and in different settings, poor access to radiology procedures and increased utilization are two contrasting issues. Inappropriate use of radiology results in increased cost and unnecessary exposure. A well-constructed and system-based improvement framework together with leadership and good teamwork will address many of the emerging challenges such as those summarized in the 2012 "International Conference on Radiation Protection in Medicine - Setting the Scene for the Next Decade". By documenting examples of leadership and innovation, the expert panel in this publication facilitates discussions and contributes to the development and implementation of tailored solutions to meet the needs of individual practitioners, radiology facilities, and healthcare systems in different settings. Under a globalized environment with limited resources, the common objective is to maximize impact and minimize waste in the delivery of such actions.

The International Radiology Quality Network, a Commission of the International Society of Radiology, looks forward to collaborating with other stakeholders and contributing to actions to improve quality and appropriate use of radiology, radiation safety, and patient-centered care.

Lawrence Lau, MBBS, FACR (Hon), FRANZCR, FRCR, FAMS (Hon), DDR, DDU
Chairman
International Radiology Quality Network

The International Society of Radiographers and Radiological Technologists (ISRRT) is very pleased to offer support to this outstanding publication. The ISRRT is the international voice for radiographers and radiological technologists and is recognized by the United Nations as an official non-governmental organization and has official status with the World Health Organization (WHO).

It is very timely that *Radiological Safety and Quality – Paradigms in Leadership and Innovation* has been written. The need for leadership, innovation, multi-disciplinary and inter-sectorial collaboration, and stakeholder participation in actions to improve radiological safety and quality is imperative for today's environment.

The mission of ISRRT includes "improving the standards of delivery and practice of medical imaging and radiation therapy throughout the world..." and our vision includes "... representing the practice of medical imaging and radiation therapy technology by promoting the highest achievable standards of patient care and profession practice." This wonderful publication fits perfectly with the mission and vision of the ISRRT and promotes the necessity of global solidarity and our common interest to improve radiation safety and radiological quality.

Michael D. Ward, PhD, RTR, FASRT
President
International Society of Radiographers and Radiological Technologists

In the era of personalized medicine, appropriate and properly performed diagnostic imaging procedures play a central role in the management of most diseases. However, an open access to information from varying sources can at times confuse the public and even lead to false perceptions and mistrust of diagnostic professionals and safety of procedures. Radiation safety is an emotional subject, and is becoming an important consideration for many patients.

Radiologists and other professionals involved in diagnostic procedures are the gatekeepers, not only of the appropriateness of procedures but also the safety and quality aspects. It is our job to inform and assure the public of ongoing efforts to improve safety and quality when delivering these services.

In establishing a "safety and quality" culture, collective leadership across the continents and regions with differing resources is needed. This must be achieved by international collaboration in areas such as: justification, appropriateness, safety standards, and standards of interpretation and reporting. Innovation is needed in making "affordable" decision support tools available. Educational efforts must be coordinated, and it must be emphasized that different needs exist in different areas of the world, necessitating the adoption of more tailored safety and quality solutions.

In this respect, the International Society of Radiology is establishing an "International Commission for Radiological Safety and Quality":

- To be the Society's representative responsible for radiological safety and quality issues
- To bring together leaders and experts in the field to formulate strategies for adopting and improving a "safety and quality" culture
- To coordinate educational efforts in the field
- To collaborate with international agencies to implement the strategies in different settings

The publication of this book is therefore a welcome step in the collective efforts to enhance the focus on safety and quality in diagnostic imaging and radiation medicine.

Jan Labuscagne, MB ChB, M Med (Rad D), FRANZCR
President
International Society of Radiology

The leadership of the Image Gently Campaign is delighted to note the creation of this pioneering, interdisciplinary, and international textbook on quality and safety in the field of radiology. The discussions of radiation protection and quality improvement could not be more timely and the use of an international and collaborative approach provide us with a model for how we all may share information and harmonize best practices to aid patient care worldwide.

What is vital today is to share our knowledge across disciplines and to critically evaluate the limited resources we have to help our patients achieve improved health. Drs. Lau and Ng have gathered together international leaders in quality and radiation safety to provide solutions to the difficult problems we face today. Their model includes the engagement of all stakeholders to not only create and implement these solutions but also to adapt them to local environments.

We believe in such a model. The Image Gently Campaign is an initiative of the Alliance for Radiation Safety in Pediatric Imaging. The campaign goal is to change practice by increasing awareness of the opportunities to promote radiation protection in the imaging of children.

In 2007, the Alliance for Radiation Safety in Pediatric Imaging[1] created an international coalition, representing more than 70 medical organizations that are dedicated to radiation protection for children. Our efforts both within the medical community and with patients and parents have provided open source access to all of our education and advocacy materials. Both Drs. Lau and Ng and many of the chapter authors have been leaders in collaborations with our Alliance to promote changing the practice of medical imaging and decreasing radiation exposure in children worldwide.

There is no doubt that this book will be of value to the radiology community worldwide. We thank Drs. Lau and Ng for this wonderful contribution to our profession.

Kimberly E. Applegate, MD, MS, FACR
Chair, International Outreach for Image Gently
The Alliance for Radiation Safety in Paediatric Imaging

Marilyn J. Goske, MD
Chair
The Alliance for Radiation Safety in Paediatric Imaging

Medical exposure is the largest contributor to radiation exposure to the public from artificial radiation sources in most parts of the world. The United Nations Scientific Committee on the Effects of Atomic Radiation (UNSCEAR) is committed to increasing awareness and to improving knowledge with regard to the levels, effects,

---

[1] Society for Pediatric Radiology, American College of Radiology, American Society of Radiologic Technologists, and American Association of Physicists in Medicine.

and use of ionizing radiation in medicine. While strengthening the collection of data by UNSCEAR to improve its evaluations is important, there is a more general issue related to coordination of the development and implementation of evidence-based radiation protection standards, guidelines, and policies worldwide.

An international forum for consultation and harmonization in radiation protection and safety is the Inter-Agency Committee of Radiation Safety (IACRS). Its objective is to promote consistency and coordination of policies with regard to areas of common interests in radiation protection and safety such as applying principles, criteria, and standards of radiation protection and safety and translating them into regulatory terms; coordinating research and development; advancing education and training; and promoting widespread information exchange. The IACRS is composed of representatives from nine intergovernmental organizations[2] concerned with radiation protection and safety and from five non-governmental organizations[3] accorded as observers.

Many members of the IARCS have contributed to this publication, which would help to deepen the understanding among health authorities and regulators on how best to collaborate with the medical community on radiation protection and safety matters.

There is a need for cooperation between international societies representing professionals applying radiation in healthcare and the IARCS to better promote the safe and appropriate use of radiation in medicine.

Malcolm Crick, MA, MSRP, CRadP
Secretary
United Nations Scientific Committee on the Effects of Atomic Radiation

The World Health Organization (WHO) welcomes this global collaboration to produce a publication addressing two major paradigms in leadership and innovation in radiology: radiation safety and quality.

Radiation safety is an essential component of good medical practice. It has established itself as subject of interest not only for radiation safety bodies and health authorities but also for policy makers, health care providers, researchers, manufacturers, patients, and general public. This is confirmed by the fact that several of these key stakeholders are co-authors of the present publication.

---

[2] European Commission (EC); Food and Agriculture Organization of the United Nations (FAO); International Atomic Energy Agency (IAEA); International Labor Organization (ILO); Nuclear Energy Agency of the Organization for Economic Co-operation and Development (NEA/OECD); Pan American Health Organization (PAHO); United Nations Scientific Committee on the Effects of Atomic Radiation (UNSCEAR); and World Health Organization (WHO).

[3] International Commission on Radiation Units and Measurements (ICRU); International Commission on Radiological Protection (ICRP); International Electrotechnical Commission (IEC); International Radiation Protection Association (IRPA); and International Standards Organization (ISO).

The use of ionizing radiation in health care has substantially increased during the last decade. New technologies, applications, and equipment are constantly being developed to improve the safety and efficacy. At the same time, incorrect or inappropriate handling of these increasingly complex technologies can also introduce potential health hazards for patients and staff. This demands public health policies that both recognize the multiple health benefits that can be obtained and minimize health risks. The two pillars for managing such risks in radiology are justification of procedures and optimization of protection to deliver a radiation dose commensurate with the medical purpose.

Co-sponsored by eight international organizations, the International Basic Safety Standards for Protection against Ionizing Radiation and for the Safety of Radiation Sources (BSS) have been recently revised and a substantial part of the new safety requirements refer to the medical use of ionizing radiation. As one of the BSS cosponsors, WHO is fully engaged in fostering its implementation.

Quality improvement in radiology encompasses radiation safety. Improving radiation protection culture of medical practice is crucial to ensure that patients benefit from the use of radiation in medical imaging and will contribute to health systems strengthening, with a more cost-effective allocation of health resources.

Maria del Rosario Pérez, MD
Scientist, Radiation Team
Unit of Interventions for Healthy Environments
Department of Public Health and Environment
World Health Organization

Preface

Radiology and medical imaging save lives and are indispensible in patient-centered medicine. The stakeholders from developed and developing countries face many similar challenges, which could potentially jeopardize radiation safety and radiological quality. Poor access to and inappropriate use of procedures are two contrasting issues. When radiation is used in medicine, the aim is to maximize benefits and minimize risks. Inappropriate use leads to unnecessary radiation exposure and increased cost.

Good teamwork and an integrated framework underpin improvement actions for practitioners, radiological facilities and healthcare systems. A framework consisting of safety and quality measures, synergistic implementation strategies, and performance enhancements supports the development and implementation of improvement actions and enables the identification of gaps and value-adding opportunities. Each stakeholder plays a unique and complementary role. For example, in a system-based action, individual experts prepare recommendations and guidance tools, and international agencies and professional organizations advocate their adoption by national regulatory authorities and use by practitioners. The key to maximize resources and secure success is by completing the project loop. Under a globalized environment, an inclusive engagement platform promotes stakeholder collaboration, improves awareness, facilitates cross-fertilization, minimizes duplication, and enables collective resources mobilization.

Radiological Safety and Quality – Paradigms in Leadership and Innovation is a multi-disciplinary and inter-sectorial collaboration by an international team of experts and leaders from academic institutions, radiology facilities, professional organizations, regulatory authorities, UN, and international agencies. It is a privilege for us to serve as co-editors and we thank the authors for their expert and valuable contributions. We very much appreciate the foreword messages offered by leaders from international professional organizations and agencies, by reiterating our common goal and demonstrating a joint commitment to improve radiological safety and quality through collaboration.

Established principles and new concepts, challenges, and opportunities, and leadership and innovation examples are discussed to raise awareness, share experience, and facilitate facility-based or system-wide improvements in quality care, radiation safety, and appropriate use of radiology. Such collective efforts will improve daily practice, i.e. by *doing the right procedure* by justification and *doing the procedure right* by optimization of radiological protection and error minimization.

Reston, VA, USA   Lawrence Lau
Kuala Lumpur, Malaysia   Kwan-Hoong Ng

Acknowledgement

We learned from the feedback and are motivated by the comments from our patients and the public over the years about the need for the strengthening of advocacy and actions to improve radiological quality and safety. We thank our colleagues and specialists in the field for sharing experience with us so we are more informed by the published evidence and recommendations. These valuable knowledge and guidance enable us to better understand the inter-related and at times complex issues on the subject and underpin the improvement actions, implementation strategies and priorities as discussed in this publication.We thank Springer for the management and production teams' commitment and professional support. The contributions from our dedicated writing team and the supporting foreword messages from leaders of international organizations and agencies are greatly appreciated.Last but not least are the understanding, support and patience from our families over this time, which have ensured a more timely completion of this exciting collaboration.

Lawrence Lau
Kwan-Hoong Ng

Contents

Part I
Radiological Safety and Quality

Chapter 1
Radiological Safety and Quality: Paradigms in Leadership and Innovation

Lawrence Lau and Kwan-Hoong Ng

Abstract Radiation medicine and medical imaging save lives and are indispensible in patient-centered care. In many parts of the world access to these procedures is poor, while in others utilization has increased significantly. Increased utilization, appropriate or inappropriate, increases cost and population exposure. From a public health perspective, procedure use should be rational and be guided by quality, safety, and appropriateness; aiming to maximize the benefits and minimize the risks.

Appropriateness and safety, including radiation safety, are key quality elements. The stakeholders in daily practice are the patients, referrers, providers and payers. These and others have delivered many actions to improve quality in different settings, addressing healthcare system, facility and end-user needs.

This chapter outlines the emerging challenges threatening radiological quality and radiation safety and discusses the solutions for healthcare systems, facilities and end-users. Good teamwork and an integrated framework would overcome many of these challenges. An action framework consisting of quality and safety measures, synergistic implementation strategies, and performance enhancements is presented. Recommendations and guidance tools are used to improve practice by indicating the requirements and processes. Each stakeholder plays a unique and complementary role in the development, advocacy, adoption and use of recommendations and tools.

The goal is to do the *right* procedure by justification and to do the procedure *right* by optimization and error minimization in daily practice. To bridge the gap between evidence and practice requires an innovative approach, leadership from authorities, collaboration with stakeholders, and participation of end-users.

L. Lau (✉)
International Radiology Quality Network, 1891 Preston White Drive,
Reston, VA, USA
e-mail: lslau@bigpond.net.au

K.-H. Ng
Department of Biomedical Imaging, Faculty of Medicine, University of Malaya Research
Imaging Centre, Lembah Pantai, 50603 Kuala Lumpur, Malaysia
e-mail: ngkh@ummc.edu.my

L. Lau and K.-H. Ng (eds.), *Radiological Safety and Quality: Paradigms
in Leadership and Innovation*, DOI 10.1007/978-94-007-7256-4_1,
© Springer Science+Business Media Dordrecht 2014

Keywords Quality radiology • Radiation protection • Radiation safety

1 Introduction

Radiation medicine (RM) includes diagnostic radiology, interventional radiology, nuclear medicine and radiotherapy. Medical imaging (MI) covers all imaging modalities with or without using ionizing radiation. Radiation medicine and medical imaging (RMMI) are indispensible in modern medicine, improve patient-centered care, and their use has expanded worldwide. Quality improvement is an integral part of good medical practice.

Globalization, consumer sophistication, communication and technological advances, corporatization, rationalization, service outsourcing, teleradiology, workflow modularization, and commoditization are reshaping practice. From a public health perspective, the use of valuable health resources should be justified and rational.

This chapter begins with an overview of radiological quality, followed by the emerging challenges and solutions for end-users, facilities, and healthcare systems. Most solutions and actions are based on recommendations or guidance tools, by indicating the requirements and processes. An inclusive stakeholder engagement platform and teamwork facilitate action delivery. For a more comprehensive approach, an integrated framework of related actions has advantages. Successful closure of the action loop from concept to change in practice requires an innovative approach, leadership from authorities, collaboration with stakeholders, and participation of end-users.

2 Basic Concepts

2.1 Stakeholders

The stakeholders are summarized in Table 1.1. Despite their differing perspectives and needs, the stakeholders share a common objective, i.e. patient-focused care and correct and safe use of procedures.

2.2 Quality Elements

The Institute of Medicine (IOM) defines quality in healthcare as safe, timely, effective, efficient, patient-centered and equitable [26, 27]. Safety and appropriateness are key quality elements (Fig. 1.1). Safety includes radiation safety and is dependant on the actions adopted by a facility and the technique used in procedures.

Table 1.1 Stakeholders in RMMI. These include those who directly provide or use procedures and others who indirectly support the system

Consumers: patients and general public

Referrers: general practitioners, specialists and other eligible providers

Providers: radiologists, radiation oncologists, nuclear medicine specialists, radiographers, nuclear medicine technologists, medical physicists, and other eligible providers

Payers: public authorities, private insurers, social services, individuals and others

Regulators: governments, health ministries, competent authorities, other related sectors, policy and decision makers

Research and academic institutions

International organizations and UN agencies

Professional, academic and scientific organizations

Medical defence organizations, malpractice insurers

Equipment manufacturers and vendors

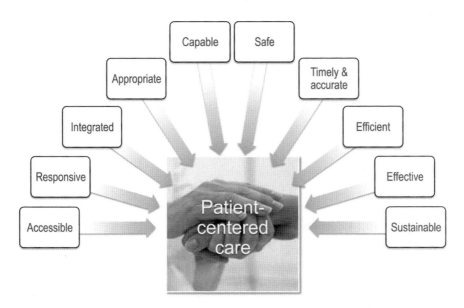

Fig. 1.1 Quality elements. For facilities this means providing responsive, appropriate, safe and patient-centered procedures by capable personnel and timely communication to referrers. For systems this means ready access to procedures and system integration, efficiency, effectiveness, and sustainability

Quality in RMMI can be characterized in many ways [5, 47]. One possibility is: 'a timely access to and delivery of integrated and appropriate procedures in a safe and responsive facility and a prompt communication of accurately interpreted results by capable personnel in an efficient, effective, and sustainable manner.'

2.3 Quality Processes

These include quality control, quality assurance and quality improvement. By using chest radiography as an example, quality control is the rejection and repeat of a non-diagnostic image. Quality assurance uses exposure charts, positioning instructions and processor settings etc. to reduce non-diagnostic images. Quality improvement is proactive and requires the analysis, development, and implementation of on-going actions for each step to achieve better diagnostic data and lower exposure.

2.4 Quality and Radiation Safety

Radiation protection issues must be addressed in facility layout; equipment selection, installation, commissioning, and maintenance; practitioner training; and the quality program [65, 93]. Justification and optimization actions, Diagnostic Reference Levels (DRLs), and individual dosimetry all contribute to radiation protection and quality in practice. However, the DRL concept is not well understood and guidance is required to improve its implementation [94]. Similarly, guidance is needed when transitioning to digital imaging [66] or introducing hybrid rooms designed for conventional surgery and fluoroscopy-guided procedures [45]. The International Commission on Radiological Protection (ICRP) publishes radiation protection recommendations and collaborates with others in radiation protection actions. In Chap. 2 Vano et al. described the positive impact of radiation protection to quality, including examples of relevant ICRP recommendations.

3 Emerging Challenges

Many challenges are similar worldwide [51], but there are local variations due to resources and settings (Table 1.2). Perhaps the most striking challenges are poor access and increased use. Increased use, appropriate or inappropriate, increases radiation exposure and cost. Increased workload increases errors. Follow-up of incidental findings further compounds these concerns. Technological advances and an aging population increase demand. Reports showing increased cancer risk from medical radiation highlight the need for a more appropriate use of procedures [67].

Inappropriate use could be due to ineffective justification, poor optimization or human errors. Poor realization of stakeholders' role and responsibilities contribute to this challenge. Some radiological equipment users have not received proper training in radiation safety and protection. In many undergraduate courses, MI, radiation protection and safety are poorly covered. Practitioners may not have much time for on-going professional development and on-going professional development teaching methodology may not be optimal for adult education. Some

Table 1.2 Emerging challenges in radiation medicine and medical imaging

Increased utilization, expenditure, and exposure
Poor access to healthcare resources and procedures
Inappropriate use of procedures
Workforce shortage
Low awareness in procedure utilization and radiation risks
Undergraduate education and practitioner competency
Poor system infrastructure
Weak policies
Action fragmentation and discontinuity
Access to and format of recommendations and guidance tools
Inertia to change and transient improvement
Volunteering
Funding and lead-time

referrers do not appreciate the differences in disease prevalence and procedure use between community and tertiary settings. Insufficient training and inexperience contribute to interpretation errors.

The workforce shortage is global and is compounded by inequitable distribution, migration and changing practice, e.g. international teleradiology. Policy change by a system or stakeholder impacts on another. Due to the shortage of radiologists, there are role extension opportunities for radiographers and these issues are complex.

Appropriate selection of equipment, equipment maintenance, and radiation protection and safety are practical issues that require careful consideration [65]. Resources vary between countries, urban and rural settings. While radiography or ultrasound can diagnose many common conditions, two-thirds of the world has no or inadequate access [99]. Poor access means radiography is used even when ultrasound is more appropriate. In others, while MRI is more appropriate, use is limited by criteria to contain cost. Access to screening mammography is age dependent. The availability and use of radiation protection and individual exposure monitoring devices vary markedly.

Poor system infrastructure and weak policies limit the implementation of recommendations. It is becoming challenging for some authorities to implement timely policy updates. For example, teleradiology potentially threatens communication and disrupts teamwork in quality and safety actions. Regulations should be in place to ensure practice is safe, e.g. outsourcing, teleradiology etc.

For actions involving many stakeholders, there is a risk of poor coordination resulting in fragmentation. Without good communication and collaboration, duplication and unintended complications are possibilities. Personnel and leadership changes lead to discontinuity of long-term actions. For any action, the aim is to improve and maintain change in practice. However, inertia to change and transient improvement are common.

Professional experts prepare recommendations and tools. Poor awareness and inadequate peer support threaten volunteering. The challenges for recommendations and guidance tools are their availability and presentation. Radiological quality and safety actions compete with others for funds and many system-based actions have a long lead-time. It is challenging to persevere, stay focused and maintain motivated with these long-term plans.

4 Solutions

Two solutions are suggested: good teamwork and an integrated framework (Fig. 1.2). An integrated framework puts related actions into perspective, informs and adds value to each other and minimizes waste. Each action requires contribution from different stakeholders. Good teamwork and collaboration are needed.

4.1 Measures

The three quality and safety measures used along the patient journey are justification, optimization and error minimization (Fig. 1.3). A procedure is used only if and when indicated and the technique is fully optimized.

4.2 Procedure Justification

Procedure justification is applied at three levels: society, procedure, and individual [33]. For most conditions, MI is not required. If indicated, it is important to verify if the information is not already available and if the relevant clinical, laboratory, imaging, and treatment details are provided in the referral. For high-dose or complex procedures, individual justification is particularly important and should take into account of all the available information. When appropriate and indicated, ultrasound or MRI should be chosen, especially in children. Availability, expertise and cost are other considerations.

Referral guidelines are tools to facilitate an appropriate use of procedures. While they are available in some countries and regions, their use in practice is low. A number of referral guideline-related projects are being conducted worldwide each focusing on a certain aspect: advocacy, development, pilot, implementation, and evaluation of use and impact [38, 100]. There are opportunities and synergies yet to be explored to strengthen these existing arrangements. In Chap. 6, Lau outlined the rationales, issues and improvement opportunities for and development and implementation of referral guidelines.

Fig. 1.2 Quality framework. An integrated framework consists of quality and safety measures, implementation strategies and performance enhancements for systems and end-users. Based on these elements, a suite of innovative actions is developed to improve quality and radiation safety

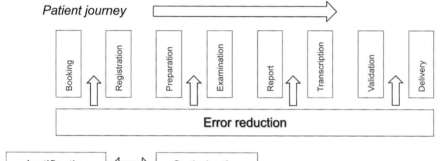

Fig. 1.3 Patient journey. Procedure justification and optimization are the pillars of radiation protection and are equally applicable to other modalities. Error reduction actions are used to reduce human errors before, during, and after a procedure. Justification, applicable at the start, is an important "gatekeeper" to eliminate unnecessary procedures

Institution-Wide Implementation

An integration of guidelines into the requesting process improves compliance. Computerized physician order entry and decision support (CPOE-DS) [76] and pre-authorization are the two alternatives. CPOE-DS is an important first step in executing a comprehensive institution-wide improvement program in procedure utilization, patient care, and dose reduction [80]. Implementation by stages works best, i.e. by piloting certain conditions in certain departments; following which actionable appropriate utilization management becomes possible. In Chap. 7 Sistrom et al. described their experience with the design, implementation, and impact of an integrated CPOE-DS system.

Justification and Optimization

Justification and optimization are related and should be considered together in some situations. For example, a pediatric CT procedure is not justified in a facility without a pediatric radiologist or staff experienced in pediatric protocols, but could be the first choice where these conditions are met. Better infrastructure such as experienced staff, equipment with dose reduction technology, and good quality program enables a facility to offer CT as the first choice for some conditions.

Procedure Efficacy

For a discussion of the efficacy of MI procedures, a hierarchical model can be used [18]. Efficacy at a lower level assures efficacy at the next higher level. For MI, Level 1 concerns with data quality as defined by image resolution, and signal-to-noise ratio etc. Level 2 relates to diagnostic accuracy, sensitivity, and specificity. Level 3 determines if the findings change the provisional diagnosis, and Level 4 the impact on patient management. A procedure adds value and affects outcome only when there is a potential to change management. Level 5 assesses patient outcome as measured by morbidity, mortality, health-related quality of life etc. Finally, Level 6 evaluates efficacy from the society's perspective, i.e. the societal costs should not exceed individual benefits.

4.3   Optimization

Optimization is applied at two levels: the design, selection, commissioning, and maintenance of equipment and the daily working procedure. The choice of actions depends on resources and will impact on exposure, radiation risk and cost. The aim is to obtain adequate diagnostic data with lowest possible dose by following the ALARA principle [33]. Education and training in radiation protection by focusing on the needs and means to apply appropriate exposure is a key component to reduce exposure and should be covered when introducing new modalities or protocols.

Optimization by Innovation

Advances in equipment design and innovations in data acquisition, processing and analysis improve the optimization of radiation protection, image quality and diagnostic data. Many recent initiatives have focused on dose reduction in pediatric CT. While each stakeholder group contributes in different ways to improve radiological quality and radiation protection of patients, the equipment manufacturers play a very important role by undertaking research and development and bringing these innovations to practice [73]. Hardware improvements include innovations in detector technology; tube design; gantry configuration; automatic exposure control; hybrid and multi-modality imaging; and digital image display. Software improvements include the use of iterative reconstruction algorithm and new processing techniques to reduce noise in low-dose scanning and to suppress streak artifacts. Computer-aided diagnosis (CAD) is being used for breast cancer detection in mammography screening and there is enormous potential in other fields. Multi-modality imaging plays a key role in the management of malignancies [102]. In Chap. 10 Gutierrez Rios and Zaidi explained the concept of image quality, outlined the equipment innovations, and discussed the challenges and opportunities affecting their use in clinical and research settings.

Optimization by Dosimetry

Dose assessment is used to improve justification and optimization, characterize radiation risks, assist quality assurance and facilitate benchmarking. Different dosimetric quantities are available for different modalities and the choice is based on the intended use. DRLs are used in facilities to monitor performance and improve optimization. Dosimetry should be an integral part of quality assurance to ensure a more effective and efficient use of procedures. In Chap. 2 Shrimpton and Ng discussed the role and use of dosimetry in radiation protection and clinical practice, including the use and effectiveness of DRLs.

4.4   Error Minimization

Errors increase with suboptimal workforce capacity and capability, high workload, workplace demands, and limited resources. An understanding of the causes and the use of controls prevent and minimize errors, risks, and adverse events. Participation in an incident reporting system is an excellent way to learn and share experience. A well-designed reporting system and a fair and just culture in facilities encourage open and honest reporting. Reporting is usually voluntary, e.g. for the participants of Radiation Oncology Safety Information System ROSIS [72]; Radiology Events Register RaER [19, 71]; Radiology Events and Discrepancies READ [81], Safety in Radiological Procedures Program SAFRAD [29] and General Radiology Improvement Database GRID [1]. However, French practitioners are legally required to report over-exposures in radiotherapy [3]. The WHO published a framework for patient safety that is currently being developed into a classification [78, 98]. In Chap. 11 Mandel and Runciman outlined the rationales, benefits, issues, and features of an incident reporting system for MI; including the challenges limiting and the solutions encouraging reporting.

4.5 Implementation Strategies

The strategies are listed in Table 1.3 and illustrated in Fig. 1.4 and are applicable to healthcare systems and end-users. Each strategy addresses a certain aspect of implementation and complements each other. Providing recommendations and tools alone does not guarantee their use. By providing up-to-date evidence-based

Table 1.3 Framework to improve radiation medicine and medical imaging

Quality and safety measures	Implementation strategies	Performance enhancements
Justification	Conduct research	Promote leadership
Optimization	Promote awareness	Engage stakeholders
Error minimization	Provide education	Strengthen communication
	Strengthen infrastructure	Identify collaborations
	Implement policies	Build teams
	Evaluate impact	Facilitate innovations
	Apply on-going improvement	Champion safety culture
		Mobilize resources
		Encourage participation
		Reward change and excellence

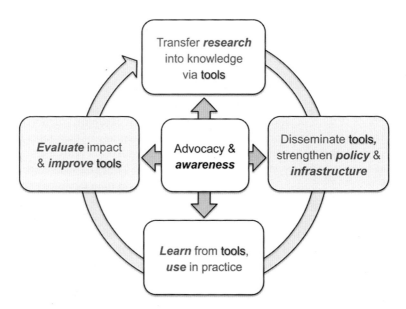

Fig. 1.4 Synergy between the implementation strategies. The common vehicles underpinning improvement actions are evidence-based recommendations and guidance tools (tools). This diagram demonstrates the synergy between the implementation strategies. The tools should be regularly evaluated to determine their effectiveness and be improved over time

contents relevant to a setting in a more user-friendly interface and encouraging adoption by the local authority and end-users improve this probability. The key is to strengthen the weakest link. Many improvement actions are inter-related and synergy should be sought. For example, based on population exposure surveys or DRL assessments, improvement actions should follow.

5   Research in Radiation Effects

Research is a systematic investigation into a particular subject to reveal the evidence and facts, which are used to underpin the development of theories and applications. In RMMI, research includes studies into radiation effects, risks, and benefits and monitoring of exposure at regional, national and facility levels (Fig. 1.5).

5.1   Radiation Effects, Benefits, and Risks

Ionizing radiation can cause stochastic and non-stochastic (deterministic) effects. Epidemiological studies in atomic bomb survivors, radiotherapy patients and occupationally exposed cohorts, showed a significant increase of cancer risk at doses *above* 100 mSv. There is a statistically detectable health risk from low doses, but this is low and difficult to detect. There are issues when attributing health effects to medical radiation exposure due to biases and uncertainties associated with risk assessment at low doses. These include dosimetric uncertainties; epidemiological and methodological uncertainties; uncertainties from low statistical power and precision in epidemiology studies; uncertainties in modeling of radiation risk data; and generalization of risk estimates across different populations [22, 40, 59, 90].

Depending on dose, ionizing radiation may cause cell death or non-lethal transformation, resulting in cancer after a long latency period. If the cell death is extensive, tissue dysfunction may be clinically evident. Such effect, e.g. skin burns, is termed tissue reaction or 'deterministic effect'; the severity increases with dose above a certain threshold. The non-lethal transformation of cells is termed probabilistic or 'stochastic effect'; the probability of occurrence is a function of dose, i.e. the smaller the dose, the smaller the probability.

To ensure informed decision, an analysis of or a statement on radiation-induced cancer risks and deaths from MI procedures should be balanced and accompanied by an account of the benefits, i.e. reduction in morbidity, mortality and cost of more invasive procedures.

Cancer Risk

Epidemiological studies indicate a higher cancer risk in children, especially for thyroid cancer, breast cancer and leukemia, as observed in atomic bomb survivors

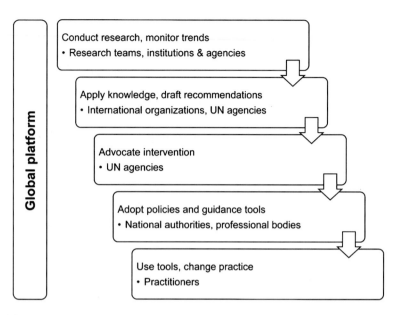

Fig. 1.5 Translating research findings to improve practice. Research provides the scientific basis to improve knowledge in radiation risks; compare utilization, exposures, and trends; underpin advocacy messages, recommendations, and guidance tools; and guide interventions to reduce risks and improve practice

and after the Chernobyl accident. For young children exposed to radiation during the atomic bombings of Hiroshima and Nagasaki there is an increased risk of childhood leukemia and other cancers later in life. The atomic bomb survivors exposed *in-utero* experienced an increased risk of childhood solid cancers and of cancers in adult life. Second malignant neoplasm, mainly breast and thyroid cancer, have been reported following pediatric radiotherapy after sufficiently long follow-up periods.

Radiation-induced cancer risks are highly dependent on age at the time of the exposure and increase in children because of organ sensitivity and life expectancy. Radiation dose is cumulative, i.e. patients undergoing multiple high dose procedures, could reach exposures at which epidemiological studies showed an increase in cancer risk. Genetic predisposition, genomic instability, deficiencies in DNA repair, regulatory processes of cell proliferation and endocrine factors all seem to be involved in the age dependence of cancer risk following exposure.

Risk Assessment

Radiation risk assessment in public health is complex. In a recent WHO report on the Fukushima incident, four steps were used to assess public health impact by: (1) identifying the specific ionizing radiation source and exposure pathway; (2) applying known radiation dose and biological effects data; (3) estimating doses including lifetime organ doses; and (4) characterizing lifetime cancer risks according to age at exposure, sex and cancer type [101].

Global Research Platform

A common platform enables stakeholder engagement; collaboration and synergy; larger study cohorts; knowledge and experience sharing; gap analysis and action prioritization; and joint resource mobilization. The shaping and promotion of integrated radiation risk research is facilitated by strengthening advocacy; building national research capacity; and adopting a strategy guided by quality, impact, sustainability and tangible deliverables [68]. The agenda would cover: basic science at molecular, cellular, tissue, and animal levels; experimental, epidemiological, and clinical elements; and population exposure assessments. The promotion of radiation risk research is one of the priorities in 'Bonn Call-for-Action', an international conference organized by the IAEA [30].

5.2 Exposure Monitoring

Exposure monitoring for healthcare systems, facilities and individuals facilitates improvements in radiation protection and optimization [32, 33]. Exposure monitoring for staff is an integral part of practice. Exposure monitoring for individuals is gaining support and is of particular importance to those requiring regular procedures.

Global and National Monitoring

The United Nations Scientific Committee on the Effects of Atomic Radiation (UNSCEAR) regularly conducts national surveys, analyzes trends and shares this data with other international agencies and organizations on global levels, effects and risks of ionizing radiation, and the use of radiation procedures. This contributes to the collaborative development and implementation of global radiation safety recommendations, norms, and standards. UNSCEAR has developed a strategy with IAEA and WHO to improve survey participation, data analysis and trend estimation [9]. An account of UNSCEAR's work and their impact on radiation safety, the challenges and innovative solutions is provided by Shannoun and Crick in Chap. 5.

Local Monitoring

The ICRP recommends the use of DRLs to benchmark performance and to ensure procedure doses do not deviate significantly from peer facilities unless there is a known, relevant, and acceptable reason [34]. The use of DRLs is explained in Chap. 2 by Vano et al.

6   Advocacy and Communication

The aim is to raise awareness and provide trustworthy information on the immediate benefits and potential risks of procedures and radiation safety; and to advocate more appropriate use of procedures and risk prevention by guidance tools.

6.1   Stakeholder Communication

Risk communication is a dialogue between those responsible for assessing, influencing and controlling radiation risks and those who may be affected. Evidence-based advice facilitates decision-making by improving stakeholder understanding of the risks and benefits of procedures. The WHO is developing a toolkit to improve risk communication in pediatric imaging.

6.2   Evidence-Based Advocacy

Organizations and agencies provide evidence-based information and education material on the benefits and risks of radiation procedures to the public, patients and practitioners. Examples include: Image Gently [24], Image Wisely [25], InsideRadiology [84], RadiologyInfo [2], Radiation Protection of Patients [31], and Virtual Departments [87]. The IAEA Radiation Protection of Patients is an indispensible website for radiation protection resources. Image Gently aims to change practice by lowering dose in pediatric imaging and its freely available CT protocols are size-based and independent of manufacturer or machine. The ICRP has prepared recommendations for 'Radiological Protection in Pediatric Diagnostic and Interventional Radiology' [36].

6.3   Social Media

This new technology enables an innovative platform for real-time and interactive communication. Organizations are exploring social media to strengthen existing communication strategies and to explore innovative applications. The effectiveness of this approach in RM is not well known at present. However, by keeping the contents factual, not too technical and being conversational contribute to a success-ful campaign. In Chap. 12 Berris and Rehani presented the IAEA Radiation Protection of Patients Unit's early experience in the use of social media and outlined the pros and cons of this innovative approach.

7 Education and Training

Providing education and training is perhaps the most commonly used and more tangible strategy to improve knowledge, competency, capability and capacity. Referrers and providers of RM must be properly trained in radiation protection [35]. The 'train-the-trainer' approach is an effective option to improve capacity and enables training delivery in local settings. The vision of the learning institution, accountability of the teaching faculty and appropriateness of the teaching resources all benefit from an institution-wide on-going improvement process. Useful training topics include: practice improvement strategies, quality processes, safety culture, procedure justification, optimization of protection and error reduction etc.

7.1 Applying Praxis to Enhance Education

Postgraduate education requires a different approach to undergraduate teaching. Success hinges on matching the contents to the recipients and delivering this in a user-friendly format and media. By adopting good teamwork between clinical, educational, and information technology experts and committing to quality teaching, the faculty will deliver the teaching programs successfully. In Chap. 13 Ros et al. explained how quality assurance, praxis and total quality management are successfully applied to assure the quality and sustainability of curriculum development and program implementation.

8 Infrastructure Strengthening

There is a need to strengthen workforce capacity and physical infrastructure to build teams and implement actions. Some national and regional examples are outlined. Social entrepreneurship is an alternative to bridge the radiological healthcare gap. Some examples employed to strengthen infrastructure and improve quality include the use of: clinical audit or practice accreditation; Lean or Six Sigma; a radiation safety culture; and individual exposure monitoring.

8.1 National and Regional Examples

China

Some of the emerging challenges are utilization, equipment, practitioners, workload, and system infrastructure issues. Awareness is low and infrastructure suboptimal for radiation protection and quality control. Radiation protection devices may not be available or used. Led by the Ministry of Health, regional Departments of Health, Chinese Society of Radiology (CSR) and others, improvements are being progressively introduced through: education and training, Regional Radiation Quality Control Centers, research, awareness raising, and team building in quality control and radiation safety in facilities. In Chap. 21 Song et al. provided an account of the current status, issues, resources, regulations, laws and solutions for RMMI and radiation protection in China.

Korea

The awareness of radiation issues amongst practitioners and public is low. The Ministry of Education, Science and Technology is responsible for nuclear medicine and radiation therapy policies and Ministry of Health and Welfare for diagnostic radiology. The Korea Food and Drug Administration (KFDA) is working to reduce population exposure by establishing DRLs. The DRL values for most radiographic procedures in Korea are similar to the United Kingdom [82]. Inter-organization networks such as the Korean Alliance for Radiation Safety and Culture in Medicine (KARSM) plays a key role in reducing exposure through awareness and education. In Chap. 22, Sung and Choi outlined examples of radiation protection regulations, findings of national procedure utilization and exposure surveys, current issues and solutions under a dualistic regulatory system.

Africa

Many urban communities have good access to healthcare; but not so in remote communities that occupy 80 % of Africa. The challenges include workforce shortage, workload, population exposure, equipment, budget and awareness in radiation safety and protection. Collective efforts by individuals, institutions and organizations are needed to strengthen education and training. For example, the Ernest Cook Ultrasound Research and Education Institute (ECUREI) was established to provide ultrasound education in the region and to lead regional outreach programs [43]. The Forum of National Regulatory Bodies in Africa (FNRBA) was formed to strengthen the effectiveness and sustainability of national radiation protection infrastructure. More experienced countries assist others by facilitating training and establishing Training Centres of Excellence in radiation safety. In Chap. 20 Kawooya et al. discussed these challenges and outlined innovative actions.

8.2   Social Entrepreneurship

Social entrepreneurship is a positive, collaborative, and pervasive approach to improve sustainable access to procedures and radiation protection through individual and organizational leadership, end-user participation and innovative program delivery [23]. Some examples include Physicians Ultrasound in Rwanda Education Initiative [69], RAD-AID International [70], and the Malaysian College of Radiology's Value Added Mammogram Program [7] etc. Access to procedures depends on infrastructure, i.e. physical, professional, equipment, technological, public health, regulatory, and financial etc. Sustainable solutions might target one, some, or all of these elements. An integrated solution is important, e.g. a breast cancer-screening program is incomplete without surgical and oncological support. In Chap. 24 Ho explained the concept, provided examples and advocated the use of social entrepreneurship as an innovative approach to improve access in developing countries.

8.3 Radiation Safety Culture

A radiation safety culture promotes awareness of safety and radiation risks; encourages the sharing of responsibilities among stakeholders; maintains the heritage for the next generation; and improves the quality and effectiveness of a safety program. It supports the implementation of radiation protection and works best in a safety conscious environment where individuals feel free to raise concerns without fear of retaliation or discrimination [92]. The International Radiation Protection Association (IRPA) has initiated a process to promote a safety culture [37]. In Chap. 15 Classic et al. described and discussed organizational culture; safety culture; and the features, benefits and implementations of radiation safety culture.

8.4 Clinical Audit and Accreditation

Clinical audit and practice accreditation are multi-disciplinary processes used to improve practice. They are guided by standards and assessment processes. The standards usually cover professional, technical and administrative components [49]. The two processes differ in flexibility and are complementary. Clinical audit is a systematic assessment, which can be internal or external, partial or comprehensive, and at single or multi-levels [41, 42]. There is no penalty to facilities that fail to meet the criteria. The focus is on improving care, resource use, and professional education. While mandatory in the Europe Union (EU), each country can determine its own audit process [13]. Accreditation is an evaluation of a facility's competence to perform certain procedures. It is formal, structured and external. Image review and onsite inspection are the norm. Accreditation could be granted, maintained, or denied. Accreditation programs are available in Australia, Korea, New Zealand, United Kingdom and United States [10, 49]. In some countries, service reimbursement is linked to participation.

Through advocacy, desktop audits, reciprocate auditor arrangements, random site visits, and collaborative development of requirements and assessment processes, some of the challenges such as poor awareness of the benefits; demand for resources; auditor training; and standard and assessment issues etc. can be overcome. In Chap. 17 Järvinen and Wilcox explained the features, regulatory environments, challenges and possible solutions to improve uptake.

8.5   Lean and Six Sigma

The two proven strategies in process improvement are Six Sigma and Lean, which have been used extensively in manufacturing but only recently applied to healthcare [56]. The Six Sigma objective is high performance; reliability and value such that only at six standard deviations away from the mean will one encounter a defect. The five phases of Six Sigma are: define, measure, analyze, improve and control (DMAIC). The Lean methodology focuses on process flow and considers any activity that does not add value as waste [95]. It uses visual tools such as process mapping, value stream mapping, and flow-charting to understand the processes. The Lean principles include zero defects, mistake proofing, waste minimization, and one-piece flow without batching. Successful facility-wide implementation requires the involvement of the whole organization and teamwork in learning and applying this process. In Chap. 16 Glenn and Blackmore explained how the Lean process contributes to continuous quality and process improvements in patient care and the use of clinical decision support in MI [6].

8.6   Individual Dose Monitoring

In digital systems, it is possible to easily collect and archive an individual's dosimetric data. Regulatory and governmental authorities, international agencies, professional organizations, and patient advocacy groups are increasingly interested in the recording, monitoring and potentially reporting of such data for individual patients [61–63, 88, 91]. The IAEA has developed a template for dose tracking and monitoring under the SmartRadTrack Program [74]. The National Academies of Sciences sponsored the Beebe Symposium focusing on dose estimation, recording, monitoring and reporting [64].

There are clear stakeholder benefits, but what and how to measure are the challenges [75], e.g. absence of an agreed measurement of exposure that is common to all modalities. The decision of what and how to measure and the implementation approach depends on resources. The 4 A's process, covering awareness, accountability, ability and action is applicable [17]. In Chap. 4 Frush et al. discussed the rationales, approaches, benefits, challenges, and solutions for dose recording, monitoring and potentially reporting for individuals.

9   Policy Implementation

Authorities monitor technological advances, review scientific data, assess policies and legislations for effectiveness, and update recommendations and guidance on a regular basis. Examples of collaborations between regulators, health authorities, and professional organizations are discussed, e.g. national implementation of justification; regional radiation safety initiatives; international basic safety and clinical teleradiology standards; and programs to ensure competency. A system-based framework facilitates the development of integrated improvement actions.

9.1   National Implementation of Procedure Justification

Public health policymakers consider guidelines as means to encourage more effective use of resources, reduce unnecessary exposure, and ensure good medicine. Preauthorization by using referral guidelines has resulted in a significant reduction in utilization, exposure and cost [4, 53]. Nation-wide implementation of procedure justification is complex requiring leadership, inter-sectorial consultation and collaboration. Three synergistic pathways are required: practitioner education and training, organizational infrastructure, and information technology tools. It is essential to sustain implementation by regular updates based on clinical impact and stakeholder feedback. In Chap. 8 Luxenburg et al. detailed the processes underpinning a successful national implementation of referral guidelines.

9.2   European Initiatives

In the EU radiation safety standards are proclaimed in the Euratom Treaty and the secondary legislations, e.g. the Euratom Directives [8]. Member States transpose these into national regulations, thus ensuring consistency. The European Commission (EC) publishes guidelines, information and implementation guidance tools for a more effective and consistent national implementation of policies and actions. The EC Directorate-General for Energy is undertaking a series of initiatives to strengthen justification and optimization, practitioner education and training, radiation risk research, awareness raising, radiation safety culture, and population and individual exposure monitoring. The European Medical ALARA Network (EMAN) [14], Heads of European Radiological Protection Competent Authorities (HERCA), Medical Radiation Protection Education and Training (MEDRAPET) [54], and Multidisciplinary European Low Dose Initiative (MELODI) [55] are effective platforms for stakeholder engagement and collaboration. In Chap. 23 Simeonov et al. documented these initiatives and explained their roles towards the strengthening of radiation protection in Europe.

9.3   International Basic Safety Standards

A regulatory framework documenting the radiation safety and protection requirements for patients, staff and public ensures a more responsible use of radiation in medicine. The requirements should be the same worldwide and the latest 'International Basic Safety Standards' (BSS) was updated in 2011 [28]. The BSS is an excellent example of successful collaboration among leading stakeholders over the past decades towards international harmonization of norms and standards. However, these standards are not mandatory. Their uptake by the health sector is low and the engagement of health authorities in BSS application could be strengthened. In Chap. 18 Le Heron and Borras explained the development, current status and use of the BSS.

9.4   Clinical Teleradiology Standards

Technological advances, clinical needs, 24/7 cover, workforce shortage, increased utilization, rationalization and outsourcing etc. are driving the use of clinical teleradiology. The issues and impact on practice were previously reported [11, 20, 48, 57]. Consideration is required to balance stakeholder needs, patient care and radiation safety. Authorities and organizations have published good practice guides focusing on legal, organizational, technological and scientific elements [13, 15, 39, 44, 57]. In Chap. 19 Dixon and Moore discussed the issues and solutions relating to teleradiology practice from the point of view of the facility, management, practitioner, patient, profession, and medical community.

9.5   Practitioner and System Competency

Competency is an ability to perform a task properly. The competency of practitioners, facilities and healthcare systems underpin quality and safety. In a competent system, quality and radiation protection actions are followed, monitored and improved continuously. Programs focusing on practitioners include certification, recertification, revalidation, continuing medical education (CME), continuing professional development (CPD), maintenance of professional standards (MOPS), maintenance of certification (MOC), and credentialing etc. Many organizations apply the CanMEDS framework [77] in the design of competency-based training, continuing education and assessment programs [85]. To ensure competency within facilities, professional standards are developed. For healthcare systems, it is a challenge to develop and implement comprehensive actions to tackle the inter-relating issues concerning appropriate use and radiation safety. The relationship between the profession and the regulator is complex and to share a common vision is of great value. In Chap. 14 Kenny discussed collaborations to ensure competency and to achieve benefits across the healthcare system.

9.6   System-Based Improvement Framework

The stakeholders have delivered many individual actions to improve quality and safety in different settings. However, a prospectively designed and system-based initiative covering a comprehensive range of integrated actions is less common. Examples of a holistic approach are the 'World Health Organization's Global Initiative on Radiation Safety in Health Care Settings' [97] and the 'Quality Use of Diagnostic Imaging Program' [50, 86]. In a system-based initiative, there are more stakeholders and more issues involved, resulting in wider impact. However, the lead-time is usually longer before any tangible outcome is achieved. In Chap. 26 Lau outlined the features, challenges and solutions for a system-based quality improvement initiative. The design and implementation of actions based on a range of radiological quality and safety measures, implementation strategies and performance enhancements under a comprehensive framework was explained.

10   Factors Influencing Performance

A number of other factors influence the outcome of these improvement actions. A set of performance enhancements is available to improve the uptake of recommendations and guidance tools (Table 1.3).

10.1   Influencing Factors

Societal, legal and organizational factors and values [58], and individual technical and non-technical skills [16] influence an action's outcome (Fig. 1.6). Some societal and

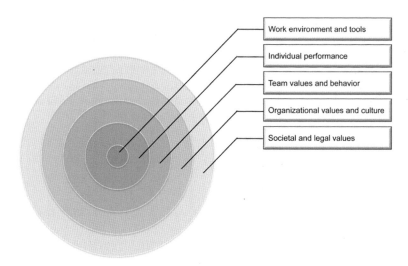

Fig. 1.6   Factors influencing outcome of improvement actions

legal factors are: timing; awareness; community values, needs, norms, expectations and demands; political priorities; economic pressures; rules, laws, regulations, and policy implementation; and partnership barriers. The organizational factors include: management values, goal, and style; authority hierarchy; safety policy and culture; and commitment to and involvement in quality and safety improvement actions.

Team performance depends on group value; goodwill; communication; decision-making; coordination; cooperation; collaboration; bias; cohesiveness; personnel turnover; and perception of responsibility. Training, knowledge, skill, competency, perception, awareness, interest, and participation are contributing factors for individuals. A facility's physical environment, setting and layout, and the availability of equipment and tools are other local factors.

10.2 Leadership and Participation

While individuals advocate for a worthy cause, a good idea must be sponsored by an organization or agency for implementation. To bring innovations to practice, strong leadership is vital by explaining the rationale to the end-users. End-user participation is needed to implement guidance tools in facilities.

10.3 Teamwork

Teams must have a sense of cohesiveness [46] and good teamwork improves performance and outcome. In facilities, teams use quality maps, performance indicators and clinical audits to mitigate potential risks along the patient journey, thus supporting the Hippocratic Oath '... to do no harm ...' [83]. In healthcare systems, teams work together to improve equipment design; to develop practice standards; to implement regulations; to promote audit and accreditation; to coordinate and streamline research; to champion a safety culture; and to strengthen practitioner education and training etc. The focus, expertise and resources of organizations and agencies differ. However, each step of an action requires the contribution from different stakeholders who play unique and complementary roles (Fig. 1.7). In Chap. 9, George et al. discussed the contribution of teamwork to improvement actions in facilities and healthcare systems.

10.4 Reward Excellence

The traditional reimbursement by volume model could have contributed to an unsustainable escalation of healthcare expenditure in some countries. The stakeholders are exploring alternatives to contain cost, improve quality, and reward

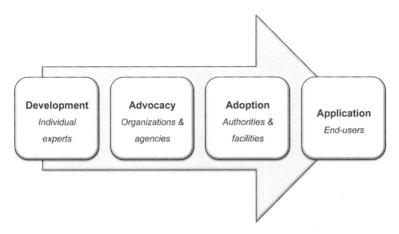

Fig. 1.7 Teamwork. Each step of an action requires the contribution from different stakeholders who play unique and complementary roles. For examples, professional experts prepare evidence-based recommendations and guidance tools, organizations and agencies advocate their adoption by regulatory authorities and use by the practitioners

performance, e.g. by using pay-for-performance (P4P) or value-based purchasing models [12]. P4P is tied to achievement in pre-defined goals such as appropriateness, service, satisfaction, quality, and outcomes. The challenges include: (1) defining quality metrics [79]; (2) meeting various P4P criteria; and (3) ensuring reward is greater than cost. More work and reform are needed to overcome operational, regulatory, political, and cultural barriers. In Chap. 25 Duszak Jr and Silva III discussed the reimbursement systems in the United States, the evolving trends, challenges and innovations for P4P programs.

11   Discussion

11.1   Access and Use

Most countries share similar challenges. However, resources differ between countries and between regions within a country, thus accounting for the variations in the access to and use of procedures. Access is an issue for resource poor countries and the remote or rural regions of better-resourced states. The annual per capita healthcare expenditure between high and low gross national income (GNI) countries [96] differs by a factor of 700 times. Seventy percent of radiation procedures are consumed by 25 % of the world's population [89]. The annual global collective effective dose has increased by 75 % over the two latest UNSCEAR reports [89]. In the United States the annual exposure from medical use of radiation is equal to background exposure [60].

11.2   Benefits and Risks

The system of radiological protection in medicine aims to balance radiation risks without unduly limiting the potential benefits for individuals and society. The ICRP recommends that procedures using ionizing radiation should be undertaken only if they will contribute to patient management or population health improvement. Thus, any potential procedural risk should be less than the risk of missing a treatable disease. Further, the benefit to an individual must be balanced with potential cancer induction to the community.

11.3   Setting the Agenda

A suite of improvement actions could be constructed and discussed under an integrated framework of quality and safety measures, synergistic implementation strategies, and performance enhancements [52] (Table 1.3). Recommendations and guidance tools are developed for the healthcare systems and end-users that are appropriate to the local setting by choosing appropriate elements from this framework. The framework enables identification of possible synergies and value-adding opportunities between actions. However, resources limit the development of a full gamut of actions. Many actions are independently planned and usually reactive to a certain issue. In such instances, it is useful to be aware of other related actions to achieve some synergy and minimize duplication.

11.4   Closing the Loop

Perhaps the most important component of an action plan is to narrow the gaps between evidence and practice and close the action loop (Fig. 1.8). At the start the stakeholders should be conscious of these steps and allocate sufficient resources to pilot, implement, evaluate and improve the recommendations and tools and to complete this loop. An action half completed and not implemented has no impact and is a waste of resources.

11.5   Multi-Disciplinary and Inter-Sectorial Collaboration

Quality and radiation safety involve many disciplines and sectors. Improvement actions require participation and expertise from different groups. For example, actions within facilities require input and collaboration from radiologists, radiographers, medical physicists and administrators by providing guidance to

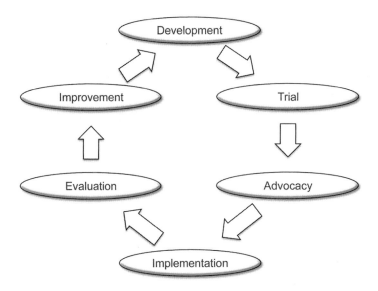

Fig. 1.8 Closing the action loop. The stakeholders should allocate sufficient resources to complete the action loop

clinical priorities, technical considerations, radiation protection and resource allocation respectively. A similar approach is applicable to interactions between other RM stakeholders and policymakers when making decisions on system-based actions.

From a national and system-wide perspective, health and non-health authorities, i.e. economic, education, environment and industry, from time to time have to tackle overlapping issues and develop policies with potential health, economic or social implications, which could be beyond their competence. Better communication and collaboration between sectors improve awareness, streamline action coordination, prevent unintended consequence and achieve better outcome.

11.6 Global Leadership and Local Implementation

Globalization applies to RMMI as to other sectors. Good communication and an inclusive, multi-disciplinary, and inter-sectorial platform facilitates stakeholder engagement and collaboration; improves awareness, facilitates cross-fertilization, minimize duplication; and enables collective resources mobilization. Professional organizations provide expert advice and advocacy at international level in global action development and facilitate local implementation. UN agencies, with their independency, credibility, infrastructure, and links to organizations and authorities, play a leading role by facilitating national infrastructure strengthening and policies implementation.

A range of issues and actions were identified in the 'International Conference on Radiation Protection in Medicine – Setting the Scene for the Next Decade' [30]. The priorities documented in 'Bonn Call-for-Action' are to: shape and promote a research agenda; provide information on medical exposure and foster radiation benefit-risk dialogue; strengthen justification, optimization and radiation incident prevention; engage manufacturers to improve equipment safety; promote radiation protection culture; strengthen radiation safety regulations and policies; and audit compliance. These recommendations guide the design of global improvement actions in the coming years. A global, comprehensive, and system-based framework and good teamwork contribute to the delivery of this diversified range of objectives.

A system-based approach is more integrated and has the potential to achieve wider improvements across the entire healthcare system. To achieve the desired outcome for system-based improvement, it is important to inform the stakeholders and end-users by the most effective way, i.e. by engaging the health ministries and radiation regulatory authorities. One of the key strengths of the UN agencies is their strong global network of health ministries, regulatory authorities, professional organizations, and research institutions. The IAEA counterparts are mainly the radiation protection regulatory authorities while the WHO counterparts are the health ministries, which in many countries are also responsible for radiation protection matters. Cooperation between the health ministries and radiation safety authorities has to be fostered.

12   Conclusion

Through an integrated framework, good teamwork, strong leadership and innovative approaches many of the emerging radiological quality and safety challenges are addressed. Each stakeholder plays a unique and complementary role. Collaboration is strength. A good concept alone will not deliver system-wide improvement. Combining actions in research, awareness, education, infrastructure building and policy implementation will.

The objective in everyday practice is to do the *right* procedure by justification and to do the procedure *right* by optimization and error minimization. This is indeed most tangible and will directly benefit every patient.

It is a great privilege for us to serve as co-editors of this publication. We appreciate the strong support from a team of experts who are leaders in their fields and willing to share their experience. By adopting a multi-disciplinary and inter-sectorial approach and by discussing (1) the established principles and new concepts; (2) the challenges and improvement opportunities, and (3) the examples of leadership and innovations in a common forum, our aim is to raise awareness, share experience, and facilitate collaborations towards more coordinated system-wide improvement in quality radiology and radiation safety.

References

1. American College of Radiology (2012) General radiology improvement database (GRID). http://www.acr.org/SecondaryMainMenuCategories/quality_safety. Accessed 1 Mar 2013
2. American College of Radiology and Radiological Society of North America (2013) RadiologyInfo. http://www.radiologyinfo.org/. Accessed 1 Mar 2013
3. Autorité de sûreté nucléaire (2012) http://www.french-nuclear-safety.fr. Accessed 1 Mar 2013
4. Blachar A, Tal S, Mandel A et al (2006) Preauthorization of CT and MRI examinations: assessment of a managed care preauthorization program based on the ACR Appropriateness Criteria and the Royal College of radiology guidelines. J Am Coll Radiol 3(11):851–859
5. Blackmore CC (2007) Defining quality in radiology. J Am Coll Radiol 4(4):217–223
6. Blackmore CC, Mecklenburg RS, Kaplan GS (2011) Effectiveness of clinical decision support in controlling inappropriate imaging. J Am Coll Radiol 8(1):19–25
7. College of Radiology, Academy of Medicine of Malaysia (2013) The CoR Mammogram Program (CMP). http://www.radiologymalaysia.org/breasthealth/smp/. Accessed 1 Mar 2013
8. Council of the European Union (2010) The high contracting parties of EURATOM (1957), consolidated version of the treaty establishing the European Atomic Energy Community. Off J Eur Union (OJEU) 2010/C 84/01
9. Crick M, Shannoun F, Le Heron J et al (2010) Opportunities to improve the global assessment of medical radiation exposures. In: International conference on radiation protection in medicine, Varna, Bulgaria
10. Destouet JM, Bassett LW, Yaffe MJ et al (2005) The ACR's mammography accreditation program: ten years of experience since MQSA. J Am Coll Radiol 2(7):585–594
11. Dixon AK, Fitzgerald R (2008) Outsourcing and teleradiology: potential benefits, risks and solutions from a UK/European perspective. J Am Coll Radiol 5(1):12–18
12. Duszak R (2009) P4P: pragmatic for practice. J Am Coll Radiol 6:477–478
13. European Commission (2009) Radiation protection no 159: European commission guidelines on clinical audit for medical radiological practices (Diagnostic radiology, nuclear medicine and radiotherapy). Publication Office of the European Union, Luxembourg
14. European Medical ALARA Network (2013) http://www.eman-network.eu/. Accessed 1 Mar 2013
15. European Society of Radiology (2010) ESR Position on the proposal for a directive of the European Parliament and of the council on the application of patients' rights in cross-border healthcare (Spanish Presidency compromise paper). http://www.myesr.org/html/img/pool/20100713_ESR_Position_Cross_border_healthcare_directive_Final.pdf. Accessed 1 Mar 2013
16. Flin R, O'Connor P, Crichton M (2008) Safety at the sharp end: a guide to non-technical skills. Ashgate, Farnham
17. Frush DP, Denham CR, Goske KJ et al (2012) Radiation protection and dose monitoring in medical imaging: a journey from awareness, through accountability, ability and action…but what is the destination. J Patient Saf 8, pre published ahead of print
18. Fryback DG, Thornbury JR (1991) The efficacy of diagnostic imaging. Med Decis Making 11(2):88–94
19. Galloway H, Hibbert P, Agar A (2009) Radiology events register progress report – second phase. http://www.ranzcr.edu.au/quality-a-safety/qudi/past-projects/quality-services-accredited-providers. Accessed 1 Mar 2013
20. Golding SJ, Webster P, Dixon AK (2010) Impact of a clinical governance programme on reporting standards in MRI: findings of the national service in outsourced MRI. In: Proceedings of UK radiological congress 2010, BJR congress series. http://bjr.birjournals.org/site/misc/Proceed_2010.pdf. Accessed 1 Mar 2013
21. Heads of European Radiological protection Competent Authorities (2013) http://www.herca.org/. Accessed 1 Mar 2013

22. Hendee WR (2013) Risk of medical imaging. Med Phys 40(4):040401. doi:10.1118/1.4794923
23. Ho E (2012) Social radiology: where to now? Biomed Imaging Interv J 8(1):e9. doi:10.2349/biij.8.1.e9
24. Image Gently (2012) The alliance for radiation safety in paediatric imaging. http://www.pedrad.org/associations/5364/ig/. Accessed 1 Mar 2013
25. Image Wisely (2012) Radiation safety in adult medical imaging. http://www.imagewisely.org. Accessed 1 Mar 2013
26. Institute of Medicine (2000) To err is human: building a safer health system. National Academies Press, Washington, DC
27. Institute of Medicine (2001) Crossing the quality chasm: a new health system for the 21st century. National Academies Press, Washington, DC
28. International Atomic Energy Agency (2011) Radiation protection and safety of radiation sources: international basic safety standards – Interim Edition. IAEA safety standards series GSR part 3 (Interim), IAEA, Vienna
29. International Atomic Energy Agency (2012a) Safety in radiological procedures (SAFRAD). https://rpop.iaea.org/SAFRAD/About.aspx. Accessed 1 Mar 2013
30. International Atomic Energy Agency (2012b) International conference on radiation protection in medicine – setting the scene for the next Decade. http://www.iaea.org/newscenter/news/2012/workerprotection.html. Accessed 1 Mar 2013
31. International Atomic Energy Agency (2013) Radiation protection of patients. https://rpop.iaea.org/RPoP/RPoP/Content/index.htm. Accessed 1 Mar 2013
32. International Commission on Radiological Protection (2001) Diagnostic reference levels in medical imaging: review and additional advice. Ann ICRP 31(4):33–52
33. International Commission on Radiological Protection (2007) ICRP publication 105. Radiation protection in medicine. Ann ICRP 37(6):1–63
34. International Commission on Radiological Protection (2007) ICRP publication 103. The 2007 recommendations of the International Commission on radiological protection. Ann ICRP 37(2–4):1–332
35. International Commission on Radiological Protection (2009) ICRP publication 113. Education and training in radiological protection for diagnostic and interventional procedures. Ann ICRP 39(5):7–68
36. International Commission on Radiological Protection (2013) ICRP publication 121. Radiological protection in pediatric diagnostic and interventional radiology. Ann ICRP 42(2):1–63
37. International Radiation Protection Association (2012) Radiation protection culture. http://www.irpa.net/index.php/radiation-protection-culture.html. Accessed 1 Mar 2013
38. International Radiology Quality Network (2010) Referral guidelines project. http://www.irqn.org/work/referral-guidelines.htm. Accessed 1 Mar 2013
39. International Radiology Quality Network (2012) Principles of international clinical teleradiology http://www.irqn.org/work/teleradiology.htm. Accessed 1 Mar 2013
40. Jacob P (2012) Uncertainties of cancer risks estimates for applications of ionizing radiation in medicine. In: International conference on radiation protection in medicine – setting the scene for the next decade, IAEA, Bonn
41. Järvinen H (2011) Clinical auditing and quality assurance. A refresher course lecture. In: Proceedings of the third European IRPA congress 2010. http://www.irpa2010europe.com/proceedings.htm. Accessed 1 Mar 2013
42. Järvinen H, Alanen A, Ahonen A et al (2011) Guidance on internal audits and self-assessments: support to external clinical audits. In: Proceedings of the conference of the Nordic society of radiation protection (NSFS). http://nsfs.org/NSFS-2011/documents/session-07/S7-O1.pdf. Accessed 1 Mar 2013
43. Kawooya MG, Goldberg BB, De Groot W et al (2010) Evaluation of US training for the past 6 years at ECUREI, the World Federation for Ultrasound in Medicine and Biology (WFUMB) Centre of Excellence, Kampala. Uganda Acad Radiol 17(3):392–398

44. Kenny LM, Lau LSW (2008) Editorial: clinical teleradiology – the purpose of principles. Med J Aust 188(4):197–198
45. Klein LW, Miller DL, Goldstein J, on behalf of the Multispecialty Occupational Health Group et al (2011) The catheterization laboratory and interventional vascular suite of the future: anticipating innovations in design and function. Catheter Cardiovasc Interv 77 (3):447–455
46. Kopans DB (2007) Breast imaging. Lippincott Williams and Wilkins, Philadelphia
47. Lau LSW (2006) A continuum of quality improvement in radiology. J Am Coll Radiol 3 (4):233–239
48. Lau LSW (2006) Clinical teleradiology in Australia: practice and standards. J Am Coll Radiol 3(5):377–381
49. Lau LS (2007) The design and implementation of the RANZCR/NATA accreditation program for Australian radiology practices. J Am Coll Radiol 4(10):730–738
50. Lau LSW (2007) The Australian national quality program in diagnostic imaging and interventional radiology. J Am Coll Radiol 4(11):849–855
51. Lau LSW, Perez MR, Applegate KE et al (2011) Global quality imaging: emerging issues. J Am Coll Radiol 8(7):508–512
52. Lau LSW, Perez MR, Applegate KE et al (2011) Global quality imaging: improvement actions. J Am Coll Radiol 8(5):330–334
53. Luxenburg O, Vaknin S, Polak G et al (2004) Annual research report. The Israel National Institute for Health Policy and Health Service Research (Hebrew)
54. Medical Radiation Protection Education and Training MEDRAPET. http://www.medrapet.eu/. Accessed 1 Mar 2013
55. MELODI Multidisciplinary European Low Dose Initiative. http://www.melodi-online.eu/. Accessed 1 Mar 2013
56. Miller D (ed) (2005) Innovation series 2005: going lean in health care. Institute for Healthcare Improvement, Cambridge, MA
57. Moore VA, Allen B Jr, Campbell SC et al (2005) Report of the ACR task force on international teleradiology. J Am Coll Radiol 2(2):121–125
58. Moray N (2000) Culture, politics and ergonomics. Ergonomics 43(7):858–868
59. Müller W (2012) Can we attribute health effects to medical radiation exposure? In: International conference on radiation protection in medicine – setting the scene for the next Decade. IAEA, Bonn
60. National Council on Radiation Protection and Measurements (2009) Report no.160 Ionizing radiation exposure of the population of the United States. NCRP, Bethesda
61. National Institute of Health (2009) New diagnostic imaging devices at the NIH clinical center to automatically record radiation exposure. http://www.nih.gov/news/health/aug2009/cc-17.htm. Accessed 1 Mar 2013
62. National Quality Forum (2010a) Safe practice 34: pediatric imaging. In: Safe practices for better healthcare – 2010 Update: a consensus report. Washington, DC
63. National Quality Forum (2010b) Safe practices for better healthcare – 2010 Update: a consensus report. Washington, DC
64. National Research Council (2012) Tracking radiation exposure from medical diagnostic procedures: workshop report. Committee on tracking radiation doses from medical diagnostic procedures; nuclear and radiation studies board; Division on Earth and Life Studies. National Academies Press, Washington, DC
65. Ng KH, Mclean ID (2011) Diagnostic radiology in the tropics: technical considerations. Semin Musculoskelet Radiol 15(5):441–445
66. Ng KH, Rehani MM (2006) X ray imaging goes digital. BMJ 333(7572):765–766
67. Pearce MS, Salotti JA, Little MP et al (2012) Radiation exposure from CT scans in childhood and subsequent risk of leukaemia and brain tumours: a retrospective cohort study. Lancet 390 (9840):499–505

68. Perez MR, Lau LSW (2010) WHO workshop on radiation risk assessment in paediatric health care. J Radiol Prot 30(1):105–110
69. Physicians Ultrasound in Rwanda Education Initiative PURE (2013) https://sites.google.com/site/rwandaultrasound/home. Accessed 1 Mar 2013
70. RAD-AID International (2013) http://www.rad-aid.org. Accessed 1 Mar 2013
71. Radiology Events Register (2012) http://www.raer.org/index.htm. Accessed 1 Mar 2013
72. Radiation Oncology Safety Information System (2012) http://www.rosis.info/. Accessed 1 Mar 2013
73. Rehani MM (2013) Challenges in radiation protection of patients for the 21st century. Am J Roentgenol 200(4):762–764
74. Rehani MM, Frush DP (2011) Patient exposure tracking: the IAEA smart card project. Radiat Prot Dosimetry 147(1–2):314–316
75. Rehani MM, Frush DP, Berris T et al (2012) Patient radiation exposure tracking: worldwide programs and needs – results from the first IAEA survey. Eur J Radiol 81(10):e968–e976
76. Rosenthal DI, Weilburg JB, Schultz T et al (2006) Radiology order entry with decision support: initial clinical experience. J Am Coll Radiol 3(10):799–806
77. Royal College of Physicians and Surgeons of Canada (2005) CanMEDS framework. http://www.royalcollege.ca/public/canmeds/framework. Accessed 1 Mar 2013
78. Runciman W, Hibbert P, Thomson R et al (2009) Towards an international classification for patient safety: key concepts and terms. Int J Qual Saf Health Care 21(1):18–26
79. Silva E III (2010) PQRI: from bonus to penalty. J Am Coll Radiol 7(11):835–836
80. Sistrom CL, Dang PA, Weilburg JB et al (2009) Effect of computerized order entry with integrated decision support on the growth of outpatient procedure volumes: seven-year time series analysis. Radiology 251(1):147–155
81. Spencer PA (2012) Shared learning can help minimize errors. http://www.auntminnieeurope.com/index.aspx?sec=nws&sub=rad&pag=dis&ItemID=607379. Accessed 1 Mar 2013
82. Sung DW (2011) Investigation of patient dose for diagnostic reference levels (DRL) in radiographic examination: national survey in Korea. Final report. Korea Food and Drug Administration, Seoul
83. The Hippocratic Oath. http://www.nlm.nih.gov/hmd/greek/greek_oath.html. Accessed 1 Mar 2013
84. The Royal Australian and New Zealand College of Radiologists (2013a) InsideRadiology. http://www.insideradiology.com.au/. Accessed 1 Mar 2013
85. The Royal Australian and New Zealand College of Radiologists (2013b) CPD overview. http://www.ranzcr.edu.au/cpd/overview. Accessed 1 Mar 2013
86. The Royal Australian and New Zealand College of Radiologists (2013c) The quality use of diagnostic program. http://www.ranzcr.edu.au/quality-a-safety/qudi. Accessed 1 Mar 2013
87. The Royal College of Radiologists (2012) Virtual departments. http://www.goingfora.com/index.html Accessed 1 Mar 2013
88. The State of California SB 1237 (2010) http://www.leginfo.ca.gov/pub/09-10/bill/sen/sb_1201-1250/sb_1237_bill_20100929_chaptered.html. Accessed 1 Mar 2013
89. United Nations Scientific Committee on the Effects of Atomic Radiation (2010) Sources and effects of ionizing radiation. UNSCEAR 2008 report to the general assembly with scientific annexes. United Nations, New York
90. United Nations Scientific Committee on the Effects of Atomic Radiation (2013) UNSCEAR 2012 report to the general assembly, with annexes. United Nations, New York (in press)
91. United States Food and Drug Administration (2010) Initiative to reduce unnecessary radiation exposure from medical imaging. http://www.fda.gov/Radiation-EmittingProducts/RadiationSafety/RadiationDoseReduction/ucm199994.htm. Accessed 1 Mar 2013
92. United States Nuclear Regulatory Commission (2010) Safety culture. http://www.nrc.gov/about-nrc/regulatory/enforcement/safety-culture.html. Accessed 1 Mar 2013
93. Vano E (2011) Global view on radiation protection in medicine. Radiat Prot Dosimetry 147(1–2):3–7

94. Vano E, Gonzalez L (2001) Approaches to establishing reference levels in interventional radiology. Radiat Prot Dosimetry 94(1–2):109–112
95. Womack JP, Jones DT (2003) Lean thinking: banish waste and create wealth in your corporation. Free Press, New York
96. World Bank (2012) Total health expenditure per capita. http://data.worldbank.org/indicator/SH.XPD.PCAP. Accessed 1 Mar 2013
97. World Health Organization (2008) WHO global initiative on radiation safety in health care settings technical meeting report. WHO, Geneva
98. World Health Organization (2009) Conceptual framework for the international classification for patient safety version 1.1. WHO, Geneva
99. World Health Organization (2010a) World Health Report – health systems financing: the path to universal coverage. http://www.who.int/whr/2010/en/index.html. Accessed 1 Mar 2013
100. World Health Organization (2010b) Medical imaging specialists call for global referral guidelines. http://www.who.int/ionizing_radiation/medical_exposure/referral_guidelines.pdf. Accessed 1 Mar 2013
101. World Health Organization (2013) Health risk assessment from the nuclear accident after the 2011 Great East Japan earthquake and tsunami based on a preliminary dose estimation. http://www.who.int/ionizing_radiation/pub_meet/fukushima_dose_assessment/en/index.html. Accessed 1 Mar 2013
102. Zaidi H, Del Guerra A (2011) An outlook on future design of hybrid PET/MRI systems. Med Phys 38(10):5667–5689

Chapter 2
How Radiation Protection Influences Quality in Radiology

Eliseo Vano, Kwan-Hoong Ng, and Lawrence Lau

Abstract Radiation safety is a key quality element in medical imaging and interventional radiology. Radiologists, referrers and other practitioners involved in the use of radiation medicine must be properly trained in radiation protection (RP) to ensure quality care and patient safety. Workforce shortage, workload increase, workplace changes, and budget challenges are emerging issues around the world, which could place quality at risk. Selecting the right procedure by justification, using the right dose and choosing adequate imaging data by optimization, and preventing errors along the patient journey must be considered and applied in practice. This chapter describes the positive impact of RP to quality in imaging and intervention through justification and optimization actions, including the use of Diagnostic Reference Levels (DRL) and individual patient dose recording and tracking. The relevant International Commission on Radiological Protection (ICRP) recommendations are highlighted together with a discussion on a wider cooperation between relevant professional groups, industry and other stakeholders. A set of actions to improve RP in quality programs for medical imaging and interventional radiology is suggested. It is concluded that RP and radiation management are integral elements for quality in imaging and intervention. Radiation safety topics should be covered in education and training programs and research projects in radiation medicine. While an individual action addresses a certain aspect of RP, collectively these actions will improve quality in medical imaging and

E. Vano (✉)
Radiology Department, School of Medicine, Complutense University and San Carlos Hospital, 28040 Madrid, Spain
e-mail: eliseov@med.ucm.es

K.-H. Ng
Department of Biomedical Imaging, Faculty of Medicine, University of Malaya Research Imaging Centre, Lembah Pantai, 50603 Kuala Lumpur, Malaysia

L. Lau
International Radiology Quality Network, 1891 Preston White Drive, Reston, VA 20191-4326, USA

L. Lau and K.-H. Ng (eds.), *Radiological Safety and Quality: Paradigms in Leadership and Innovation*, DOI 10.1007/978-94-007-7256-4_2,
© Springer Science+Business Media Dordrecht 2014

interventional radiology. Leadership and on-going collaboration by developing and implementing innovative actions will ensure RP and quality objectives are attained in the not too distant future.

Keywords Diagnostic reference levels (DRL) • Optimization of protection • Procedure justification • Quality improvement • Quality program • Quality radiology • Radiation protection • Radiation safety • Radiology errors

1 Introduction

1.1 Quality and Safety

Quality medical imaging has been defined as 'a timely access to and delivery of integrated and appropriate procedures, in a safe and responsive practice, and a prompt delivery of an accurately interpreted report by capable personnel in an efficient, effective, and sustainable manner' [28]. Radiation safety is a key quality element when ionizing radiation is used in imaging and intervention. Medical imaging and interventional radiology save lives, are indispensable in healthcare, and their use has greatly expanded worldwide. Radiologists and other eligible operators of radiological equipment, simple or sophisticated, must be properly trained in radiation safety and RP to ensure quality care and patient safety. Budget constraint is a critical issue that may influence the level of quality and impact on basic radiation safety. Sometimes, new medical imaging technology is introduced without due consideration for the economic and human resources required to ensure adequate staff training, including RP, or to implement a basic quality control program, including patient dosimetry.

1.2 Issues

Workforce shortage, workload increase, workplace changes, and budget challenges are emerging issues around the world, which could place quality medical imaging and intervention radiology at risk [28]. It is important for the stakeholders to collaborate and jointly address these issues in order to ensure patient safety and improve the quality of care. Several strategies have been proposed to improve quality medical imaging by conducting research, promoting awareness, providing education and training, strengthening infrastructure, and implementing policies. In practice, these actions will result in selecting the right procedure (justification), using the right dose (optimization), and preventing errors along the patient journey [29].

The growth in medical imaging and interventional radiology over the past two decades has yielded undisputable benefits to patients in terms of better quality of life and longer expectancy. This growth reflects new technologies and applications, including new imaging modalities. However, part of this growth can be attributed to overutilization. In 2009, the American Board of Radiology Foundation hosted a summit to discuss the causes and effects of overutilization of medical imaging [11]. The key elements contributing to overutilization were: payment mechanisms and financial incentives; practice behavior of some referring physicians; self-referral, including referral for additional medical imaging procedures; defensive medicine; missed educational opportunities when inappropriate procedures were requested; patient expectation; and duplicating procedures. A range of actions to reduce over-utilization was proposed: national collaboration to develop evidence-based appropri-ateness criteria; wider use of practice guidelines in the request and delivery of procedures; decision support at the point of care; education of referring physicians, patients, and the public; accreditation of facilities; management of self-referral and defensive medicine; and payment reform.

A quality program for medical imaging and interventional radiology should maximize the benefits and minimize the risks of radiation exposure. Radiation risk should be considered and quantified by regular systematic radiation dose audits. The increased use of high dose procedures, e.g. CT and interventional procedures; the duplication or repeat of these procedures in certain group of patients; and the age of some patients, i.e. paediatric and young adults, could substantially increase radiation risk.

1.3   Principles

The driving principles of a quality system for medical imaging and interventional radiology are procedure justification and optimization of diagnostic information, including image quality. When ionizing radiation is used in imaging and interven-tion, the proper management of radiation exposure for the patients and staff has to be considered. Of course, optimization should include error prevention, covering unintended or over exposures.

In relation to the management of radiation exposure, it is helpful to use the term 'diagnostic information' instead of 'image quality' because the new imaging modalities can combine several acquisition modes, e.g. fluoroscopy, cine, digital subtraction, rotational acquisition, cone beam CT, etc., and can result in differences in exposure and image quality. Sometimes, 'noisy images' may be acceptable and diagnostic if the saving in radiation dose is significant. The objective is to obtain 'adequate diagnostic information' for the clinical task, rather than 'best image quality' especially if the latter could involve much higher exposure.

Sometimes quality medical imaging is judged only by the accuracy and timeli-ness of a report, especially in clinical teleradiology, without radiation risk being taken into account [56]. Consistent with conventional practice, providers of clinical teleradiology are encouraged to apply good practice principles and guidelines to underpin quality practice and radiation safety [24, 25, 27]. With digital imaging, it is possible to obtain good or even excellent image quality and at times too much diagnostic information while the patient could be inadvertently over-irradiated. Therefore, a good and timely report is not necessarily a guarantee that the procedure has been carried out with the most optimal radiation dose management.

The successful development and implementation of quality system in medical imaging and interventional radiology requires leadership and collaboration between radiologists, other clinicians, radiographers, medical physicists, and nursing professionals. Active participation of and good communication between the referrers and other healthcare managers at all levels are essential to ensure coordinated and continuity of care and the requested procedures are justified and not duplicated. Patient information on the benefits and risks of procedures should be available to facilitate informed decision-making and patient satisfaction. For example, some procedures are not recommended if and when the benefits do not outweigh the radiation risks, as the case for some paediatric CT procedures.

Interventional radiology is an area in which a quality program, including radiation safety, has been developed including considerations for the pre-procedural, procedural and post-procedural steps along the patient journey [30–32]. The aim is to prevent radiation-induced injuries relating to these complex procedures and to address the deficiencies due to the absence of a proper radiation management program.

1.4   Individual Exposure Recording and Monitoring

Reiner suggested the elements for consideration when quantifying radiation safety and quality in medical imaging [36]. These included an automated recording, tracking, and analysis of quality data. Very few data on radiation dose are being prospectively collected, tracked, or analyzed. These individual patient dose and image quality data can be used for education and training, certification, research, practice guidelines development, and new technology development.

In 2012, an international collaboration on patient radiation safety has led to the publication of a 'Joint position statement on the IAEA patient radiation exposure tracking' [13], supported by the Conference of Radiation Control Program Directors USA (CRCPD), European Society of Radiology (ESR), Food and Drug Administration USA (FDA), IAEA, International Organization for Medical Physics (IOMP), International Society of Radiographers and Radiological Technologists (ISRRT) and World Health Organization (WHO). The scope includes the recording, reporting and tracking of radiation doses of all imaging and interventional procedures employing ionizing radiation, including radiography, fluoroscopy, CT and nuclear medicine procedures.

This chapter focuses on the impact of RP on quality in medical imaging and interventional radiology and describes the role of RP by justification through referral criteria and guidelines and by optimization of procedures. The relevant recommendations of the ICRP are highlighted. There is a need to improve the participation of and the wider cooperation between all stakeholders, e.g. radiologists, other clinicians, radiographers, medical physicists and other stakeholders to advance and strengthen radiation safety programs including those dealing with RP training and patient dose recording and tracking.

2 Justification and Referral Guidance

All medical imaging and interventional radiology procedures must be justified for a good clinical reason and the radiation dose involved in some of these procedures should also be considered. If available and when appropriate, imaging or intervention by other modalities not using ionizing radiation should be selected.

2.1 Referral Guidelines and Radiation Protection

Some international organizations and national professional societies have published and promoted referral guidelines [2, 6, 44, 55]. For these guidelines to remain relevant and to reflect rapid technological advances and changing practice, timely updates are needed, requiring both human and financial resources. In Europe, the Council Directive on Medical Exposures stated in one of its articles: 'Member States of the European Union shall ensure that recommendations concerning referral criteria for medical exposure, including radiation doses, are available to the prescribers of medical exposure' [3]. In addition to the existing national guidelines, the European Commission published 'Radiation Protection 118. Referral guidelines for imaging' in 2000. This document included discussions on RP and provided a list of the typical radiation dose of the different radiological procedures [6]. However, the majority of the content was not related to RP. Due to limited resources these European guidelines have not been updated regularly, but it is important to recognize the initial RP recommendations are still currently valid. Further, an update of the typical radiation doses for diagnostic radiological and nuclear medicine procedures have been published as part of the DOSE DATAMED and SENTINEL programs [7, 8]. Currently, a new version of DOSE DATAMED (DOSE DATAMED 2) is being implemented [10].

2.2 Means to Improve Care and Minimize Exposure

The following is an extract from the European Guidelines [6] highlighting the scenarios where inappropriate use could be avoided and RP enhanced:

1. Repeating investigations which have already been done e.g. at another hospital, in an outpatient department, or in the accident and emergency department. HAS

IT BEEN DONE ALREADY? Every attempt should be made to get the previous films (images). The transfer of digital data through electronic links may assist in the future.

2. Investigation when results are unlikely to affect patient management, e.g. when the anticipated 'positive' finding is usually irrelevant, e.g. degenerative spinal disease (as 'normal' as hair turning grey with age) or because a positive finding is so unlikely. DO I NEED IT?
3. Investigating too often, i.e. before the disease could have progressed or resolved or before the results could influence treatment. DO I NEED IT NOW?
4. Doing the wrong investigation. Imaging techniques are developing rapidly. It is often helpful to discuss an investigation with a specialist in clinical radiology or nuclear medicine before it is requested. IS THIS THE BEST INVESTIGATION?
5. Failing to provide appropriate clinical information and questions that the imaging investigation should answer. Deficiencies here may lead to the wrong technique being used, e.g. the omission of an essential view. HAVE I EXPLAINED THE PROBLEM?
6. Over-investigating. Some clinicians tend to rely on investigations more than others. Some patients take comfort in being investigated. ARE TOO MANY INVESTIGATIONS BEING PERFORMED?

The typical effective dose for some common diagnostic radiology procedures can vary by a factor of 1,000, e.g. from the equivalent of 1 or 2 days of natural background radiation (0.02 mSv for a chest radiograph) to 4.5 years (for a CT of the abdomen). However, there is a substantial variation in the level of background radiation between and within countries. These general recommendations should be applied and other specific advice should also be considered, e.g. when transitioning from conventional film-screen to digital imaging [16, 35].

These six basic and clear advices on RP published in the 2000 EC guidelines are still valid today and probably will be in the coming years. A regular update of referral criteria should be made by the professional organizations after taking into account of RP issues and having RP specialists participating in the working groups. Amongst other updates, this process enables the incorporation of radiation dose for the various procedures as reported in the most recent literature. It is important to recognize that while a certain procedure is generally justified for a particular condition, its appropriate use for a certain patient requires a careful consideration of an individual's specific circumstances.

2.3 An ICRP Perspective

Justification, as stated by the ICRP, shall be applied at three levels [18]. In principle, the decision to adopt or continue an activity involves a consideration of the benefits and risks of the available options. In medical imaging and interventional radiology, this consideration usually leads to a number of alternatives that will do more good than harm. The decision process could be complex but is necessary. The harm, more strictly speaking the detriment, is not confined to radiation but includes others as well as economic and societal costs. Often, radiation risk is only a small part of the total consideration.

For these reasons, the Commission uses the term 'justification' to the first of the three levels, i.e. it only requires that the net benefit for the procedure be positive. Selection of the best available option is usually a task beyond the responsibility of RP organizations.

In some healthcare systems, commercial interest may encourage more referrals to medical imaging, since these procedures may be a major source of income to the parties concern. This reason, with or without utilization-based incentives, could lead to increased use of procedures well beyond the norm for good medical practice. The Commission disapproves referrals that confer unjustified risk to patients, and that are inconsistent with medical ethics and the principles of RP.

The three levels of justification of a radiological practice in medicine are:

- The first and most general, the proper use of radiation in medicine is accepted as doing more good than harm to society. This general level of justification is now taken for granted and is not discussed further.
- The second level, a specified procedure with a specified objective is defined and justified, e.g. a chest x-ray for a patient with relevant symptom(s), or a group of individuals at risk for a condition that can be detected and treated. The aim of this second level of justification is to determine whether the radiological procedure will improve the diagnosis, or will provide necessary information about the exposed individuals. The justification of a radiological procedure at this level is a matter for national and international professional bodies, in conjunction with national health and radiological protection authorities, and the corresponding international organisations. The total benefits from a procedure include not only the direct health benefits to the patient, but also the benefits to the patient's family and to society. Justification at this level should be reviewed from time to time, as more information becomes available about the risks and effectiveness of the existing procedure and the availability of new procedures.
- The third level, the application of a procedure to an individual patient should be justified, i.e. the particular application should be judged to do more good than harm to an individual patient. Hence all individual medical exposures should be justified in advance, taking into account of the specific objectives of the exposure and the characteristics of the individual involved. Justification of individual exposures should include checking that the required information is not already available. For high-dose procedures, such as complex diagnostic and interventional procedures, individual justification by the practitioner is particularly important and should take account of all the available information. It is often possible to speed up the procedure by defining referral criteria and patient categories in advance.

Although the principal aim of medical exposures is to do more good than harm to an individual patient, the Commission recommends consideration of radiation exposure to staff and other individuals should be part of this justification process.

2.4 How Optimization Affects Justification

Justification and optimization are closely linked and these two principles should be considered jointly in some cases. For example, due to technological advance but limited by resources, between and within countries, a procedure such as paediatric CT could not be justified in a facility without a paediatric radiologist or experience in paediatric low dose protocols, but could be the first option in another facilities where these conditions are met. The availability of better infrastructure such as experienced paediatric radiologist, CT scanner with dose reduction technology, and good quality program enables a facility to offer CT as the first choice for some conditions.

Therefore, if and when RP is fully optimized, certain procedures could be justified and appropriate even though these are usually not when such provisos are not met. A similar consideration could be extended to some adult procedures. The choice between invasive cardiac catheterization and CT coronary angiography is a good example illustrating how local experience and expertise, equipment availability, and use of dose reduction strategies could influence the selection of an imaging modality [39, 40].

3 Optimization of Radiation Protection and Data

3.1 Diagnostic Reference Levels

To improve the optimization of RP of patient in diagnostic procedures, the ICRP recommends the use of Diagnostic Reference Levels (DRLs) to compare procedures, which is applicable to groups of similar patients rather than individuals. DRLs are used to ensure the doses do not deviate significantly from those achieved at peer departments for that procedure unless there is a known, relevant, and acceptable reason for this deviation [18].

However, the DRL concept is not well understood by many practitioners and referrers. The following provides some helpful hints relating to the use of DRLs [49]:

- DRLs are not dose limits, they should be used as investigation levels;
- DRLs are not applicable to individual patients;
- Comparison with DRLs shall be made using mean or median values of a sample of patient doses;
- Quantities used as DRLs should be easily measured;
- The use of DRLs should be made in conjunction with the evaluation of image quality or diagnostic information;
- DRLs should be applied with certain flexibility, i.e. allowing tolerances for patient size, condition, etc.;

- DRLs are not differentiators for good or bad practice;
- Values that are UNDER the DRLs may not necessarily be optimised values;
- Values that are OVER the DRLs should require an investigation and optimization of the x-ray system or operational protocols;
- DRLs should be used in a dynamic and continuous process of optimization;
- The goal in using DRLs is not to reduce patient doses if image quality or diagnostic information is compromised; and
- Compliance or faults with DRLs should be discussed with the staff of the imaging department.

Optimization could be a challenge when introducing new imaging modalities and acquisition protocols. However, equipment manufacturers have significantly improved the hardware and post-processing tools in recent years, especially for interventional procedures and CT. The end-users contribute by undertaking clinical evaluation of these systems and acquisition protocols. The combined efforts have resulted in the successful use of dose-saving strategies, e.g. substantial patient doses reduction in CT coronary angiography [37].

3.2   Patient Dosimetry

In the past, patient dosimetry in interventional radiology was made over a small sample of procedures to calculate the mean or median values as part of clinical audits and in the use of DRLs. With the introduction of digital systems, it is possible to easily collect and archive dosimetric and demographic data for these procedures together with images, as part of the Digital Imaging and Communication in Medicine (DICOM) header or other DICOM services, e.g. Modality Performed Procedure Step (MPPS) or Radiation Dose Structured Reports (RDSR), and to manage an individual patient's dose data [51, 52].

The analysis of this data is subjected to quality control and should include: (a) a periodic calibration of patient dose quantities as reported by the x-ray system, (b) an automatic detection and identification of high individual dose values, (c) a periodic statistical analysis of the local DRLs and comparison with national or regional DRLs, and (d) corrective actions when indicated to meet the requirements of quality assurance and clinical audit programs.

An example of an automatic patient dosimetry management system is the 'Dose on Line for Interventional Radiology (DOLIR)' used to analyze, monitor, audit and archive individual patient dosimetry for fluoroscopy-guided cardiology and interventional radiology procedures in seven catheterization laboratories at the San Carlos Hospital, Madrid [42]. Figure 2.1 shows the data in Dose Area Product (DAP) and Fig. 2.2 in cumulative air kerma. Comparative benchmarks such as median value, local reference level and trigger level could be individually or collectively selected and displayed. Programs such as this enable a ready identification of those procedures associated with high exposure and alert prompt intervention to those with potential radiation-induced skin injuries.

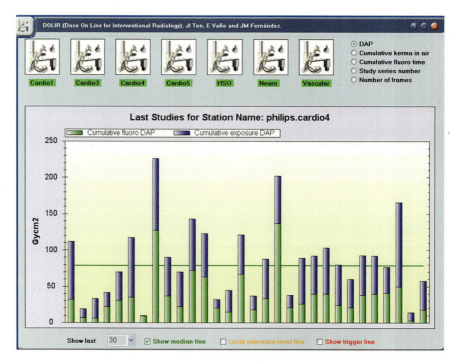

Fig. 2.1 The Dose On Line for Interventional Radiology (*DOLIR*) Program displaying Dose Area Product (*DAP*) values. This example shows the exposures for the last 30 patients (*far left of lower screen*) in DAP (*right of upper screen*) expressed in Gy cm^2 including a breakdown of the fluoroscopic and cine acquisition exposures. The median DAP was chosen (*centre left of lower screen*) and represented by the *green horizontal line*. This program enables the operators to readily identify those procedures with higher exposure

The North American based Society of Interventional Radiology (SIR) Standards of Practice Committee has recently published a document on 'Quality improvement guidelines for recording patient radiation dose in the medical record for fluoroscopically guided procedures' [30]. This document stated that ideally all available patient radiation dose data should be recorded. The society also recognizes that this may become mandatory in the future, as the FDA has already expressed an intention to establish requirements for CT and fluoroscopic devices to provide radiation dose data for incorporation into an individual's medical record or a radiation dose registry. The guidelines suggested an adequate recording of the different dose metrics, including skin dose mapping, for all fluoroscopy-guided interventional procedures is needed. The establishment of threshold levels to enable prompt analysis was also suggested.

3.3 Radiation Protection for Patients and Staff

The most important changes introduced by the ICRP in 2007 concerning the optimization of diagnostic imaging, were the application of DRLs for interventional

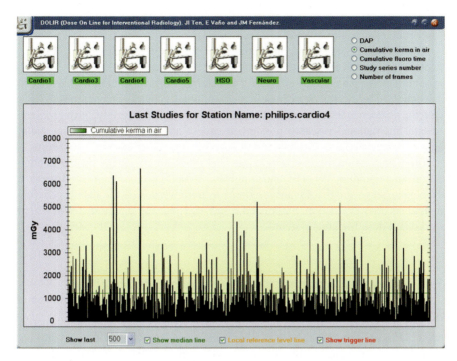

Fig. 2.2 Another example of the DOLIR Program displaying exposures in cumulative air kerma. The parameter chosen here (*right of upper screen*) was cumulative air kerma at the entrance reference point given in mGy for the last 500 cardiology procedures (*far left of lower screen*). The reference benchmarks selected (*centre of lower screen*) were: median value (*green line*, at approximately 1,200 mGy for this laboratory), local reference level (*orange line*, at 2,000 mGy) and trigger level (*red line*, at 5,000 mGy). The DOLIR Program enables ready identification of those procedures resulted in high exposure and prompt action on those with potential radiation-induced skin injuries

procedures [15, 18] and to consider the exposure to staff as part of optimization [17, 18]. The latter is particularly relevant to staff involved in fluoroscopy-guided interventional radiology procedures, especially in the context of the ICRP Statement on tissue reactions (deterministic effects), and the recommended changes to the occupational dose limits for the lens of the eye and the need to improve optimization for some high dose procedures [21].

The Commission drew attention to recent epidemiological evidence, which suggested that tissue reactions could occur when the threshold doses are at or might be lower than previously considered, i.e. 0.5 Gy for the lens of the eyes (radiation induced opacities) and 0.5 Gy for circulatory disease to the heart or brain. Doses of this magnitude to staff (lens of the eyes) and to patients could be reached in some complex interventional procedures and the Commission recommended particular emphasis on optimisation in these circumstances. Therefore, when discussing optimization, it is important to consider the RP of patients and staff together. The Commission has recommended the reduction of the dose limit to the lens of the eyes from 150 to 20 mSv/yr.

Other medical specialties in addition to radiology and cardiology also use fluoroscopy-guided interventional procedures as alternative to surgery, especially in some elderly patients unsuitable for general anesthesia or due to other clinical constraints. The increasing interest in these minimally invasive techniques together with the updated international recommendations on radiation safety, have promoted several international research projects in patient and staff dosimetry that should help to improve radiation safety and quality when performing these procedures.

The Commission's recommendations and other improvements in the pending European regulation [9] will encourage the radiology community and users to develop strategies and software programs to improve dosimetric data processing, individual dose evaluation, automated result analysis, and data transfer to patient record and dose tracking system.

3.4   Education, Training and Optimization Strategies

Education and training in RP by focusing on the need and means to apply appropriate exposure in radiation procedures is a key component to reduce dose to patients and staff. The Commission published a set of recommendations on 'Education and training in radiological protection for diagnostic and interventional procedures' [19].

In recent years several scientific and professional societies have produced guidelines on radiation safety, including patient dosimetry for interventional radiology. Some of these guidelines have been adopted simultaneously by the American and European societies of interventional radiology [4, 30, 32]. Other recommendations have been produced by expert groups and later endorsed by the professional societies [5]. The role of the European Commission in the publication of guidelines and reports [7, 8] is particularly important in the optimization of interventional procedures.

DRLs are still a challenge especially for interventional radiology. The Commission proposed their application to interventional radiology in 2001 and 2007 [15, 18] but it is still a long way from their effective application. The National Council on Radiation Protection and Measurements (NCRP) in USA has recently published a document on this issue [33]. The ICRP formed a working party in 2012 to provide more specific advice on the use of DRLs in interventional procedures and new imaging techniques.

Radiation protection and optimization are very important in paediatric practice especially in those procedures using high doses such as CT and interventional procedures. The Commission has prepared recommendations for 'Radiological Protection in Paediatric Diagnostic and Interventional Radiology' [23]. The IAEA is promoting programs on RP in paediatric interventional radiology in Latin America [54] and several relevant papers with results on patient and staff doses have recently been published [45–47]. Image Gently, conducted by The Alliance for Radiation

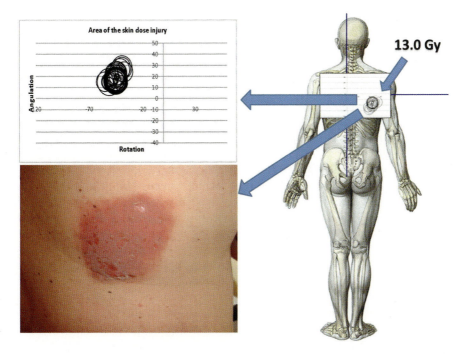

Fig. 2.3 A case of radiodermatitis. This occurred following a complex interventional cardiology procedure resulted in a peak skin dose of 13 Gy [50]. The DOLIR system enables immediate clinical intervention and the skin lesion resolved within a few months [42]. Patient dose audit and clinical follow-up of potential skin injury should be part of a quality improvement program for all interventional radiology and cardiology procedures (Reproduced after modification from Vano et al. (2012) Importance of a patient dosimetry and clinical follow-up program in the detection of radiodermatitis after long percutaneous coronary interventions. Cardiovasc Intervent Radiol. With kind permission from Springer Science + Business Media)

Safety in Pediatric Imaging, has formed a paediatric interventional radiology group and recently launched the 'Step Lightly' Campaign [43]. The European Commission is expected to launch a program to review the use of DRLs in Europe for pediatrics including their application in interventional procedures in 2013.

The follow-up of patients suspected of potential radiation injuries arising from complex interventional or CT procedures is a performance indicator that should be incorporated into a quality program. Two recent papers reported the incidence of and criteria for patient inclusion in the follow-up protocol. Applying the SIR-CIRSE recommendations [38], the first paper on cardiac procedures reported a follow-up rate of 0.31 % and a skin injury rate of 0.03 % [50]. An example of radiodermatitis after a complex interventional cardiology procedure is shown in Fig. 2.3. The second paper was on neuroradiology procedures performed in the same hospital. Following optimization and applying the peak skin dose criteria of >3 Gy as stated in the CIRSE and SIR guidelines, a patient follow-up rate of 1 % was reported [53].

3.5   Built-in Optimization

Hybrid rooms catering for conventional surgery and fluoroscopy-guided procedures are expected to increase in the future and RP is an important consideration. Some of the critical RP issues to be considered include: appropriate shielding, structure required for ceiling-suspended screens, protective garments for staff, patient dose monitoring and recording, appropriate personal dose monitoring, RP training for staff, designation of an individual responsible for RP, and support from qualified medical physicist.

The Multispecialty Occupational Health Group (MSOHG) has prepared recommendations on innovative designs for these rooms [26]. The involvement of medical, paramedical, engineering, construction, equipment, management, and other stakeholders is required to comprehensively consider various economic and stakeholder needs. A team approach, involving the specialists working together rather than in competition, will most likely lead to better patient outcome. Collectively, solutions will be developed to underpin the quality and safety requirements for: structure, space, ventilation, cooling, infection control, supporting infrastructure, and medical imaging (including hardware, operating table, software, imaging protocols, data archive and supporting equipment) etc. The safety needs of patients and staff, including radiation safety, should be one of the priorities.

4   Impact of Radiation Protection Recommendations

4.1   ICRP Recommendations

The ICRP revised the risk factors for stochastic effects in 2007 [17] and there are no substantial changes to the overall cancer risk coefficient since the 1990 report [14]. However, relevant changes were proposed for some organs such as breast and lung. The Commission emphasized the higher radiation risk for children and the need for caution when applying effective dose in medical exposure. Some refinements were made to the medical use of radiation by recommending the use of DRLs for interventional radiology and the use of staff doses as part of justification and optimization [18].

For tissue reactions, i.e. deterministic effects, the most relevant changes proposed in 2011 are new threshold doses for radiation opacities (cataracts) in the lens of the eyes and for circulatory disease to the heart and brain. These changes, especially that on lens opacities, should have significant impact on the radiation safety of professionals involved in fluoroscopy-guided interventional procedures and on the requirements in quality programs. The ICRP released its statement in April 2011 and included an update on the dose limit for the lens of the eye for occupationally exposed persons [21]. The immediate consequence was the adoption and incorporation of this change in the International Basic Safety Standards [12] and European Basic Safety Standards [9].

The Commission has recently produced three documents detailing the recommendations for diagnostic and interventional radiology: Publication 117 on 'Radiological protection in fluoroscopically guided procedures performed outside the imaging department' [20]; Publication 120 on 'Radiological protection in cardiology' [22]; and Publication 121 on 'Radiological protection in paediatric diagnostic and interventional radiology' [23].

4.2 European Commission Actions

The European regulations are quite strict on quality programs for medical imaging procedures. In fact, many requirements are included as part on the current Medical Exposure Directive [3], which will be further improved in the upcoming Directive on Basic Safety Standards [9]. Quality assurance programs, clinical audits and inspections by competent authorities are required. These quality programs include the quality control of the imaging equipment, patient dosimetry and involvement of Medical Physics Experts. Training of professionals involved in medical exposures, including RP training, is another aspect of the European Directive.

Quality assurance is defined in the European Directive [3] as all those planned and systematic actions necessary to provide adequate confidence that a structure, system, component, or procedure will perform satisfactorily and comply with the agreed standards. Quality control is part of quality assurance. The set of operations (programming, coordinating, implementing) intended to maintain or to improve quality. It covers monitoring, evaluation and maintenance at required levels of all characteristics of performance of equipment that can be defined, measured, and controlled.

In the new European Basic Safety Standards [9] optimization includes the selection of equipment, the consistent production of adequate diagnostic information or therapeutic outcomes, the practical aspects of medical exposure procedures, quality assurance, and the assessment and evaluation of patient and staff doses. Accidental and unintended exposures shall be part of the quality assurance program. Member States of the European Union 'shall implement a system for the registration and analysis of events involving or potentially involving accidental or unintended exposures'.

5 Impact of Radiation Protection on Quality Programs

5.1 Improve Quality by Radiation Protection

Radiation protection and radiation safety are key elements for quality program for medical imaging and interventional radiology. Radiation protection must be considered in: (a) the design of a facility, i.e. x-ray room, waiting area, patients and staff flow, shielding, equipment selection, informatics infrastructure, and personnel training etc.; (b) equipment installation and commissioning, i.e. acceptance testing, connectivity between modalities with RIS and PACS, and staff training etc. and (c) equipment use, i.e. routine daily practice, quality assurance program, quality clinical audits including patient and staff doses and image quality, and on-going training in quality and safety etc.

Good management in a medical imaging and interventional radiology facility includes leadership in radiation safety and RP by advocating the importance of RP to the health professionals and ensuring RP is a substantial part of a quality management system in practice [48].

Scientific and professional societies and organizations, agencies, competent authorities, industry, and standard organizations etc. have contributed to improvements in radiation safety in imaging and intervention. The best outcome will be achieved when all the stakeholders, i.e. medical doctors, radiographers, medical physicists, other health professionals, regulators, health authorities and industry are working together.

5.2   Radiation Protection Actions in Quality Programs

The following is a list of RP actions, which could form part of a medical imaging quality program [41, 48]:

- To improve the RP competences of medical doctors, medical physicists, engineers, radiographers, technicians, nurses, etc. by education and training;
- To promote a closer working relationship between medical physicists and other health professionals, e.g. collaboration between medical physicists and cardiologists is still uncommon;
- To promote a closer working relationship between radiologists, radiographers and medical physicists to improve justification and optimization and to reduce errors;
- To provide adequate infrastructure in a medical imaging facility, including radiation dose management and radiation safety for patients and staff [34];
- To improve interdisciplinary cooperation in research projects in radiation health effect and radiation health risk, especially with epidemiologists and radiobiologists;
- To define staffing requirements for RP and radiation safety, e.g. RP may be compromised by inadequate or unqualified staff;
- To integrate patient and staff RP into medical practice, e.g. interventional radiology;
- To improve the integration of RP into clinical quality programs;
- To address safety issues and to prevent incidents and radiation injuries when introducing new technologies and techniques, especially in interventional radiology;
- To improve RP in paediatric imaging, CT and interventional radiology;

- To improve the measurement, recording, analysis and archival of individual patient dose, which will impact on procedure justification and optimization [1];
- To refine criteria for the justification of radiological procedures in asymptomatic individuals after taking into account of the radiation dose;
- To optimize the use of new imaging technology, e.g. flat detectors, PET/CT etc.;
- To collaborate with equipment designers and manufacturers in RP, image quality improvement and dose reduction;
- To improve the automated collection of patient and staff doses and data transfer to individual and population exposure databases;
- To promote the proper use of DRLs in diagnostic and interventional procedures;
- To improve the communication to patients on radiological risks and to minimize self-referral for certain procedures; and
- To increase the support to medical physics in medical imaging facilities.

6   Conclusion

Quality in medical imaging and interventional radiology needs radiation safety on board. Radiological protection and good radiation management are integral elements for quality imaging and intervention. Successful implementation of justification and optimization requires a close cooperation between radiologists, radiographers, medical physicists and other stakeholders. Patient dose estimation, recording and audit are challenges and solutions are needed. Radiation safety for staff is a priority for some procedures. Radiation safety should be covered in education and training programs and in research projects dealing with radiological imaging and intervention. While an individual action addresses a certain aspect of RP, collectively they will improve quality in medical imaging, interventional radiology and patient care. Many stakeholders contribute to these improvements worldwide through RP actions. The ICRP leads by developing and publishing RP recommendations and collaborates with other stakeholders by facilitating the implementation of innovative RP actions. These combined efforts will ensure the common RP and quality objectives are reached in the not too distant future.

References

1. Abdullah A, Sun Z, Pongnapang N et al (2012) Comparison of computed tomography dose reporting software. Radiat Prot Dosimetry 151(1):153–157
2. American College of Radiology (ACR 2012) ACR appropriateness criteria. Available at: http://www.acr.org/Quality-Safety/Appropriateness-Criteria. Accessed 28 Aug 2012
3. Council of the European Union (1997) Council Directive 97/43/Euratom of 30 June 1997 on health protection of individuals against the dangers of ionizing radiation in relation to medical exposure, and repealing Directive 84/466/EURATOM. OJEC 1997/L 180:22–27

4. Dauer LT, Thornton RH, Miller DL et al on behalf of the Society of Interventional Radiology Safety and Health Committee and the Cardiovascular and Interventional Radiology Society of Europe Standards of Practice Committee Society of Interventional Radiology Safety and Health Committee (2012) Radiation management for interventions using fluoroscopic or computed tomographic guidance during pregnancy: a joint guideline of the Society of Interventional Radiology and the Cardiovascular and Interventional Radiological Society of Europe with Endorsement by the Canadian Interventional Radiology Association. J Vasc Intervent Radiol 23(1):19–32

5. Durán A, Hian SK, Miller DL et al (2013) A summary of recommendations for occupational radiation protection in interventional cardiology. Catheter Cardiovasc Interv 81(3):562–567

6. European Commission (2000) Radiation Protection 118. Referral guidelines for imaging. European Communities, Luxembourg

7. European Commission (2008) Radiation Protection 154. European guidance on estimating population doses from medical x-ray procedures. EU Publications Office. Available at: http://ec.europa.eu/energy/nuclear/radiation_protection/doc/publication/154.zip. Accessed 22 Aug 2012

8. European Commission (2009) Safety and efficacy for new techniques and imaging using new equipment to support European legislation (SENTINEL): supporting digital medicine. Available at: http://ec.europa.eu/research/energy/print.cfm?file=/comm/research/energy/fi/fi_cpa/rpr/article_3854_en.htm. Accessed 18 Aug 2012

9. European Commission (2012a) Proposal for a Council Directive laying down basic safety standards for protection against the dangers arising from exposure to ionizing radiation. Available at: http://ec.europa.eu/energy/nuclear/radiation_protection/doc/2012_com_242.pdf. Accessed 21 Aug 2012

10. European Commission (2012b) Dose DATAMED II program. European program. Available at: http://ddmed.eu/. Accessed 22 Aug 2012

11. Hendee WR, Becker GJ, Borgstede JP et al (2010) Addressing overutilization in medical imaging. Radiology 257(1):240–245

12. International Atomic Energy Agency (2011) Radiation protection and safety of radiation sources: international basic safety standards – Interim edition, IAEA Safety standards series GSR Part 3 (Interim). IAEA, Vienna

13. International Atomic Energy Agency (2012) Joint position statement on the IAEA Patient Radiation Exposure Tracking. Available at: https://rpop.iaea.org/RPOP/RPoP/Content/News/position-statement-IAEA-exposure-tracking.htm. Accessed 18 Aug 2012

14. International Commission on Radiological Protection (1991) Recommendations of the International Commission on Radiological Protection. ICRP Publication 60. Ann ICRP 21 (1–3):1–201

15. International Commission on Radiological Protection (2001) Diagnostic reference levels in medical imaging: review and additional advice. Ann ICRP 31(4):33–52

16. International Commission on Radiological Protection (2004) ICRP Publication 93. Managing patient dose in digital radiology. Ann ICRP 34(1):1–73

17. International Commission on Radiological Protection (2007) ICRP Publication 103. The 2007 recommendations of the International Commission on Radiological Protection. Ann ICRP 37 (2–4):1–332

18. International Commission on Radiological Protection (2007) ICRP Publication 105. Radiation protection in medicine. Ann ICRP 37(6):1–63

19. International Commission on Radiological Protection (2009) ICRP Publication 113. Education and training in radiological protection for diagnostic and interventional procedures. Ann ICRP 39(5):7–68

20. International Commission on Radiological Protection (2010) ICRP Publication 117. Radiological protection in fluoroscopically guided procedures performed outside the imaging department. Ann ICRP 40(6):1–102

21. International Commission on Radiological Protection (2011) ICRP statement on tissue reactions. Approved by the Commission on 21 Apr 2011. Available at: http://www.icrp.org/docs/ICRP%20Statement%20on%20Tissue%20Reactions.pdf. Accessed on 28 Aug 2012. See also: ICRP Publication 118. ICRP statement on tissue reactions and early and late effects of radiation in normal tissues and organs-threshold doses for tissue reactions in a radiation protection context. Ann ICRP 41(1–2):1–322
22. International Commission on Radiological Protection (2013) ICRP Publication 120. Radiological protection in cardiology. Ann ICRP 42(1):1–125
23. International Commission on Radiological Protection (2013) ICRP Publication 121. Radiological protection in paediatric diagnostic and interventional radiology. Ann ICRP 42(2):1–63
24. International Radiology Quality Network (2009) Top 10 principles of international clinical teleradiology. Available at: http://www.irqn.org/work/teleradiology.htm. Accessed 6 Sept 2012
25. Kenny LM, Lau LS (2008) Clinical teleradiology – the purpose of principles. Med J Aust 188 (4):197–198
26. Klein LW, Miller DL, Goldstein J et al on behalf of the Multispecialty Occupational Health Group (2011) The catheterization laboratory and interventional vascular suite of the future: anticipating innovations in design and function. Catheter Cardiovasc Intervent 77(3):447–55
27. Lau LS (2006) Clinical teleradiology in Australia: practice and standards. J Am Coll Radiol 3 (5):377–381
28. Lau LS, Pérez MR, Applegate KE et al (2011) Global quality imaging: emerging issues. J Am Coll Radiol 8(7):508–512
29. Lau LS, Pérez MR, Applegate KE et al (2011) Global quality imaging: improvement actions. J Am Coll Radiol 8(5):330–334
30. Miller DL, Balter S, Dixon RG et al on behalf of the Society of Interventional Radiology Standards of Practice Committee (2012) Quality improvement guidelines for recording patient radiation dose in the medical record for fluoroscopically guided procedures. J Vasc Intervent Radiol 23(1):11–18
31. Miller DL, Balter S, Schueler BA et al (2010b) Clinical radiation management for fluoroscopically guided interventional procedures. Radiology 257(2):321–332
32. Miller DL, Vannó E, Bartal G et al on behalf of the Cardiovascular and Interventional Radiology Society of Europe and the Society of Interventional Radiology (2010b) Occupational radiation protection in interventional radiology: a joint guideline of the Cardiovascular and Interventional Radiology Society of Europe and the Society of Interventional Radiology. Cardiovasc Intervent Radiol 33(2):230–239
33. National Council on Radiation Protection and Measurements (2012) NCRP report 172. Reference levels and achievable doses in medical and dental imaging: recommendations for the United States. NCRP, Bethesda
34. Ng KH, McLean ID (2011) Diagnostic radiology in the tropics: technical considerations. Semin Musculoskelet Radiol 15(5):441–445
35. Ng KH, Rehani MM (2006) X ray imaging goes digital. BMJ 333(7572):765–766, 14
36. Reiner BI (2009) Quantifying radiation safety and quality in medical imaging, part 3: the quality assurance scorecard. J Am Coll Radiol 6(10):694–700
37. Sabarudin A, Sun Z, Ng KH (2012) A systematic review of radiation dose associated with different generations of multidetector CT coronary angiography. J Med Imaging Radiat Oncol 56(1):5–17
38. Stecker MS, Balter S, Towbin RB et al on behalf of the Society of Interventional Radiology Safety and Health Committee and the Cardiovascular and Interventional Radiology Society of Europe Standards of Practice Committee (2009) Guidelines for patient radiation dose management. J Vasc Intervent Radiol 20(7 Suppl):S263–S273
39. Sun Z, Ng KH (2012) Diagnostic value of coronary CT angiography with prospective ECG-gating in the diagnosis of coronary artery disease: a systematic review and meta-analysis. Int J Cardiovasc Imaging 28(8):2109–2119

40. Sun Z, Choo GH, Ng KH (2012) Coronary CT angiography: current status and continuing challenges. Br J Radiol 85(1013):495–510
41. Sun Z, Ng KH, Ramli N (2011) Biomedical imaging research: a fast-emerging area for interdisciplinary collaboration. Biomed Imaging Intervent J 7(3):e21, Epub 2011 July 1
42. Ten JI, Fernandez JM, Vaño E (2011) Automatic management system for dose parameters in interventional radiology and cardiology. Radiat Prot Dosimetry 147(1–2):325–328
43. The Alliance for Radiation Safety in Pediatric Imaging (2012) Image gently campaign: step lightly for interventional radiology. Available at: http://www.pedrad.org/associations/5364/ig/?page=584. Accessed 23 Aug 2012
44. The Royal College of Radiologists (2012) RCR referral guidelines (iRefer). Available at: http://www.rcr.ac.uk/content.aspx?PageID=995. Accessed 28 Aug 2012
45. Ubeda C, Vano E, Gonzalez L et al (2010) Scatter and staff dose levels in paediatric interventional cardiology: a multicentre study. Radiat Prot Dosimetry 140(1):67–74
46. Ubeda C, Vano E, Miranda P et al (2011) Radiation dose and image quality for paediatric interventional cardiology systems. A national survey in Chile. Radiat Prot Dosimetry 147(3):429–438
47. Ubeda C, Vano E, Miranda P et al (2012) Pilot program on patient dosimetry in pediatric interventional cardiology in Chile. Med Phys 39(5):2424–2430
48. Vano E (2011) Global view on radiation protection in medicine. Radiat Prot Dosimetry 147(1–2):3–7
49. Vano E, Gonzalez L (2001) Approaches to establishing reference levels in interventional radiology. Radiat Prot Dosimetry 94(1–2):109–112
50. Vano E, Escaned J, Vano-Galvan S et al (2013) Importance of a patient dosimetry and clinical follow-up program in the detection of radiodermatitis after long percutaneous coronary interventions. Cardiovasc Intervent Radiol 36(2):330–337
51. Vano E, Fernandez JM, Ten JI et al (2005) Patient dosimetry and image quality in digital radiology from online audit of the X-ray system. Radiat Prot Dosimetry 117(1–3):199–203
52. Vano E, Fernández JM, Ten JI et al (2007) Transition from screen-film to digital radiography: evolution of patient radiation doses at projection radiography. Radiology 243(2):461–466
53. Vano E, Fernandez JM, Sanchez RM et al (2013) Patient radiation dose management in the follow-up of potential skin injuries in neuroradiology. AJNR Am J Neuroradiol 34(2):277–282
54. Vano E, Ubeda C, Miranda P et al (2011) Radiation protection in pediatric interventional cardiology: An IAEA pilot program in Latin America. Health Phys 101(3):233–237
55. World Health Organization (2008) WHO global initiative on radiation safety in health care settings technical meeting report. WHO, Geneva
56. Wong WS, Roubal I, Jackson DB et al (2005) Outsourced teleradiology imaging services: an analysis of discordant interpretation in 124,870 cases. J Am Coll Radiol 2(6):478–484

Chapter 3
Dose Assessment in the Management of Patient Protection in Diagnostic and Interventional Radiology

Paul C. Shrimpton and Kwan-Hoong Ng

Abstract The increasingly widespread use of x-rays in diagnostic and interventional radiology provides not only enormous benefits to patients, but also significant radiation exposure for populations. The protection of patients against potential radiation harm requires the elimination of all unnecessary and unintended x-ray exposure in relation to effective clinical diagnosis. Dosimetry is an essential management tool for patient safety by allowing the assessment of typical radiation risks in support of the justification of procedures, and the routine monitoring and comparison of typical doses in pursuit of the optimization of patient protection. Periodic assessment of patient doses should form an integral part of quality assurance in x-ray facilities and is best based on practical measurements that provide useful characterization of patient exposure, such as entrance surface dose, dose-area product and, for CT, weighted CT dose index and dose-length product. In addition, the monitoring of cumulative dose during complex interventional procedures underpins the effective management of potential tissue reactions. Mean values determined for practical dose monitoring quantities in a facility for each type of procedure and patient group can form the basis not only for estimates of typical organ and effective doses utilizing appropriate coefficients, but also local diagnostic reference levels (DRLs). DRLs represent a pragmatic mechanism for promoting continuing improvement in performance by facilitating comparison with national values and practice elsewhere. The application of DRLs in the UK over the last 25 years, within a coherent framework for patient protection, has successfully helped to reduce unnecessary exposures, with national DRLs for many procedures falling by a factor of two.

P.C. Shrimpton (✉)
Medical Dosimetry Group, Centre for Radiation, Chemical and Environmental Hazards, Public Health England, Chilton, Didcot, Oxon OX11 0RQ, UK
e-mail: paul.shrimpton@phe.gov.uk

K.-H. Ng
Department of Biomedical Imaging, Faculty of Medicine, University of Malaya Research Imaging Centre, Lembah Pantai, 50603 Kuala Lumpur, Malaysia
e-mail: ngkh@ummc.edu.my

L. Lau and K.-H. Ng (eds.), *Radiological Safety and Quality: Paradigms in Leadership and Innovation*, DOI 10.1007/978-94-007-7256-4_3,
© Crown Copyright 2014

Keywords Dose monitoring • Diagnostic reference levels (DRLs) • Patient dose • Patient protection • Radiation risk • Trends in UK practice

1   Introduction

Continuing advances in the technology of x-ray imaging and its increasing application in clinical practice have ensured that diagnostic and interventional radiology remain essential tools in the support of healthcare. When properly conducted, x-ray procedures can provide enormous benefits to patients, in terms of the diagnosis and management of disease. Periodic surveys of the worldwide use of radiation in medicine conducted by the United Nations Scientific Committee on the Effects of Atomic Radiation (UNSCEAR) have highlighted the widespread utilization of x-ray procedures and steady growth in their global number. Between global reviews published for 1988 and 2008 [31], the annual total number of medical (excluding dental) x-ray procedures rose from 1,380 million to 3,143 million (corresponding to an increase in annual frequency from 280 to 488 per 1,000 population).

Notwithstanding the undoubted usefulness of diagnostic and interventional radiological procedures, there is also a potential hazard to patients from their exposure to x-rays, with risk in the shorter term of tissue reactions (such as skin erythema) when specific doses exceed thresholds, and in the longer term of stochastic effects such as cancer and, in relation to their offspring, hereditary effects [15]. Whereas these risks remain relatively low in comparison with many other risks for patients [32], the present consensus for prudence is based on the avoidance of all unnecessary and unintended x-ray exposures through active management of radiation protection for patients. The scale of the concomitant population dose from practice in diagnostic radiology is demonstrated by the UNSCEAR surveys [31], which report an increase in global collective dose from medical x-rays from 1.8 million man Sv for 1988 to 4.0 million man Sv for 2008, corresponding to annual per caput doses of 0.35 mSv and 0.62 mSv, respectively. Whereas on a global scale medical x-rays presently account for 20 % of the total dose from all sources of ionizing radiation, this proportion varies between individual countries depending on differing patterns of utilization, being, for example, 35 % in the USA [24] and 15 % in the UK [11]. In addition, computed tomography (CT) procedures are the dominant source of population dose from diagnostic x-rays in many developed countries [9] with, for example, contributions of around 70 % in both USA [24] and UK [11]. This pattern identifies CT as a particular focus for patient protection initiatives, including its use in relation to asymptomatic individuals [4].

Such broad analyses of collective dose from diagnostic and interventional radiology serve no other purpose in radiological protection than to highlight potential scope for improvement towards optimized practice and to monitor trends. They also underpin the need for particular attention to the protection of patients through the assessment of patient dose.

2   Framework for Patient Protection

The effective protection of patients against potential harm from x-rays requires the close control of all exposures with a view to preventing tissue reactions and reducing the probability of stochastic effects, whilst maintaining intended benefits from the procedure. The essence of patient protection in diagnostic and interventional radiology is therefore the elimination of all x-ray exposure that is unnecessary and unintended in relation to effective clinical diagnosis. This component of practice includes procedures that are unlikely to provide worthwhile information for patient management, repeats of procedures where information has been lost or was incomplete, all those exposures where doses could reasonably have been reduced by changes in technique or equipment without loss of clinical efficacy, and exposures due to human error.

The framework for eliminating unnecessary medical exposures has been established by the International Commission on Radiological Protection (ICRP), most recently in its 2007 recommendations [15]. The key principles for protecting patients undergoing x-ray procedures are as follows:

- Justification – do more good than harm by ensuring a positive net benefit for every procedure;
- Optimization of protection – maximize the benefit by keeping doses as low as reasonably achievable (ALARA) to meet the specific clinical purpose of the procedure.

The further application of strict limits on the doses to individual patients is inappropriate, since this could also limit benefits. However, the application of diagnostic reference levels (DRLs) is recommended as a practical way of promoting optimization of protection [13], as discussed further below. These fundamental principles for patient protection are enshrined in international [5] and national [6] legislation. Dosimetry provides essential support to these crucial elements in the framework for patient protection.

3   Dosimetry for Purpose

There are several valid and different reasons for wanting to know dose in support of the justification of x-ray procedures and the optimization of patient protection. This necessarily leads to the use in practice of a range of dose quantities and approaches in order to match most closely each particular purpose, as discussed below.

3.1   Assessment of Radiation Risk

Knowledge of mean absorbed doses to all the radiosensitive organs provides the most complete description of patient exposure that is required for the assessment of risk. However, these doses are in general difficult to measure directly in view of the location of many such organs and tissues relative to the complex partial-body exposures resulting from x-ray procedures. Organ doses can only be assessed by using dose coefficients determined for reference patients by measurement in physical phantoms or calculation by using Monte Carlo techniques and virtual phantoms [12, 18, 31]. These doses can be expressed relative to simpler quantities that can be more easily measured in routine practice, such as entrance surface dose (ESD) per radiograph, dose-area product (DAP) per complete radiographic or fluoroscopic procedure and, for CT, volume weighted CT dose index ($CTDI_{vol}$). Effective dose (E), defined by ICRP [15] as the weighted sum of doses to particular organs, can provide a convenient way of comparing practice between different radiology facilities or types of medical exposure, although it was not originally intended for this purpose [15]. Estimates of organ and effective doses depend on the specific dosimetry models assumed, in relation, for example, to bone and also to the size and anatomy of the reference patient, including whether it is mathematical- or voxel-based, and representative for adults or children [12, 18, 21, 22, 31].

The long-term risk of cancer following exposure to x-rays not only varies between organs, but it also depends on the sex and age of the exposed person [15, 23, 30]. According to models developed by ICRP [15], the lifetime risk of cancer incidence following uniform whole-body irradiation is higher at all ages for females than males, and decreases steadily with age by about a factor of 2 for every 30 years (Fig. 3.1). Patterns of risk vary between types of x-ray procedure owing to significant differences in organ doses and typical risks have been derived in relation to UK radiology practice using the ICRP risk models and typical organ doses from national surveys [32]. Typical lifetime risks of cancer incidence range from less than 1 in a billion ($<10^{-9}$) for any patient having an x-ray procedure of the knee or foot, to over 1 in 1,000 ($>10^{-3}$) for CT of the whole trunk of a young girl. For practical purposes such typical risks can usefully be classified into broad categories spanning 'negligible' risk to 'moderate' risk [32].

Contrary to advice from ICRP [15], it has become common practice to convert estimates of E for particular x-ray procedures into radiation risks using the ICRP nominal probability coefficients for fatal cancer or aggregated detriment. Such an approach has serious limitations in view, for example, of the average nature of the ICRP risk coefficients over whole populations. Detailed analyses of risk per unit E [32] have demonstrated that lifetime cancer incidence risks for a particular age band, sex and procedure can differ from those using the ICRP coefficient for detriment-adjusted cancer (5.5 % per Sv) by up to a factor of ten. Whereas the risk coefficients (% per Sv) for whole body exposure shown in Fig. 3.1 can provide estimates of risk for most x-ray procedures within ±50 % of more detailed assessments, series of age- and sex-specific coefficients derived for selected regions of procedure allow improved estimates of risk from E (within ±30 %) (Table 3.1). Such estimates of risk remain typical for a type of procedure and patient group (age and sex), and do not, of course, relate to individuals. Risk of radiation-induced heritable effects can be assessed on the basis of typical doses to the gonads [32].

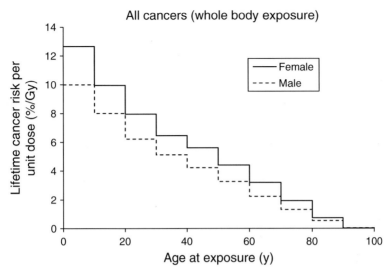

Fig. 3.1 Lifetime risk of cancer incidence by age and sex for all cancers, following uniform whole-body irradiation

Table 3.1 Proposed total lifetime cancer risk per unit effective dose (% per Sv) as a function of age at exposure and sex for x-ray procedures conducted in different of regions of anatomy

Anatomical region	Age group (years)									
	0–9	10–19	20–29	30–39	40–49	50–59	60–69	70–79	80–89	90–99
MALE										
Head	18	13	9.1	6.8	5.2	3.6	2.2	1.2	0.5	0.1
Neck	9.1	6.2	4.1	2.8	2.0	1.3	0.8	0.4	0.2	0.0
Chest	8.3	7.0	5.8	5.1	4.6	4.0	3.0	1.9	0.8	0.0
Abdomen & pelvis	12	9.7	7.5	6.0	4.7	3.4	2.2	1.1	0.4	0.0
Whole body (uniform)	**10**	**8.0**	**6.2**	**5.1**	**4.2**	**3.3**	**2.2**	**1.3**	**0.6**	**0.04**
FEMALE										
Head	15	11	7.6	5.5	4.6	3.0	1.7	0.9	0.3	0.0
Neck	20	12	7.2	4.2	2.6	1.6	1.0	0.5	0.2	0.0
Chest	14	12	10	8.8	8.3	7.1	5.4	3.3	1.3	0.0
Abdomen & pelvis	10	8.3	6.6	5.2	4.4	3.2	2.0	1.1	0.4	0.0
Whole body (uniform)	**14**	**11**	**8.5**	**6.8**	**5.8**	**4.4**	**3.1**	**1.8**	**0.7**	**0.02**

3.2 Tissue Reactions

The ICRP 2007 recommendations included consideration of the detriment arising from non-cancer effects of radiation on health. These effects, previously called deterministic effects, are now referred to as tissue reactions because it is increasingly recognized that some of these effects are not determined solely at the time of irradiation but can be modified after radiation exposure [16]. Complex interventional procedures involving prolonged fluoroscopy and extended use of radiography have potential, depending on factors relating to both the equipment and technique, for high, localized doses to patients that are above thresholds for tissue reactions [14].

Radiation-induced skin reactions range from mild and transient to severe and clinically devastating [3]. Damage can be evident in the epidermis, the dermis and the subcutaneous tissues. Although commonly referred to as skin injuries, severe radiation injuries can also extend into subcutaneous fat and muscle. The entrance skin dose, the interval between irradiations and the size of the skin area being irradiated, all affect the severity of the radiation injury. In addition a variety of physical and patient-related factors also influence the severity. The timing of the appearance of various specific tissue responses depends on intrinsic biological factors. The dose threshold due to biological variability can be quite low for the most sensitive patients relative to that for an average patient.

ICRP [16] has recommended that there are some tissue reactions, particularly those with very late manifestation, where threshold doses are, or might be, lower than previously considered. For the lens of the eye, the threshold absorbed dose is now considered to be 0.5 Gy (compared with the previous estimate of 2 Gy). In consequence, ICRP [16] has recommended a change in the dose limit for the eye in relation to occupational exposures, with implications for managing the radiation safety of practitioners involved in fluoroscopically-guided interventional (FGI) procedures. Medical practitioners should also be aware that the absorbed dose threshold for circulatory disease may be as low as 0.5 Gy to the heart or brain. Doses to patients of this magnitude could be reached during some complex interventional procedures and therefore particular emphasis should be placed on optimization in these circumstances.

Dose monitoring is essential for individual patients undergoing FGI procedures using established metrics that include fluoroscopy time, number of radiographic or fluoroscopic frames, DAP and air kerma at the reference point ($K_{a,r}$). Institutions should set trigger levels for DAP and $K_{a,r}$ so as to identify patients for clinical follow-up, with a view to the early detection and management of skin injuries [17, 25].

3.3 Monitoring and Comparison of Performance

In addition to knowledge of typical risks for x-ray procedures, there is also a need for routine monitoring of dose in support of achieving and maintaining optimization of patient protection. Such dosimetry is best based on practical dose quantities that can easily be assessed by measurement or calculation, in order to provide suitable characterization of patient exposure during a procedure for the purposes of comparison of performance, both over time and between different facilities. These monitoring quantities include, as mentioned above, ESD, DAP and $CTDI_{vol}$, together also with dose-length product (DLP) and size-specific dose estimate (SSDE) [1] in relation to CT. Their periodic monitoring for samples of patients undergoing a range of procedures should form an integral part of quality assurance and performance testing in each x-ray facility (see for example [20]). This dose monitoring data will also provide the basis for more detailed assessments of patient exposure (such as typical organ and effective doses), as describe above.

Wide scale surveys have long since demonstrated wide variations in dosimetric performance between facilities for similar types of x-ray procedure [29, 31]. Such patterns suggest lack of uniform implementation of optimization of patient protection and have led to the development of the concept of procedure-specific diagnostic reference levels (DRLs) in order to promote improvements in practice.

4   Diagnostic Reference Levels (DRLs)

4.1   Application

DRLs represent a quality improvement tool within the framework for quality assurance and control initiatives in relation to radiation safety for patients. These doses are specified in terms of the practical dose quantities ESD, DAP, $CTDI_{vol}$ and DLP, as dose indictors (rather than more complete measures of patient dose) in relation to comparison of typical practice between different x-ray rooms for each type of procedure and patient group. DRLs are not intended for application to individual patients, although action is required when they are consistently being exceeded. The DRL concept is now widely established as a useful way of promoting improvements in patient protection [13, 15] through its operation at two levels: national and local [19].

National DRLs are commonly set on the basis of third quartile values for the distributions of mean doses observed from a large sample of x-ray rooms during wide scale dose surveys [13, 19]. As such, they are clearly not optimum doses, but merely a practical trigger to help identify facilities with unusually high doses (in the top 25 % of national practice) and where review and remedial action might be most urgently needed for improved practice. For continuing effectiveness, national DRLs should be periodically reviewed and revised on the basis of updated survey data reflecting changes in national practice and technology.

Local DRLs are set by individual x-ray facilities on the basis of their mean doses observed for samples of patients for each patient group and type of procedure. Mean values above the national DRL should be investigated and either justified as being necessary or reduced through appropriate changes in practice to improve patient protection. These locally determined mean doses (as characterizing typical practice) can be set as local DRLs for subsequent comparison with practice elsewhere. Local DRLs should be reviewed annually and revised as necessary following periodic local dose audit to monitor trends, being conducted, for example, every 3 years or on significant change of equipment or technique [20].

4.2   Effectiveness: The UK Experience

National DRLs were first established (under the term *suggested guideline doses*) in the UK in 1989 following a national dose survey in England in the mid-1980s [29]. Subsequent UK initiatives include the agreement of a national protocol for patient dose measurements in diagnostic radiology [27] and the setting up of a national patient dose database for conventional x-ray procedures at the then National Radiological Protection Board (subsequently part of the Health Protection Agency and, from 1 April 2013, now incorporated into Public Health England). Quinquennial reviews of this database have led to updated national DRLs for 1995, 2000, 2005 and 2010 [10], published as NRPB/HPA national reference doses that inform national DRLs set by the Department of Health [7]. National DRLs have also been recommended for CT on the basis of national dose surveys for 1989 and 2003 [28], with the results from a third UK review for 2011 in publication.

Over the last 25 years, UK national DRLs for conventional x-ray procedures have in general fallen by a factor of 2 [10], as summarized in Tables 3.2 and 3.3. Significant reductions have also been observed in relation to national DRLs for paediatric patients (Table 3.4), although changes for CT have, thus far, been rather less (Table 3.5). These trends reflect significant changes in practice to improve

Table 3.2 Trends in UK national diagnostic reference levels for some radiographic procedures on adult patients.

Radiographic procedure	UK National DRL for entrance surface dose (mGy)				
	1985	1995	2000	2005	2010
Skull AP/PA	5	4	3	2	1.8
Skull LAT	3	2	1.5	1.3	1.1
Chest PA	0.3	0.2	0.2	0.15	0.15
Chest LAT	1.5	0.7	1	0.6	0.5
Thoracic spine AP	7	5	3.5	4	3.5
Thoracic spine LAT	20	16	10	7	7
Lumbar spine AP	10	7	6	5	5.7
Lumbar spine LAT	30	20	14	11	10
Lumbar spine LSJ	40	35	26	26	–
Abdomen AP	10	7	6	4	4
Pelvis AP	10	5	4	4	4

AP Antero-posterior, *PA* Postero-anterior, *LAT* Lateral, *LSJ* Lumbo-sacral joint

Table 3.3 Trends in UK national diagnostic reference levels for some fluoroscopic procedures on adult patients

| Procedure | UK National DRL for dose-area product (Gy cm^2) | | | | |
	1985	1995	2000	2005	2010
Intravenous urography	40	23	16	14	14
Barium meal	25	17	13	14	12
Barium enema	60	32	31	24	21

Table 3.4 Trends in UK national diagnostic reference levels for some fluoroscopic procedures on paediatric patients

| Procedure | Age group (years) | UK National DRL for dose-area product (Gy cm^2) | | |
		2000	2005	2010
Micturating cystourethography	0	0.4	0.3	0.1
	1	1.0	0.8	0.3
	5	1.0	0.8	0.3
	10	2.1	1.5	0.4
	15	4.7	2.5	0.9
Barium meal	0	0.7	0.4	0.1
	1	2.0	1.2	0.2
	5	2.0	1.2	0.2
	10	4.5	2.4	0.7
	15	7.2	6.4	2.0
Barium swallow	0	0.8	0.4	0.2
	1	1.5	1.3	0.4
	5	1.5	1.3	0.5
	10	2.7	2.9	1.8
	15	4.6	3.5	3.0

Table 3.5 Trends in UK national diagnostic reference levels for some CT procedures on adult patients

| Procedure | UK National DRL for dose-length product (mGy cm) | |
	1989	2003
Head[a]	1,050	930
Chest[b]	650	580
Abdomen[b]	780	470
Abdomen & pelvis[b]	–	560
Chest, abdomen & pelvis[b]	–	940

[a]Dose values refer to the 16 cm diameter standard CT dosimetry phantom
[b]Dose values refer to the 32 cm diameter standard CT dosimetry phantom

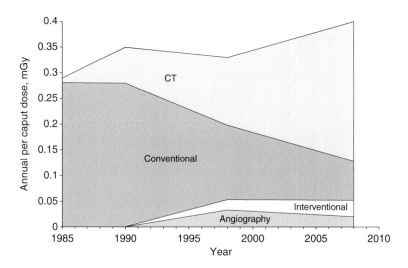

Fig. 3.2 Trends in contributions to UK annual per caput dose from diagnostic and interventional radiology by category of x-ray procedure

patient protection through the reduction of unnecessary patient exposures. The application of DRLs, as an essential element within a coherent framework for managing patient dose, has undoubtedly helped both to raise awareness of levels of dose in routine practice and also to promote and facilitate improvements towards optimization of protection.

In consequence, the average annual dose per head of population in the UK from x-ray procedures has remained relatively constant over the last 25 years, at between 0.3 and 0.4 mSv [11]. The total number of medical and dental x-ray procedures has increased steadily by nearly 30 % during this period, with much of this growth relating to higher dose procedures, such as CT. As illustrated in Fig. 3.2, however, the growth in UK population dose from CT has largely been offset (thus far) by a reduction in the contribution from conventional x-ray procedures. In stark contrast, similar data for the USA demonstrate a significant rise (by a factor 6) in the per caput dose from x-rays between the early 1980s (0.4 mSv) and 2006 (2.3 mSv) [24].

DRLs are now gaining wider use elsewhere, particularly in Europe [8] and, more recently, the USA [2, 26], as a valuable practical tool for promoting reduction of unnecessary exposures.

5 Conclusions

Dosimetry provides essential support to the effective management of patient protection in diagnostic and interventional radiology by facilitating the assessment of radiation risks (in relation to justification) and the comparison of performance (in relation to optimization of practice). It also underpins the management of tissue reactions during complex fluoroscopically-guided interventions. Periodic monitoring of patient doses should form an integral part of quality assurance in x-ray facilities and is best based on practical measurements that provide useful characterization of patient exposure. Such monitoring quantities would include entrance surface dose, dose-area product and, for CT, weighted CT dose index and dose-length product. Mean values determined in a facility for each type of procedure and patient group form the basis not only for estimates of typical organ and effective doses utilizing appropriate coefficients, but also local diagnostic reference levels (DRLs) for comparison with national values and practice elsewhere as a pragmatic mechanism for promoting continuing improvement in performance. The application of DRLs in the UK over the last 25 years, within a coherent framework for patient protection, has successfully helped to reduce unnecessary exposures, with national DRLs for many procedures falling by a factor of two.

Continuing developments in technology and clinical practice dictate that routine dosimetry will become increasingly important in order to help ensure the effective and efficient use of x-rays for the continuing benefit of patients.

References

1. American Association of Physicists in Medicine (2011) Report No 204: size-specific dose estimates (SSDE) in pediatric and adult body CT examinations. American Association of Physicists in Medicine, College Park
2. American College of Radiology (2008) ACR practice guidelines for diagnostic reference levels in medical x-ray imaging. American College of Radiology, Reston
3. Balter S, Hopewell JW, Miller DL et al (2010) Fluoroscopically guided interventional procedures: a review of radiation effects on patients' skin and hair. Radiology 254(2):326–341
4. Committee on Medical Aspects of Radiation in the Environment (2007) 12th Report: The impact of personally initiated x-ray computed tomography scanning for the health assessment of asymptomatic individuals. Health Protection Agency, Chilton
5. Council of the European Union (1997) Council Directive 97/43/Euratom of 30 June 1997 on health protection of individuals against the dangers of ionizing radiation in relation to medical exposure. Off J Eur Commun L180:22
6. Department of Health (2000) The ionizing radiation [Medical Exposure] regulations 2000 (IR(M) ER). https://www.gov.uk/government/organisations/department-of-health. Accessed 26 May 2012
7. Department of Health (2007) Guidance on the establishment and use of diagnostic reference levels (DRLs). https://www.gov.uk/government/organisations/department-of-health. Accessed 26 May 2012
8. European ALARA Network (2007) The diagnostic reference levels (DRLs) in Europe. http://www.eu-alara.net/. Accessed 26 May 2012
9. European Commission (2008) Radiation protection 154. European guidance on estimating population doses from medical X-ray procedures. EU publications office. http://ddmed.eu/_media/background_of_ddm1:rp154.pdf. Accessed 26 May 2012
10. Hart D, Hillier MC, Shrimpton PC (2012) Report HPA-CRCE-034: doses to patients from radiographic and fluoroscopic x-ray imaging procedures in the UK – 2010 review. Health Protection Agency, Chilton

11. Hart D, Wall BF, Hillier MC et al (2010) Report HPA-CRCE-012: frequency and collective dose for medical and dental x-ray examinations in the UK, 2008. Health Protection Agency, Chilton

12. International Atomic Energy Agency (2007) Technical reports series 457: dosimetry in diagnostic radiology: an international code of practice. International Atomic Energy Agency, Vienna

13. International Commission on Radiological Protection (1996) ICRP publication 73: radiological protection and safety in medicine. Ann ICRP 26(2):P31

14. International Commission on Radiological Protection (2000) ICRP publication 85: avoidance of radiation injuries from medical interventional procedures. Ann ICRP 30(2):P63

15. International Commission on Radiological Protection (2007) ICRP publication 103: the 2007 recommendations of the International Commission on radiological protection. Ann ICRP 37(2–4):1–332

16. International Commission on Radiological Protection (2012) ICRP publication 118: ICRP statement on tissue reactions and early and late effects of radiation in normal tissues and organs (threshold doses for tissue reactions in a radiation protection context). Ann ICRP 41(1–2):1–322

17. International Commission on Radiological Protection (2013) ICRP publication 120: radiological protection in cardiology. Ann ICRP 42(1):1–125

18. International Commission on Radiation Units and Measurements (2005) ICRU Report 74: patient dosimetry for x rays used in medical imaging. J ICRU 5(2):P113

19. Institute of Physics and Engineering in Medicine (2004) Report 88: guidance on the establishment and use of diagnostic reference levels for medical x-ray examinations. Institute of Physics and Engineering in Medicine, York

20. Institute of Physics and Engineering in Medicine (2005) Report 91: recommended standards for the routine performance testing of diagnostic x-ray imaging systems. Institute of Physics and Engineering in Medicine, York

21. Jansen JTM, Shrimpton PC (2009) Calculation of normalized organ doses for pediatric patients undergoing CT examinations on four types of CT scanner. In: Dössel O, Schlegel WC (eds) Proceedings of world congress on medical physics and biomedical engineering (Munich, Sept 7–12 2009). IFMBE Proceedings 25(3):116–119

22. Jansen JTM, Shrimpton PC (2011) Calculation of normalized organ and effective doses to adult reference computational phantoms from contemporary computed tomography scanners. Prog Nuc Sci Technol 2:165–171

23. National Research Council of the National Academies (2006) Health risks from exposure to low levels of ionizing radiation. Biological effects of ionizing radiation (BEIR) VII phase 2. The National Academies Press, Washington, DC

24. National Council on Radiation Protection and Measurements (2009) Report 160: ionizing radiation exposure of the population of the United States. National Council on Radiation Protection and Measurements, Bethesda

25. National Council on Radiation Protection and Measurements (2010) Report 168: radiation dose management for fluoroscopically-guided interventional medical procedures. National Council on Radiation Protection and Measurements, Bethesda

26. National Council on Radiation Protection and Measurements (2012) Report 172: reference levels and achievable doses in medical and dental imaging: recommendations for the United States. National Council on Radiation Protection and Measurements, Bethesda

27. National Radiological Protection Board (1992) IPSM/NRPB/CoR national protocol for patient dose measurements in diagnostic radiology. National Radiological Protection Board, Chilton

28. Shrimpton PC, Hillier MC, Lewis MA et al (2005) Report NRPB-W67: doses from computed tomography (CT) examinations in the UK – 2003 review. Health Protection Agency, Chilton

29. Shrimpton PC, Wall BF and Hillier MC (1989) Suggested guideline doses for medical x-ray examinations. Radiation protection – theory and practice. In: Goldfinch EP (ed) Proceedings of 4th international symposium of the society for radiological protection (Malvern, 4–9 June 1989), Institute of Physics, Bristol

30. United Nations Scientific Committee on the Effects of Atomic Radiation (2008) Effects of ionizing radiation – volume 1. UNSCEAR 2006 report to the general assembly with scientific annexes. United Nations, New York
31. United Nations Scientific Committee on the Effects of Atomic Radiation (2010) Sources and effects of ionizing radiation – volume 1. UNSCEAR 2008 report to the general assembly with scientific annexes. United Nations, New York
32. Wall BF, Haylock R, Jansen JTM et al (2011) Report HPA-CRCE-028: radiation risks from medical x-ray examinations as a function of the age and sex of the patient. Health Protection Agency, Chilton

Chapter 4
Monitoring of Medical Radiation Exposure for Individuals

Donald P. Frush, Richard Morin, and Madan M. Rehani

Abstract Radiation exposure and the potential risks from medical imaging continue to be highly visible issues, especially for the medical imaging community. There has been progress with dose management, particularly dose reduction, due to both technical and application advancements. There are various existing actions focusing on the collection and analysis of population exposure data for radiological procedures. However, there is an increasing mandate, particularly in the United States but globally as well, for dose recording, monitoring and reporting of medical radiation exposure for individual patients. There are many benefits to radiation monitoring and the strategies for implementation will vary, partly depending on resources. This action supports tenants of medical professionalism, including transparency and the primacy of patients. While challenges remain, the medical imaging community must take ownership of the innovative actions related to the recording and monitoring of cumulative exposures from medical radiation procedures. A successful outcome requires vision, leadership, and collaborations from the different stakeholders in medical imaging to overcome challenges and ensure implementation of actions to further improve quality in radiology and radiation safety.

Keywords Dose monitoring • Dose recording • Dose reporting • Ionizing radiation • Medical imaging procedures • Radiation dose

D.P. Frush (✉)
Pediatric Radiology, 1905 Children's Health Centre, Box 3808 DUMC, Durham, NC 27710, USA
e-mail: donald.frush@duke.edu

R. Morin
Department of Radiology, Mayo Clinic Jacksonville, Jacksonville, FL 32224, USA

M.M. Rehani
Radiation Protection of Patients Unit, RSM, NSRW, International Atomic Energy Agency, PO Box 100, A 1400 Vienna, Austria

L. Lau and K.-H. Ng (eds.), *Radiological Safety and Quality: Paradigms in Leadership and Innovation*, DOI 10.1007/978-94-007-7256-4_4,
© Springer Science+Business Media Dordrecht 2014

1   Introduction/Background

The dangers of radiation from medical imaging were recognized shortly after the discovery of x-ray [21]. Despite this, the management of radiation exposure and the associated radiation risk is still one of the most significant issues currently confronting radiology, and is increasingly recognized in the broader medical community. This is exemplified by the scrutiny of improper use, including overutilization of CT and the resultant risks of cancer related to exposure, in particular. This has promoted and has been subsequently paralleled by a profound investment in numerous technical advancements for radiation dose reduction such as beam filtration and digital technology (i.e. no lost films and better remote transmittal of studies obviating repeat procedures) for radiography, pulsed technology in fluoroscopy, and iterative reconstruction algorithms for CT, as examples.

These advancements in both technology, as well as improvement in performance and application, have been supplemented by educational endeavors such as the Image Gently Campaign, from the Alliance for Radiation and Safety in Pediatric Imaging [16, 17], and the Image Wisely initiative [8]. Many of these efforts have been ongoing for more than a decade. However, there has only relatively recently been an evolving and global mandate for radiation dose recording, monitoring, and reporting in medical imaging [36, 41]. A monitoring program has tangible benefits for both the patient as well as a facility. Awareness of its dose profiles for the different procedures using ionizing radiation, can be compared with those from other similar local or international facilities. The successful implementation of this innovative action has some significant challenges but these difficulties should not deflect effort since the call to record and monitor radiation dose or dose indices for medical imaging is increasingly prevalent. There is a fundamental and sufficient justification to assure the safety and welfare of the patients we, as a medical imaging community, care for. Embedded in this mandate of accountability and responsibility is that this process of medical radiation recording, monitoring, and potentially reporting is one for which we must also actively take ownership. This evolution of radiation protection of patients over the last decade or so follows the model of 4 As (Table 4.1): awareness, accountability, ability and action [13, 15]. This model is entirely applicable to the issues of radiation dose recording, monitoring, and reporting. . .but we are just at the initial stages of awareness.

1.1   Driving Forces for the Monitoring of Medical Radiation Exposure for Individuals

How did we get to this sense of urgency? Two fundamental principles of radiation protection of personnel are justification and optimization [25]. These principles have come into sharper focus for medical radiation recently for several reasons. First, the utilization of medical imaging procedures employing ionizing radiation

Table 4.1 The model of 4 As of how leaders can implement the adoption of new technologies

Awareness: The leaders must be aware of the performance gaps before anything can be achieved. Awareness requires that adequate information be provided to leaders at all levels. The facility requires that structures and systems be in place to provide a continuous flow of information to leaders from multiple sources about the causes and possible solutions for the risks, hazards, and performance gaps that contribute to patient safety. Leaders at any level need a clear under-standing of the performance shortfalls in order to act

Accountability: Accountability of the leaders to closing performance gaps is a key success factor – someone needs to "own" the changes that must be made to processes, systems, and expectations of staff. Due to the slow but critical transformation from the legacy "command and control" accountability structures to "team-based" approaches, few leaders are directly accountable for specific and measurable patient safety performance gaps. High-performing organizations have seen the light and have teamed clinical with administrative leaders to work towards joint goals

Ability: A team or unit may be aware of the gaps and may be accountable for those gaps. However, if they are not able to make the necessary changes, change will not occur. Worse, "learned helplessness" can set in, galvanizing the troops to the status quo. The dimension of ability may be measured as the capacity for change. It requires investment in knowledge and skills, compensated staff time, and the "dark green dollars" of line item budget allocations. Prelimi-nary results from the Texas Medical Institute of Technology Research Test Bed, which is studying the impact of patient safety practices and solutions in hundreds of community hospitals, indicate that few hospitals have made adequate investments in patient safety

Action: Finally, to accelerate the adoption of innovations, organizations need to take explicit actions toward line-of-sight targets that close the performance gaps, that can be easily measured, and that can generate early wins. Multiple objectives that can be achieved by direct actions must be integrated into an improvement program, to ensure improvement that can be easily scored

Adapted from [13, 15]

has increased. In the USA, there has been a nearly sevenfold increase in the amount of radiation from medical imaging in the past 30 years, in large part due to the use of nuclear imaging and CT [32]. This increase use in CT is not limited to adults. In one review over a 3-year period of more than 350,000 children enrolled in healthcare networks, 42.5 % had an imaging procedure that used ionizing radiation. Nearly 8 % of all children had at least one CT during this period; given the fact that some children had more than one CT, CT accounted for 11.9 % of all procedures [9]. Currently it is estimated that 85 million CT procedures are performed per year in the USA [23].

In addition, the media have continued to draw public attention to radiation exposure from both justifiable and improper uses of medical imaging, including the risks of cancer and other biological effects such as alopecia from misuse of CT perfusion [50], erythema from equipment and personnel failures in CT [48], brain cancer from dental radiography [49], and risk from inappropriate collimation from neonatal radiography [51]. Moreover, the medical literature has highlighted the potential connections between low-level radiation exposures, such as seen with medical imaging and cancer for CT in several opinion pieces and editorials [7, 26, 39], as well as in recent scientific investigations. Examples include the association of a few chest radiographs and leukemia [5], CT and cancer [6, 44], and most recently the first direct association between CT performed in childhood with leukemia and brain cancer [37]. While debatable and often based on cascading assumptions and dose estimates (including cumulative dose) and risk, these medical publications continue to prompt, and often provoke, the medical imaging community in particular to justify the practice of medical imaging.

This issue of radiation protection and increasing use of radiation in medicine is not unique to the United States. Currently, there are nearly four billion worldwide medical imaging procedures that use ionizing radiation per year [53], equating to more than one exposure per person every other year, globally. Both the International Action Plan on Radiation Protection of Patients of the International Atomic Energy Agency (IAEA) [43] and the Global Initiative on Radiation Safety in Health Care Settings of the World Health Organization (WHO) have recognized the potential radiation risks from the growing medical use and the need for collective actions from the stakeholders [41, 55].

The ubiquity of this type of medical imaging, the increasing connection between low level radiation and cancer risk, and the promulgation of these observations in the media, confront public expectations on the responsibility of healthcare providers, and explain the increasing demand for a more comprehensive blueprint for dose management beyond day-to-day procedure strategies: a blueprint of the recording, monitoring and potentially reporting of medical radiation dose. This blueprint is based on the principles of justification and optimization. Much of the efforts to date have been focused on these two aspects [12, 19]. Certainly there are challenges to the first, outlined with possible solutions recently in a well-summarized article from the American Board of Radiology Foundation Summit [19]. This accountability for utilization is in fact a shared responsibility for all qualified practitioners involved in requesting and providing medical imaging. The second component of radiation protection, consisting of optimization, leveraged through technical advances as well as protocol improvements in applications, is clearly the responsibility of the medical imaging community. This involves the medical imaging technologists, medical physicists, radiologists and other imaging experts. Many of these efforts were noted in a recent article by Hricak et al., discussed in the context of managing radiation exposure for CT [22].

2   Existing Cumulative Radiation Dose Reporting

The medical imaging community has only recently recognized (or been called to task) [36] that there is a responsibility for a more appropriate use of medical radiation exposure and the recording, monitoring and potentially reporting of dose for radiation procedures. There are several examples that highlight this need. First, in one investigation of a patient population of 31,462 seen over 22 years in a university setting, there were 190,712 CT procedures; 15 % of patients received an estimated effective dose of more than 100 mSv with 4 % receiving between 250 and 1,375 mSv [45]. In another retrospective study, in five USA healthcare markets, nearly 69 % underwent an imaging procedure that utilized ionizing radiation in the 3 year study period; 20.5 enrollees per 1,000 (0.2 %) in a pool of 655,613 had estimated cumulative effective doses of > 20 mSv per year [11]. Additionally, cumulative estimated doses from cranial radiation procedures during hospitalization for subarachnoid hemorrhage ranged from 2.4 to 36.1 Gy with a mean of 12.8 Gy [31]. The cumulative doses have also been reported in pediatric cardiac patients [2], patients in emergency settings [18], individuals with chronic illnesses [46] and pediatric oncology patients (with an estimated effective dose of up to 642 mSv) [1].

These reports were unusual a decade or more ago and signal a growing, although controversial [10], recognition across the medical disciplines of the need to address cumulative radiation exposure in medical imaging. To date, the largest inter-organizational scientific discussion was the Beebe Symposium sponsored by the National Academies of Sciences "Tracking Radiation Exposure from Medical Diagnostic Procedures", in December of 2011 which specifically addressed dose estimation, recording, monitoring and reporting [36]. Given these investigations and efforts, the focus has diverged from the justification and optimization elements of radiation protection, to accountability for how medical radiation exposure from these procedures is recorded, monitored, and reported.

The sectors that are actively involved include the regulatory, policy, and governmental agencies; and the international organizations and professional societies representing imaging experts, especially radiologists, technologists, health physicists, medical physicists, epidemiologists, radiation biologists, and patient advocacy groups [36]. For example, based on over-exposures from perfusion CT procedures, the California state legislature passed a bill SB 1237 mandating dose index recording on all patients for CT and nuclear imaging, which began July 2012 [52]. In September 2011, The Joint Commission published a Sentinel Alert, outlining 21 actions for radiation protection. These mandates include the following: Item 7: Institute a process for the review of all dosing protocols either annually or every 2 years to ensure that protocols adhere to the latest evidence. Item 8: Investigate patterns outside the range of appropriate doses. Track radiation doses from procedures repeated due to insufficient image quality or lack of availability of previous studies to identify the causes. Address and resolve these problems through education and other measures. Item 9: Record the dosage or exposure as part of the study's summary report of findings. Item 19: The Joint Commission endorses the creation of a national registry to track radiation doses as the start of a process [47].

The National Quality Forum approved measures for CT radiation dose recording as part of the safety measures [34, 35]. In February 2010, the Food and Drug Administration supported dose recording and reporting during a symposium on radiation protection [54]. Industry has responded with new programs offering the tracking of radiation exposures for individual patients in modalities using ionizing radiation [14, 38]. The National Institute of Health in 2009 called for the cumulative dose recording and tracking in all patients in their medical system [33]. Internationally, the International Atomic Energy Agency (IAEA) through the Radiation Protection of Patients Program (RPoP) has been developing a template for radiation dose tracking and monitoring called the SmartCard/ SmartRadTrack Program since 2008 [41]. A recent statement released by the IAEA summarized these efforts [24].

Dose recording could be accomplished through dose registries. In the United States, the American College of Radiology (ACR) has for more than a year, been accepting national and some international CT dose data into a dose index registry (DIR). The registry has collected the dose data for more than 1.4 million CT procedures from over 520 registered (252 active) sites to August 2012 (personal communication, Laura Coombs, PhD, ACR) [30]. Moreover, there are applications available for downloading on personal devices that enable dose recording for the public [4]. Finally, there are hardcopy forms or cards for radiation dose tracking available through the Food and Drug Administration and the Alliance for Radiation Safety in Pediatric Imaging, which further supports the momentum and account- ability for tracking. However, these paper records and applications for personal devices place this responsibility on the patients or parents. This is likely due to the fact that there is currently no more feasible way across the healthcare systems to track this information in the United States. However, this should not be wholly the patients' or parents' responsibility. Together, whether promoting the principle, providing a template or actually engaging in medical radiation dose monitoring program development such as through the ACR, it is irrefutable that there is increasing longitudinal accountability for radiation dose recording as part of the facility's quality system as well as to individual patients.

2.1   International Efforts Supporting Medical Radiation Exposure Monitoring for Individuals

A global survey was conducted by the IAEA to obtain information on the existence of a patient exposure tracking program; plans for future programs; perceived needs, goals of future programs; which procedures will be tracked; whether procedure tracking alone or dose tracking is planned; and which dose quantities will be tracked [42]. The responses from 76 countries, including all of the six most populous countries and 16 of the 20 most populous, showed that although no country has yet implemented a patient exposure tracking program at a national level, there is increased interest in this issue (46). Eight countries (11 %) indicated that such a program is actively being planned and three (4 %) stated that they have a program for tracking procedures only, but not for dose. Twenty-two (29 %) feel that such a program will be "Extremely useful", forty-six (60 %) "Very useful" and eight (11 %) "Moderately useful", with no respondents stating "Mildly useful" or "Not useful". Of note, 99 % of the countries indicated an interest in developing and promoting such a program.

It is clear that virtually all countries have an interest in and some have plans to achieve dose tracking in the near future. The current situation makes this possible within a few dozen hospitals covered by PACS systems and that there are nascent efforts to extend coverage nationwide. These findings can serve as a benchmark and stimulus for future medical radiation exposure and dose recording efforts, which offer the potential to create a data-rich environment enabling better global implementation of the radiation protection principles of justification, optimization, and dose limits.

3   Radiation Dose Monitoring Program

There are several clarifications regarding dose recording and monitoring that are necessary. First, "dose" used in the clinical arenas is often misconstrued as patient dose. Almost without exception, dose in clinical practice is an index and does not represent individual patient dose. Those radiation exposure indices such as the exposure index (EI) in digital radiography, dose area product (DAP) in fluoroscopy, or dose indices such as CT dose index (CTDI) in CT do not represent individual doses and hence individual risk. In addition, "effective dose" is a frequently used phrase in medical imaging dialogue. This concept of effective dose, while commonly encountered, has significant limitations when applied to medical imaging [27, 28]. In addition, cumulative doses do not necessarily translate to cumulative risk [10].

Related to the monitoring components of a program, there are two general aims: one is directed to the individual patient such as the cumulative dose index or dose history of all procedures; the other is a procedural dose profile that is decoupled from an individual patient, such as through a registry ranging from an individual facility archive to a national or multinational registry. The benefits, challenges and potential solutions will vary depending on whether the cumulative dose program is directed at the patients, the facilities, or both.

3.1   Benefits for Dose Recording and Monitoring

Recording and monitoring dose is more than a simple compliance measure; there are definite benefits that can be achieved. The benefits are outlined in Table 4.2. In addition, Miller recently categorized the major benefits of radiation monitoring based on justification, optimization, risk assessment and research and noted what information would or would not need to be included to achieve these goals. The information was divided into personal health information, facility identifiers, and dose data. For example, for the justification of a CT procedure, personal health information i.e. past medical imaging history would be necessary but not the dose data. Alternatively, for optimization, dose indices are required, but personal health information is relatively unnecessary [36]. Notably, from the patient's standpoint, a cumulative dose history, especially if accessible at the point of care, could be used in the decision process.

Table 4.2 The potential benefits from patient radiation exposure monitoring

1. Benefits to the patients
(a) Optimize radiation exposure
(b) Accountability for radiation protection by healthcare providers
(c) Provide an opportunity for informed discussions between patients and healthcare providers
2. Benefits to the healthcare providers referring patients for medical imaging or intervention
(a) Potential benefits from decision support
(b) Improve and justify resource utilization
3. Benefits to the healthcare providers delivering medical imaging or intervention
(a) Potential benefits from decision support
(b) Improve and justify resource utilization – includes assessing the patterns for the requests
(c) Realistic comparison of the facility's exposures with national diagnostic reference levels
(d) Protocol optimization and quality improvement
4. Benefits to the policymakers
(a) Quantitative tools to protect the public in health and safety
(b) Improve quantitative approach to radiation safety policymaking
(c) Manage medical imaging utilization
5. Benefits to the regulators
(a) Encourage the facilities to implement the diagnostic reference level process
(b) Improve data to assist the facilities to conduct reliable self-audits
6. Benefits to the researchers
(a) Provide radiation safety data
(b) Incorporate patient-specific radiation metrics into research studies
(c) Provide a quantitative basis for the development of best practices, guidelines, and appropriateness criteria
7. Benefits to the industry
(a) Promotes partnership with other stakeholders in establishing radiation exposure monitoring technology

Modified with permission from [42]

Whether or not the past dose history determines future decisions, it is important that this information be made available, if only to factor into the discussions between the patient and the healthcare provider about any issues that may arise regarding exposures and risks. To simply say: "I don't know" to queries and concerns about cumulative dose information is not supportable and contrary to good patient-centered care. The cumulative dose history assist with decisions on procedure selection or protocol modification, e.g. for follow-up procedures in patients who have had multiple prior procedures. Access to archive of cumulative dose data enables an individually tailored procedure.

For quality improvement, protocol optimization may follow by benchmarking a facility's dose profile for different procedures to the diagnostic reference levels from a range of facilities, regions or countries. Outliers of either excessively high or low radiation doses could be reviewed. The data would serve regulatory functions such as in hospital accreditation by documenting a dose monitoring program exists, such as through The Joint Commission, or in equipment accreditation by establishing more robust and dynamic dose estimations rather than by a single threshold estimation such as currently exist through the ACR for CT accreditation.

Pooled data, whether national or otherwise, could be reviewed for radiography, fluoroscopy, or CT and be used as benchmarks for an individual facility.

While there are benefits, there are significant obstacles and other challenges to developing, implementing, monitoring, and modifying a medical radiation exposure program. Some of these were recently addressed for nuclear imaging [29]. In listing and discussing these challenges, it is perhaps more efficient and effective to include a discussion of the existing or potential opportunities to address these difficulties.

3.2   Medical Radiation Exposure Monitoring Options

There is no universally agreed measure of radiation dose that is common to all imaging modalities, even though dose metrics for each modality are recommended in the international code of practice for dosimetry. The possibilities range from simply listing the procedure, such as a chest radiograph or brain CT, to the evolving methods for providing more accurate dose indices. For example, in the case for CT in adults and children the use of size-specific organ dose estimations (SSDE) is becoming more pervasive [3]. Moreover, we must realize that convention for dose estimations may be modified or replaced. For example, the CTDI has been most often used as an exposure index in CT. The recent AAPM offers more accurate size-based dose estimations which are reported to be within 10–20 % of the dose for the cross sectional dimensions representing the patient [3]. The International Atomic Energy Agency suggests the approaches and provides the templates for dose recording in its dose monitoring program [40]. Given the lack of standardization, it is not unexpected that various agencies and organizations may suggest or request recordings, with arguable relevance. For example, the Californian requirement for CT dose recording is based on CTDI and not SSDE, which has been advocated by AAPM. However, part of this variation may be due to differing resources for dose measure, archive and retrieval. A facility may only have the ability to track procedures, while others may be able to archive and track more detailed information through DICOM for interventional procedures or CT, for example. A decision of what to measure should be based on the available resources, and targeting those measures which could be adapted to evolving dose indices or measurements of these indices.

Another issue to consider is whether a dose index or risk index is more appropriate for reporting and communication. In support of using risk as the measure is a simple realization that dose is a quantity and only broadly understood when converted to risk. There is no "dose index-benefit" equation or ratio, as is recognized for risk-benefit. This point was recently discussed at the Beebe symposium in 2011 [36]. As with dose estimations, there are challenges with risk estimations. Risk measures such as lifetime attributable risk (LAR) from the Biological Effects of Ionizing Radiation (BEIR) VII data are estimations with large uncertainties [20]. In addition a cancer risk statement attached to a procedure report could have significant adverse impact on the overall value of CT to those patients or healthcare providers who are less informed of the benefits, potentially resulting in a patient's refusal to future procedures.

4 Challenges and Potential Solutions for Medical Radiation Dose Monitoring

How should the dose be recorded: in PACS, in the hospital information system or in the patient's report? Does the existence of patient portals with an access to healthcare information impact the decision on whether or not dose information is recorded? What does a 14.6 mGy $CTDI_{vol}$ mean to a 34-year-old female bank executive or a 60-year-old self-employed shop owner? Simply reporting dose in a report without explaining what it means arguably falls short of transparency, accountability and the primacy of patient care; is confusing and potentially disconcerting. However, the dose measures should be archived. The PACS option seems reasonable as an archive. Others will need careful consideration when developing dose recording programs.

A dose recording program must be supported by providing education to the healthcare practitioners. The practitioners are aware of the reasons for the procedures and the significance of the information obtained. These practitioners include family practitioners and pediatricians, who often discuss with their patients about these indications and use of procedures, as well as their potential risks, including the effects of ionizing radiation. The strategies and content for the communication of the potential benefits, dose, and radiation risks (especially if they are uncertain) with the patients are not well defined, especially with differing patient expectations and levels of understanding [36]. This focuses on the information available at the point of care, i.e. the provision of dose information and possible risks would assist with the joint determination of the appropriateness of the proposed procedure.

How will the dose information be accessed or used for patient care, patient choice and quality improvement? Before programs are implemented, a system for how data will be used must be developed including considerations as what data will be reviewed, how often, who is responsible, when changes should be instituted, and how to quantify the impact of these changes, among others. This is no small task even in sophisticated healthcare enterprises with well-developed teams of experts including information technology personnel and medical physicists. Simply recording data for the sake of recording is of little value, and could conceivably be harmful if in fact some external audit were to uncover values outside the designated ranges.

Another challenge is to decide what dose is too high or too low. The answers to these questions are far from complete. Suffice it to say that effort should be directed to support initiatives that provide reasonable and customary (even considering perhaps regional variations) dose across all modalities. In the United States, this is far from standard. The efforts through the ACR Dose Index Registry are a significant advancement and together with the data from other nations or regions could provide a clearer picture of what dose ranges may be appropriate. The data for digital radiography and fluoroscopy lag even behind those for CT in the United States.

Who is accountable for the doses delivered – the referrer or the provider of medical imaging? If an individual had undergone 30 CT procedures for an eventual diagnosis of a factitious disorder and was therefore felt to be at an inappropriate risk for cancer, who should assume this responsibility? Why aren't the benefits of CT in individuals tracked? Much of the current scrutiny appears to overlook the important balance between dose risks and benefits. We are simply being asked to measure only one aspect of CT performance.

If dose recording is embraced, this must apply to all those in the medical community who use the modality. For CT this could include cardiologists, urologists, and orthopedists. For fluoroscopy, cardiologists, gastroenterologists, orthopedics and urologists are among those using this modality. Dose recording must be inclusive to ensure a better and more comprehensive understanding of dose, risk, and strategies to manage dose by all providers. By the same token, the development of dose recording programs must be multidisciplinary and include all the key stakeholders such as administrators, information technologists, medical physicists, medical imaging technologists, radiologists and other imaging specialists, epidemiologists, radiation biologists, and risk managers.

Will dose information be used in decision-making? The opinions differ, ranging from those who believe the information specific to a particular patient and prior history makes no difference to those who consider such data essential and most useful in decision-making for medical imaging [18]. Whether or not dose information should be available in decision support, a discussion may be necessary and this information would be helpful.

What if dose index measures change due to technical advances? For example, iterative reconstruction algorithms for CT will likely lead to significant lower dose while retaining similar diagnostic quality. These technologies would unlikely be included in the registries based on procedures using older technology. The inclusion of procedures using dose reduction technologies will lower the dose ranges compared to the standard non-iterative reconstruction procedures. These newer dose indices may also appear significantly lower than other pooled data and on the surface be unacceptably low. Dose recording and monitoring programs either for individual or facility should recognize these developments are inevitable and be prepared for their impact on the objectives including the established reference levels. Finally, for patient dose information, data security and patient confidentially remain of utmost importance and this information has to be respected and protected as with any other healthcare information.

The specific recommendations from the Beebe Symposium for medical radiation dose monitoring were:

- Continue to track and monitor the overall trends and patterns of use of medical imaging;
- Continue with the ACR dose index registry efforts and expand them to include additional modalities, e.g. nuclear medicine, computed radiography and digital radiography, interventional radiology and other sites, particularly outpatient facilities;

- Within each facility, routinely report dose metrics performance with benchmarks;
- Create or use existing committees within and outside facilities to ensure the medical imaging protocols are being followed; create routine reports for this purpose for the technologists and radiologists;
- Work with the industry and information technology vendors to incorporate dose metrics directly into medical records; and ensure that dose metric information is attached to the images;
- Encourage the undertaking of national trials that quantify the benefits of medical imaging procedures;
- Implement decision support systems at all stages of patient care to optimize the use of procedures and to ensure only appropriate procedures are performed; and
- Implement and maintain comprehensive facility-wide safety programs and provide educational tools to promote awareness of radiation exposure, radiation protection and radiation safety.

5 Conclusion

Medical imaging procedures utilizing ionizing radiation have increased and thus radiation exposure is increasing. This is evident across healthcare and has garnered a widespread recognition outside the medical community, including governmental agencies, regulatory bodies, media, public, and patients. The medical imaging providers have not been as accountable for dose recording and monitoring when compared to other actions in radiation protection, especially in technical advances, improved performance and new applications. There is a range of strategies to archive dose and this information will have potential benefits to both patients and medical imaging facilities. However, there are significant albeit not insurmountable challenges. Finally, the responsibility for the development and implementation of dose recording programs must rest with the medical imaging community.

References

1. Ahmed BA, Connolly BL, Shroff P et al (2010) Cumulative effective doses from radiologic procedures for pediatric oncology patients. Pediatrics 126(4):851–858
2. Ait-Ali L, Andreassi MG, Foffa I et al (2010) Cumulative patient effective dose and acute radiation-induced chromosomal DNA damage in children with congenital heart disease. Heart 96(4):269–274
3. American Association of Physicists in Medicine (2011) AAPM Report No. 204. Size-specific dose estimates (SSDE) in pediatric and adult body CT examinations. Available at: http://www.aapm.org/pubs/reports/RPT_204.pdf. Accessed 18 July 2012
4. Baerlocher MO, Talanow R, Baerlocher AF (2010) Radiation passport: an iPhone and iPod touch application to track radiation dose and estimate associated cancer risks. J Am Coll Radiol 7(4):277–280

5. Bartley K, Metayer C, Selvin S et al (2010) Diagnostic x-rays and risk of childhood leukaemia. Int J Epidemiol 39(6):1628–1637
6. Berrington de Gonzalez A, Mahesh M, Kim KP et al (2009) Projected cancer risks from computed tomographic scans performed in the United States in 2007. Arch Intern Med 169(22):2071–2077
7. Brenner DJ, Hall EJ (2007) Computed tomography: an increasing source of radiation exposure. N Engl J Med 357(22):2277–2284
8. Brink JA, Amis ES (2010) Image Wisely: a campaign to increase awareness about adult radiation protection. Radiology 257(3):601–602
9. Dorfman AL, Fazel R, Einstein AJ et al (2011) Use of medical imaging procedures with ionizing radiation in children: a population-based study. Arch Pediatr Adolesc Med 165(5):458–464
10. Durand DJ, Dixon RL, Morin RL (2012) Utilization strategies for cumulative dose estimates: a review and rational approach. J Am Coll Radiol 9(7):480–485
11. Fazel R, Krumholz HM, Wang Y et al (2009) Exposure to low-dose ionizing radiation from medical imaging procedures. N Eng J Med 361(9):849–857
12. Frush DP (2011) Justification and optimization of CT in children: how are we performing? Pediatr Radiol 41(Suppl 2):467–471
13. Frush DP, Denham CR, Goske KJ et al (2012) Radiation protection and dose monitoring in medical imaging: a journey from awareness, through accountability, ability and action...but where will we arrive? J Patient Saf (in press)
14. GE Healthcare (2012) Track, report, and monitor radiation dose enabling optimization of scanning parameters and helping providers improve patient care. Available at: http://www.gehealthcare.com/dose/pdfs/DoseWatch-Sell-Sheet.pdf. Accessed 3 July 2012
15. George WW, Denham CR, Burgess LH et al (2010) Leading in crisis: lessons for safety leaders. J Patient Saf 6(1):24–30
16. Goske MJ, Applegate KE, Frush DP et al (2008) The image gently campaign: working together to change practice. Am J Roentgenol 190(2):273–274
17. Goske MJ, Applegate KE, Frush DP et al (2008) Image Gently: a national education and communication campaign in radiology using the science of social marketing. J Am Coll Radiol 5(12):1200–1205
18. Griffey RT, Sodickson A (2009) Cumulative radiation exposure and cancer risk estimates in emergency department patients undergoing repeat or multiple CT. Am J Roentgenol 192(4):887–892
19. Hendee WR, Becker GJ, Borgstede JP et al (2010) Addressing overutilization in medical imaging. Radiology 257(1):240–245
20. Hendee WR, O'Conner MK (2012) Radiation risks of medical imaging: separating fact from fancy. Radiology 264:312–321
21. Hrabak M, Padovan RS, Kralik M et al (2008) Scenes from the past: Nikola Tesla and the discovery of x-rays. Radiographics 28(4):1189–1192
22. Hricak H, Brenner DJ, Adelstein SJ et al (2011) Managing radiation use in medical imaging: a multifaceted challenge. Radiology 258(3):889–905
23. IMV (2012) Latest IMV CT survey shows hospitals seek to improve productivity to manage increased outpatient and emergency CT procedure volume. Available at: http://www.imvinfo.com/user/documents/content_documents/def_dis/2012_06_05_04_54_16_458_IMV_2012_CT_Press_Release.pdf. Accessed 3 July 2012
24. International Atomic Energy Agency (2012) Joint position statement for the IAEA Smart Card/SmartRadTrack project. Available at: https://rpop.iaea.org/RPOP/RPoP/Content/Documents/Whitepapers/iaea-smart-card-position-statement.pdf. Accessed 3 July 2012
25. International Commission on Radiological Protection (2008) ICRP Publication 105. Radiological protection in medicine. Ann ICRP 37(6):1–63. Elsevier, Oxford
26. Lauer MS (2009) Elements of danger – the case of medical imaging. N Engl J Med 361(9):841–843

27. Martin CJ (2007) Effective dose: how should it be applied to medical exposures? Br J Radiol 80(956):639–647
28. McCollough CH, Christner JA, Kofler JM (2010) How effective is effective dose as a predictor of radiation risk? Am J Roentgenol 194(4):890–896
29. Mercuri M, Rehani MM, Einstein AJ (2012) Tracking patient radiation exposure: challenges to integrating nuclear medicine with other modalities. J Nucl Cardiol 19(5):895–900
30. Morin RL, Coombs LP, Chatfield MB (2011) ACR dose index registry. J Am Coll Radiol 8(4):288–291
31. Moskowitz SI, Davros WJ, Kelly ME et al (2010) Cumulative radiation dose during hospitalization for aneurysmal subarachnoid hemorrhage. Am J Neuroradiol 31(8):1377–1382
32. National Council on Radiation Protection and Measurement (2009) NCRP Report No. 160: ionizing radiation exposure of the population of the United States. National Council on Radiation Protection and Measurement, Bethesda
33. National Institute of Health (2009) New diagnostic imaging devices at the NIH Clinical Center to automatically record radiation exposure. Available at http://www.nih.gov/news/health/aug2009/cc-17.htm. Accessed 6 July 2012
34. National Quality Forum (2010) Safe practice 34: pediatric imaging. In: Safe practices for better healthcare – 2010 update: a consensus report. National Quality Forum, Washington, DC
35. National Quality Forum (2010) Safe practices for better healthcare – 2010 update: a consensus report. National Quality Forum, Washington, DC
36. National Research Council (2012) Tracking radiation exposure from medical diagnostic procedures: workshop report. Committee on Tracking Radiation Doses from Medical Diagnostic Procedures; Nuclear and Radiation Studies Board; Division on Earth and Life Studies. National Academies Press, Washington, DC
37. Pearce MS, Salotti JA, Little MP et al (2012) Radiation exposure from CT scans in childhood and subsequent risk of leukemia and brain tumors: a retrospective cohort study. Lancet 380(9840):499–505
38. Radimetrics (2012) Available at: http://www.radimetrics.com/default/. Accessed 3 July 2012
39. Redberg RF (2009) Cancer risks and radiation exposure from computed tomographic scans: how can we be sure that the benefits outweigh the risks? Arch Intern Med 169(22):2049–2050
40. Rehani MM, Berris T (2013) Templates and existing elements and models for implementation of patient exposure tracking. Radiat Prot Dosimetry [Epub ahead of print]
41. Rehani MM, Frush DP (2011) Patient exposure tracking: the IAEA smart card project. Radiat Prot Dosimetry 147(1–2):314–316
42. Rehani MM, Frush DP, Berris et al (2012) Patient radiation exposure tracking: worldwide programs and needs – results from the first IAEA survey. Eur J Radiol 81(10):e968–976
43. Rehani MM, Holmberg O, Ortiz-Lopez P et al (2011) International action plan on the radiation protection of patients. Radiat Prot Dosimetry 147(1–2):38–42
44. Smith-Bindman R, Lipson J, Marcus R et al (2009) Radiation dose associated with common computed tomography examinations and the associated lifetime attributable risk of cancer. Arch Intern Med 169(22):2078–2086
45. Sodickson A, Baeyens PF, Andriole KP et al (2009) Recurrent CT, cumulative radiation exposure, and associated radiation-induced cancer risks from CT of adults. Radiology 251(1):175–184
46. Stein EG, Haramati LB, Bellin E et al (2010) Radiation exposure from medical imaging in patients with chronic and recurrent conditions. J Am Coll Radiol 7(5):351–359
47. The Joint Commission (2011) Sentinel event alert, Issue 47, Aug 24. Available at: http://www.jointcommission.org/assets/1/18/sea_471.pdf. Accessed 26 June 2012
48. The New York Times (2009) Radiation overdoses point up dangers of CT scans. Available at: http://www.nytimes.com/2009/10/16/us/16radiation.html. Accessed 3 July 2012
49. The New York Times (2010) Radiation worries for children in dentist's chairs. Available at: http://www.nytimes.com/2010/11/23/us/23scan.html?_r=1&pagewanted=allNY. Accessed 3 July 2012

50. The New York Times (2011) West Virginia hospital over radiated brain scan patients, records show. Available at: http://www.nytimes.com/2011/03/06/health/06radiation.html?_r=3. Accessed 3 July 2012

51. The New York Times (2011) X-rays and unshielded infants. Available at: http://www.nytimes.com/2011/02/28/health/28radiation.html?pagewanted=all. Accessed 3 July 2012

52. The State of California SB 1237 (2010) Available at: http://www.leginfo.ca.gov/pub/09-10/bill/sen/sb_1201-1250/sb_1237_bill_20100929_chaptered.html. Accessed 26 June 2012

53. United Nations Scientific Committee on the Effects of Atomic Radiation (2010) Sources and effects of ionizing radiation. UNSCEAR 2008 report to the general assembly with scientific annexes. United Nations, New York

54. US Food and Drug Administration (2010) Initiative to reduce unnecessary radiation exposure from medical imaging. Available at: http://www.fda.gov/Radiation-EmittingProducts/RadiationSafety/RadiationDoseReduction/ucm199994.htm. Accessed 7 July 2012

55. World Health Organization (2008) WHO global initiative on radiation safety in health care settings technical meeting report. WHO, Geneva

Chapter 5
UNSCEAR: Analysing Population Survey Data for Collaborative Development of Evidence-Based Radiation Protection Policies

F. Shannoun and M. Crick

Abstract The United Nations Scientific Committee on the Effects of Atomic Radiation (UNSCEAR) was established in 1955 to systematically collect, evaluate, publish and share data on the global levels and effects of ionizing radiation from natural and artificial sources. Regular surveys have been conducted on the frequencies of medical radiological procedures, levels of exposure, equipment used and staffing levels to determine evolving trends.

Two thirds of diagnostic radiological procedures, over 90 % of all nuclear medicine procedures, and 70 % of radiation therapy treatments were performed in the industrialized countries. This imbalance is also reflected in the availability of X-ray equipment and practitioners. The global average annual per caput effective dose from diagnostic radiological procedures has nearly doubled between 1988 and 2007 from 0.35 to 0.62 mSv.

UNSCEAR's role in collecting and sharing data and collaborating with international bodies such as IAEA, ICRP, ILO and WHO is crucial for the development and implementation of evidence-based radiation protection recommendations, guidance tools and policies worldwide. The main challenge for UNSCEAR besides data quantity and quality is data interpretation. Health risks due to medical radiation exposure include carcinogenesis. However, manifest increase in the incidence of such diseases cannot be reliably attributed to radiation exposures typical of average background levels. In particular, the uncertainties associated with epidemiological studies of individuals who have been exposed to low radiation doses are relatively high. Together with IAEA and WHO, UNSCEAR has developed a strategy to improve data collection and interpretation. By providing leadership, collaborating

F. Shannoun (✉) • M. Crick
United Nations, United Nations Scientific Committee on the Effects of Atomic Radiation
(UNSCEAR), Wagramerstraße 5, Vienna, Austria
e-mail: Ferid.SHANNOUN@unscear.org

L. Lau and K.-H. Ng (eds.), *Radiological Safety and Quality: Paradigms in Leadership and Innovation*, DOI 10.1007/978-94-007-7256-4_5,
© Springer Science+Business Media Dordrecht 2014

with other stakeholders and applying innovative solutions, UNSCEAR plays a pivotal role by ensuring appropriate information on medical radiation exposure and the possible health effects is readily available to other stakeholders thereby contributing to improvement in patient care worldwide.

Keywords Collective effective doses • Diagnostic X-ray • Global medical radiation survey • Medical radiation exposure • Nuclear medicine procedures • Population-based exposure evaluation • Radiation therapy • Radiological procedures • UNSCEAR

1 Introduction

Ionizing radiation used appropriately benefits patients by improving the detection, diagnosis and treatment of disease. Used inappropriately, it can be ineffective, increase the risks of radiation-induced disease, damage health or even threaten life. The community rightfully expects that any radiation exposure arising from medical procedure or treatment is justified and radiation protection measures are optimized. The introduction of new regulations or changes to daily practice aimed at improving patient care can have significant financial and operational implications, especially when resources are limited. Therefore, it is important that these interventions are based on reliable evidence and sound reasons.

In 1955, the United Nations Scientific Committee on the Effects of Atomic Radiation (UNSCEAR) was established by the United Nations General Assembly resolution 913(X) to collect and evaluate information on the levels and effects of ionizing radiation from natural and artificial sources. This Committee has systematically reviewed and evaluated the global and regional levels of medical exposure in order to determine trends, as well as exposures of the public and workers. It has also evaluated from time to time the evidence for radiation-induced health effects from studies of Japanese atomic bombing survivors and other exposed groups and has reviewed advances in the mechanisms of radiation-induced health effects.

Information on the use of radiation in medicine and the associated exposures is obtained through population-based surveys. Such information underpins radiation safety policy in medicine. These surveys are used to identify the trends in exposure and thus serve as an early warning sign of potential safety issues that might require attention, i.e. the procedures requiring further consideration by those concerned with radiation safety by virtue of doses or frequency. They can also be used to identify gaps in treatment capabilities and possible unwarranted dose variations for the same procedure. The surveys provide evidence to underpin advocacy messages and to develop and implement evidence-based radiation protection policies. This chapter describes and discusses these surveys, including the main findings and impact on radiation protection policy, the challenges involved when conducting and analyzing such surveys, the future challenges and some innovative solutions.

2 Medical Exposure Databases

Relatively few countries around the world conduct national surveys of medical radiation utilization and exposure on a regular basis. Germany, Switzerland and the United Kingdom are some examples of the countries, which employ an institutionalized set of arrangements to conduct surveys from time to time [1, 3, 7]. In addition, there are focused academic studies that sample from the overall population or study certain sets of patients and their exposure in some detail.

Member countries of the European Union are required to ensure that doses from medical procedures are determined for the population and for relevant reference groups, as requested under Article 12 of the European Council Directive 97/43/EURATOM [4]. The absence of an internationally accepted protocol for the evaluation of patient exposures from medical radiological procedures and the wide variations in population dose estimates between the European countries, prompted the European Commission (EC) in 2005 to launch the DOSE DATAMED project. The objective of this project was to review the past surveys of population exposures for medical radiological procedures in order to clarify the different methodologies used, to identify the sources of uncertainty, and to develop a harmonized and simplified process for future surveys. This led to the publication in 2008 of "European Guidance on Estimating Population Doses from Medical X-ray Procedures", which provides the basis for the development of a harmonized approach to assess and evaluate medical exposures for patients and the population [6]. This document gives guidance on the classification of X-ray procedures, the estimation of the frequency of procedures, and the use of the concept of effective dose. In 2011, the EC launched a follow-up project, DOSE DATAMED II, to study the doses to the European population from medical exposures [2]. The objective of the DOSE DATAMED II project has been to collect data on the frequencies and doses from X-ray and nuclear medicine procedures in the European Union and to facilitate the implementation of the European guidance [6].

Other relevant international initiatives include the development of the Nuclear Medicine Database (NUMDAB) and the Directory of Radiotherapy Centers (DIRAC) database by the International Atomic Energy Agency (IAEA) and the collection and publication of data on medical devices by the Nuclear Energy Agency (NEA) within the Organization for Economic Co-operation and Development (OECD) countries. DIRAC is a database of global radiotherapy resources and includes data on equipment and devices for teletherapy, brachytherapy, dosimetry, imaging, and quality assurance. The aim of NUMDAB is to gather and update information on the status of nuclear medicine practice worldwide. It includes data on nuclear medicine facilities, radionuclides and radiopharmaceuticals used, and the frequency of diagnostic and therapeutic procedures.

3   UNSCEAR's Medical Exposure Evaluation

UNSCEAR has conducted evaluations of global practice and analyses of trends on
the medical use of ionizing radiation over many years. The latest evaluation
covering global and regional levels and trends in medical exposure was completed
in 2008 [27] and some of its findings are presented here. The evaluation is based on
an analysis of the responses to its Global Survey of Medical Radiation Usage and
Exposures and a critical review of the literature and other information provided by
UN Member States and international organizations such as the IAEA and the World
Health Organization (WHO).

3.1   Objectives

The objectives of the evaluation are to estimate the global frequencies of use of
radiological procedures and levels of radiation exposure, with a breakdown by
radiological procedure, age, sex, healthcare level (HCL), and country; to determine
the evolving trends; and to collect information on equipment and staffing levels.

3.2   Historical Review

As early as its initial report [18], UNSCEAR recognized that exposures from
medical diagnostic and therapeutic procedures were a major component of the
total exposure due to artificial sources of radiation worldwide, a fact that remains
true today. In 1962, UNSCEAR provided information on medical exposure and
presented a comprehensive evaluation of the levels of ionizing radiation exposure
of the population and the possible effects of such exposure [19]. The UNSCEAR
1962 Report tabled for the first time the annual frequency data, i.e. the number of
procedures per 1,000 individuals, from 20 countries. In this report, the emphasis
was placed on gonadal dose and genetically significant dose because hereditary
effects were felt to be very important at that time. By 1972, estimation of bone
marrow dose was added [20]. In 1977, data on the annual frequency of specific
procedures was included [21]. In the UNSCEAR 1982 Report, data on the annual
frequency of specific procedures was published for 16 countries from five
continents and estimation of effective dose equivalent for various procedures was
reported for Japan and Poland [22]. This report was the first employing a survey
and developed in cooperation with WHO, to obtain information on diagnostic
radiology equipment and annual frequency of diagnostic X-ray procedures. This
was also the first UNSCEAR survey to include exposures from computed

Table 5.1 Radiation dosimetry [16]

Absorbed dose

The absorbed dose D, measured in grays (Gy) or milligray (mGy), represents the energy deposited in tissue per unit mass. This can be used for any form of radiation but does not account for the different biological effects of the different types of radiation.

Equivalent dose

The equivalent dose H, measured in sieverts (Sv) or millisieverts (mSv), equals to the absorbed dose D multiplied by the appropriate radiation-weighting factor (w_R) after taking into account of the biological effects of the type of radiation. For example, w_R is 1 for the types of radiation used in radiological diagnosis and nuclear medicine (gamma rays and X-rays).

Effective dose

The effective dose E, is used to quantify radiation exposure in individuals. Measured in sieverts (Sv) or millisieverts (mSv), it is a summation of the equivalent doses H to all tissues T and organs, after an adjustment for their varying radiosensitivity by using the tissue-weighting factor w_T. This is a tissue-weighted dose that takes into account of the radiosensitivity of different tissues and organs. Some tissues such as bone marrow are more radiosensitive and they are allocated a proportionally higher tissue-weighting factor than those like bone that are less radiosensitive.

$$E = \sum w_T H_T$$

An estimation of cancer risk due to medical exposure is complicated because invariably more than one organ is irradiated. The effective dose takes into account the type of radiation and the nature of the irradiated tissue or organ and provides a single dose estimate related to the total radiation risk, no matter how the radiation dose is distributed in the body. Effective dose offers a means to compare cancer risk (stochastic effect) due to medical exposures in radiology and nuclear medicine and to approximately compare the relative risks between procedures, if the representative patients or patient populations are similar with regard to age and sex. However, effective dose is not an appropriate quantity to use in radiation therapy and other high-dose situations such as might lead to deterministic effects.

Collective effective dose

The collective effective dose for a procedure, measured in man sieverts (man-Sv), is obtained by multiplying the mean effective dose E_e for a procedure by the number of procedures N_e. The number N_e may be deduced from the annual frequency (number of procedures per 1,000 population) and the estimated population size. The collective effective dose from all procedures S for the entire population is a summation of the effective dose from all procedures.

$$S = \sum E_e N_e$$

It is possible to use effective dose and even collective dose to compare the exposures between medical diagnostic procedures for the same or similar populations, but it would require additional consideration and significant corrections when comparing with other populations [26].

---

tomography (CT) [22]. In the UNSCEAR 1988 Report, the material on medical exposures was greatly expanded and an attempt was made to estimate global exposure in addition to country-specific data [23]. This was made possible with data received from the more populous countries such as China and countries from Latin America. This report also presented the first estimate of global dose to patients from medical diagnostic procedures. Since the UNSCEAR 2000 Report, the collective effective dose (in man Sievert) and the per caput effective dose (in mSv) were used to express population dose estimates (Table 5.1) [26].

3.3 Healthcare Level (HCL) Model

UNSCEAR surveys included data on medical and dental X-ray, nuclear medicine and radiation therapy procedures, most of the data coming from industrialized countries [22, 26, 27]. These surveys revealed an absence of data on the frequency and type of radiological procedures in more than half of the world's countries and only fragments of data from another quarter [22, 26, 27]. A method was therefore sought to extrapolate the existing data and to estimate the availability and frequency of medical use of radiation in countries where data were unavailable. Based on a good correlation between the "physician to population ratio" and the annual frequency of diagnostic radiological procedures [13], an analytical model was developed to enable the estimation of medical radiation exposure on a worldwide basis by grouping countries with similar resources under a particular healthcare level.

Under this model, a country is assigned to one of the four healthcare levels (HCL). Countries in HCL I have more than 1,000 physicians per million people; in HCL II, between 333 and 1,000; in HCL III, between 100 and 332; and in HCL IV, less than 100. This innovative approach enables an estimation to be made of the number and type of procedures for a given country where specific data were unavailable, by applying an average annual frequency of these procedures from other countries under the same HCL [23–27].

3.4 Survey Findings

Recent UNSCEAR surveys cover the types of medical exposures as defined by the International Commission on Radiological Protection (ICRP): (a) the exposure of patients as part of medical diagnosis or treatment; (b) the exposure of individuals as part of health screening programmes; and (c) the exposure of healthy individuals or patients voluntarily participating in medical, biomedical, diagnostic or therapeutic research programs [10]. These surveys cover three categories of practice employing ionizing radiation in medicine: diagnostic radiology (including imaging-guided interventional procedures), nuclear medicine and radiation therapy [26, 27].

The most recent UNSCEAR 2008 Report evaluated the global use of medical exposures from 1997 to 2007, determined the exposure from various modalities and procedures, and assessed the emerging trends [27]. According to this, approximately 3.6 billion diagnostic X-ray procedures (including approximately 0.5 billion dental procedures) were performed annually worldwide. Table 5.2 summarizes the estimated annual frequency of diagnostic X-ray procedures and the corresponding annual collective effective dose and per caput effective dose for this period globally and for each HCL [27].

Figure 5.1 presents the global evolution of medical radiological usage between 1970 and 2007. The annual frequency of diagnostic medical procedures (including

Table 5.2 Global and HCL estimations of annual frequencies of diagnostic X-ray procedures, dental procedures (*in brackets*) and collective effective doses for the period 1997–2007 [27]

Health-care level (HCL)	Population (1,000,000)	Annual frequency per 1,000 people	Annual collective effective dose (man Sv)	Annual per caput effective dose (mSv)
I	1,540	1,332 (*275*)	2,900,000 (*9,900*)	1.91
II	3,153	332 (*16*)	1,000,000 (*1,300*)	0.32
III	1,009	20 (*2.6*)	33,000 (*51*)	0.03
IV	744	20 (*2.6*)	24,000 (*38*)	0.03
Global	6,446	488 (*74*)	4,000,000 (*11,000*)	0.62

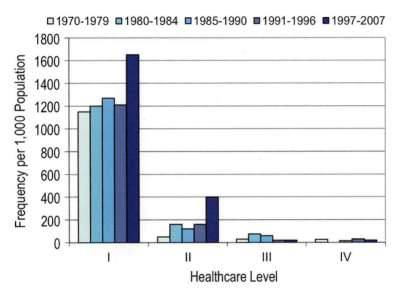

Fig. 5.1 HCL estimates of annual frequency of diagnostic X-ray procedures, including dental procedures, from five surveys between 1970 and 2007

dental procedures) in HCL I countries has increased from 1,200 per 1,000 population in the 1991–1996 period to 1,650 per 1,000 population in the 1997–2007 period [26, 27]. The annual frequency of diagnostic medical procedures in HCL III and IV countries has remained fairly constant over the same periods, although since there were limited data for these countries, there is considerable uncertainty associated with this estimate. Only 1.5 % of all worldwide diagnostic X-rays procedures were estimated to be performed in HCL III and IV countries, which together cover 27 % of the global population (Fig. 5.2). This imbalance in healthcare provision is also reflected in the availability of X-ray equipment and practitioners. New high-dose X-ray technology, particularly CT, has led to an extremely rapid growth in the annual frequency of CT procedures and a marked increase in the collective effective doses in HCL I countries. For some countries, this has resulted for the first time in history where the annual per caput effective dose of ionizing radiation from the

Fig. 5.2 Global population distribution according to the healthcare level (*HCL*) model used in the 2010 UNSCEAR report [27]

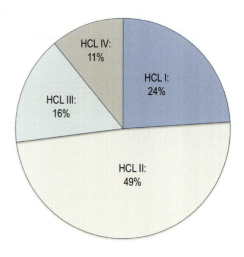

Fig. 5.3 Annual per caput effective dose in mSv for the United States in 2006 [15]

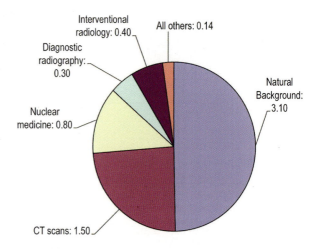

medical use of radiation is equal to or higher than the natural background exposure (Fig. 5.3).

Since the UNSCEAR 2000 Report, the total number of diagnostic X-rays procedures and per caput effective dose has increased by approximately 50 % and the total collective effective dose from medical diagnostic procedures has nearly doubled from about 2.3 million man Sv to about four million man Sv. Figures 5.4a and 5.4b summarizes the estimations of global annual frequency and collective effective dose for diagnostic X-ray procedures in the last four UNSCEAR evaluations

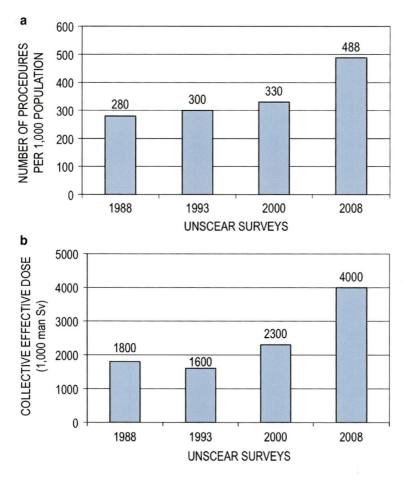

Fig. 5.4 Global estimates of diagnostic X-ray procedures in the last four UNSCEAR surveys: (**a**) annual frequency; and (**b**) annual collective effective dose

[23, 24, 26, 27]. The global average annual per caput effective dose has increased from 0.35 mSv in the 1988 to 0.62 mSv in 2007.

The utilization of nuclear medicine procedures around the world is quite uneven, with 90 % of procedures performed in HCL I countries [27]. The estimated annual frequency of diagnostic nuclear medicine procedures has not grown so dramatically over the past four survey periods (1980–1984, 1985–1990, 1991–1996 and 1997–2007), as shown in Fig. 5.5a. As presented in Fig. 5.5b, the estimated annual collective effective dose from diagnostic nuclear medicine procedures has increased by 34 %, from 150,000 to 202,000 man Sv, between the periods 1991–1996 and 1997–2007 and by a factor 2.7 compared to 74,000 man Sv in the evaluation conducted for the period 1980–1984 [23, 24, 26, 27].

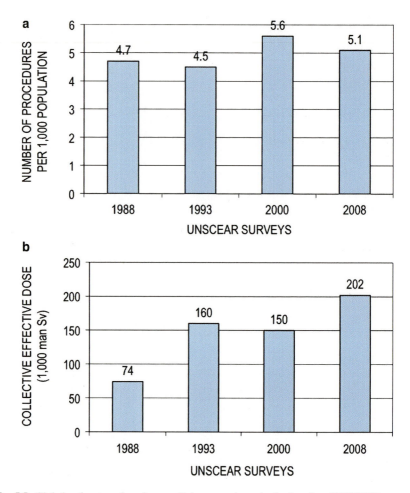

Fig. 5.5 Global estimates of nuclear medicine procedures in the last four UNSCEAR surveys: (**a**) annual frequency; and (**b**) annual collective effective dose

Figure 5.6 provides the HCL and global estimations of the annual collective dose for all diagnostic procedures, including dental, for the period 1997–2007 [27].

An estimated 5.1 million courses of radiation therapy treatment were administered annually between 1997 and 2007, up from an estimated 4.3 million in 1988. About 4.7 million treatments involved teletherapy and only 0.4 million brachytherapy. The annual frequencies of radiation therapy treatments during 1997–2007 were dominated by procedures performed in HCL I countries, accounting for about 70 % and 40 % of global teletherapy and brachytherapy, respectively [27].

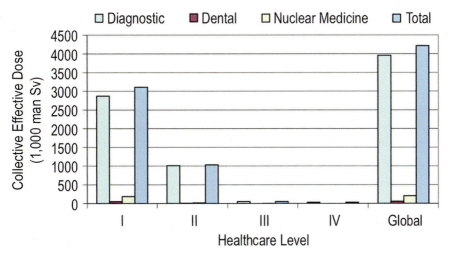

Fig. 5.6 Global and HCL estimates of annual collective effective doses for all diagnostic exposures for the period 1997–2007 [27]

4 Interpretation of Medical Exposure Data

Despite the current level of knowledge of medical radiation exposure and radiation-induced health effects, more targeted research is required. There are uncertainties in estimating the cancer risk due to ionizing radiation and in attributing health effects to and inferring risk from medical radiation exposure. An improved understanding in these areas, based on research, evidence and robust scientific debate, will lead to better application of survey data, more informed decision-making and targeted action on radiation protection and safety.

4.1 Attributing Health Effect and Inferring Risk

It is important to distinguish between a manifest "health effect" and "health risk" (likelihood for a future health effect to occur), when describing health impacts for an individual or population. When evaluating the manifest health effects and health risks from medical radiation exposure for a given population, one of the challenges is statistical fluctuation. With low-LET (linear energy transfer) radiation such as X-rays and gamma rays, the lower the dose, the higher is this uncertainty. This uncertainty increases when extrapolating the probability of incidents due to moderate dose to low and very low dose. Therefore, it is not surprising to note that a statistically significant increase in radiation-induced cancer is seen only when the exposure is 100 mSv or above [28].

A manifest health effect in an individual could be unequivocally attributed to radiation exposure only if other possible causes for an observable tissue reaction such as skin burns (deterministic effect) were excluded. Malignancies (stochastic effects) cannot be unequivocally attributed to radiation exposure because radiation is not the only possible cause and there are at present no known biomarkers that are specific to radiation exposure. A manifest increased incidence of stochastic effects in a population could be attributed to radiation exposure through epidemiological analysis provided the increased incidence is sufficient to overcome the inherent statistical uncertainties. Although demonstrated in animal studies, a manifest increase in incidence of hereditary effects in human cannot presently be attributed to radiation exposure [14, 28].

In general, a manifest increased incidence of health effects in a population cannot *reliably* be attributed to radiation exposures at levels that are typical of the global average background levels of radiation. The reasons include: (1) the uncertainties associated with risk evaluation at low doses, (2) the absence of radiation-specific biomarkers for health effects and (3) the insufficient statistical power of epidemiological studies [14, 28].

When estimating radiation-induced health effects in a population exposed to incremental doses at levels equivalent to or below natural background, it is not generally recommended to do this simply by multiplying the very low doses to a large number of individuals. However, UNSCEAR recognizes there is a need for estimations to facilitate health authorities to allocate resources accordingly and to compare the incidence of health effect from different hazardous sources. This is only valid *provided that* it is applied consistently, the uncertainties in the estimations are fully taken into account, and the projected health effects are notional [14, 28].

4.2   Uncertainties in Cancer Risk Estimation

One of the challenges relating to the interpretation, analysis and use of radiation exposure data of population is uncertainty when attributing cancer risk to ionizing radiation exposure. In epidemiological surveys of populations exposed to radiation, there are statistical fluctuations and uncertainties due to selection and information bias; exposure and dose evaluation; and model assumptions used when evaluating data. In addition, transferring the risk estimate based on data from an epidemiological study to a population of interest may not be entirely valid because of the differences in: location, setting, data collection period, age and gender profile, dose or dose rate, type of radiation and acute versus protracted exposures. Further, knowledge on individual radiosensitivity is limited [12, 28].

The uncertainty of cancer risk after exposure to ionizing radiation is often underestimated. For solid cancers after an exposure of 100 mSv, the uncertainty in the risk estimate could differ by a factor of 2–3 and the uncertainty of excess risk for a specific cancer type is considerably higher than for all solid cancers. These uncertainties in estimating cancer risk from ionizing radiation exposure need to be addressed to underpin evidence-based recommendations and enable informed decision-making [12, 28].

5 Discussion

5.1 Emerging Trends

Medical exposure remains by far the largest artificial source of exposure and continues to grow significantly. The distribution of medical exposures is uneven between countries and regions. A quarter of the world's population lives in HCL I countries, uses 66 % of diagnostic radiology procedures, 90 % of nuclear medicine procedures, and 70 % of radiation therapy treatments. In HCL I and II countries, where 75 % of the world's population resides, medical uses of radiation have increased from year to year as the benefits of the procedures have become more widely known. While there are limited data on the annual frequency of procedures in HCL III and IV countries, the utilization of radiation in medicine has remained fairly constant and at a very low level.

For diagnostic dental procedures, the annual frequency has remained constant for HCL I and II countries, but has substantially increased for HCL III and IV. The collective effective dose from diagnostic radiology has doubled since 1988 due to increased use of CT mainly in HCL I countries with a tendency for further increases in the rest of the world. Although the annual frequency of diagnostic nuclear medicine procedures has remained fairly constant since 1988, its contribution to collective effective dose has tripled, due to the introduction of high-dose cardiac procedures and an increase in hybrid imaging, i.e. PET/CT and SPECT/CT.

Increasing urbanization together with a gradual improvement in living standards means more individuals can access better healthcare. As a result, the population dose due to medical exposures has continuously increased across all healthcare levels.

5.2 Survey Challenges

The UNSCEAR surveys revealed a range of issues relating to participation, survey process, data quality and analysis. For example, limited resources and infrastructure to conduct national surveys and requests for the same data from different international agencies are contributing factors limiting national participation. The structure, consistency, clarity, complexity and, to a less extent, the language used in the questionnaires are also barriers to data collection. These and other factors such as the significant differences in equipment and resources, healthcare workforce and system, have resulted in difficulties in obtaining data from many countries and wide variations in the reported annual frequencies in earlier UNSCEAR surveys.

While data are available from HCL I countries on which UNSCEAR's evaluations are mainly based, data from HCL II to IV countries are scarce.

5.3 Improvement Strategies

In collaboration with the IAEA and WHO, an improvement strategy was developed to address deficiencies in data quality and collection and to improve participation to future UNSCEAR surveys [5]. Elements included in this strategy are an improvement in the questionnaires' structure, consistency and clarity by standardizing taxonomy and terminology, separating the section concerned with frequency of use of procedures from that concerned with dosimetric data, providing questionnaires in other UN languages other than English, developing electronic methods for data collection, establishing a small standing expert group on medical exposure, and reviewing the HCL model for better data extrapolation. In addition, UNSCEAR will focus on those procedures with significant contribution to the collective dose similar to the methodology used by the European DOSE DATAMED project [6] and will further obtain relevant data, in a cooperative and sustainable manner, from other reliable sources such as IAEA and WHO [5].

As previously mentioned, the comparison and trend analysis of population exposure data require the grouping of countries by their healthcare systems. For its future work, one of the improvement strategies is a review of the current HCL model, which is based on the "physician/population ratio", with a view to adopting the World Bank classification, which is used by WHO. This classification places countries into four income groups according to the gross national income (GNI) per capita. For example, the GNI per capita in low-income countries is US$1,000 or less, in low- to middle-income countries is from US$1,000 to 4,000, upper- to middle-income countries is from $4,000 to 12,000, and high-income countries is US$12,000 or above. This classification might better reflect the impact of a country's economic situation to healthcare expenditure and explain the variations in medical exposures. Further, this classification would enable better correlation and analysis of medical exposure data with other related and relevant data, such as healthcare resources and expenditure, between the countries within each group [17].

5.4 Impact on Radiation Safety

UNSCEAR has no direct mandate in radiation protection but it is more focused on the collection, collation, analysis and publication of medical radiation exposure data, covering the issues, levels, impact and trends. The information provided by UNSCEAR are of importance to others, such as IAEA, WHO and the International

Labor Organization (ILO), who have responsibility for the preparation of radiation safety and protection recommendations, guidance, standards, and programmes.

UNSCEAR surveys and evaluations have provided the scientific foundation to improve our understanding of the levels of radiation to which individuals are exposed and of radiation-induced health effects. This facilitates the development of evidence-based advocacy messages, recommendations and policies on radiation safety and protection. These scientific data have been used for example by the ICRP to develop recommendations on radiation protection [10, 11] and by other UN agencies to formulate international radiation protection frameworks, such as the IAEA International Action Plan for the Radiological Protection of Patients [8], and the WHO Global Initiative on Radiation Safety in Health Care Settings [29]. Further, the International Basic Safety Standards (BSS) [9] uses as its starting points UNSCEAR's scientific evaluations and ICRP's recommendations [11].

The IAEA, WHO and other UN agencies are working with their regional and national counterparts to facilitate the incorporation and implementation of the BSS into policies by the national authorities thus ensuring radiation protection of the community. UNSCEAR reports are cited in numerous publications, presentations, recommendations, and policies focusing on radiation safety and protection.

6 Conclusion

UNSCEAR's evaluations of global medical exposures are mainly based on responses from HCL I countries. Based on the lessons learned from past surveys, UNSCEAR, IAEA and WHO have developed innovative solutions to improve future collection and analysis of information on diagnostic radiology, nuclear medicine and radiation therapy. For example, under the WHO Global Initiative on Radiation Safety in Health Care Settings, collaboration between the two organizations has been strengthened to improve the collection of data on the frequency of medical radiological procedures, particularly in developing countries where this information is scarce.

Increasing medical exposure worldwide is likely associated with increased health benefits to the population as patients receive a direct benefit from their exposures. Explicit comparison of doses resulting from medical exposures with other sources is therefore inappropriate. In addition, the age distribution of patients receiving medical radiological exposure is normally older than the general population. In contrast, the introduction of new technologies has in some instances resulted in an increased use of medical radiation in children. While the magnitude of medical exposures can be assessed, it is very difficult to estimate the health risks from such use. Despites the global increase in demand for diagnostic radiology procedures; their appropriate use should be underpinned by evidence-based referral guidelines and optimization of radiation protection.

Other stakeholders apply UNSCEAR data to develop and update evidence-based radiation protection and radiation safety recommendations and guidance. To ensure data quality and relevance, the support of national authorities and their participation in surveys are required. The advocacy by and contribution from other stakeholders including national as well as international professional organizations and radiological societies in data collection is highly encouraged.

UNSCEAR's mandate has not changed over the last decades but its role as the principal focal point for international information exchange has evolved as a result of the revolution in information technology. While communication advances have facilitated data dissemination and information sharing, UNSCEAR's coordinating role and experience in global radiation exposure evaluation are most invaluable. There is a need to review and synthesize the volume of relevant data and to establish scientific consensus so policy-makers and other stakeholders could use them accordingly. The role of UNSCEAR in providing authoritative scientific information continues to be crucial to international bodies such as the IAEA, ILO, WHO and ICRP in the development of radiation protection standards, recommendations, guidance tools and policies dealing with the rapid technological advances and changing practice trends. By providing leadership, collaborating with other stakeholders and applying innovative solutions, UNSCEAR plays a pivotal role to ensure quality patient care and to improve radiation protection worldwide.

References

1. Aroua A, Burnand B, Decka I et al (2002) Nation-wide survey on radiation doses in diagnostic and interventional radiology in Switzerland in 1998. Health Phys 83(1):46–55
2. Bly R, Jahnen A, Jarvinen H et al (2011) European population dose from radiodiagnostic procedures – early results of Dose Datamed 2. In: Proceedings of the XVI NSFS conference – current challenges in radiation protection. Nordic Society for Radiation Protection, Reykjavík
3. Brix G, Nekolla E, Griebel J (2005) Radiation exposure of patients from diagnostic and interventional X-ray procedures. Facts, assessment and trends. Radiologe 45(4):340–349
4. Council of the European Union (1997) Council Directive 97/43/Euratom of 30 June 1997 on health protection of individuals against the dangers of ionizing radiation in relation to medical exposure, and repealing Directive 84/466/EURATOM. OJEC 1997/L 180/22
5. Crick M, Shannoun F, Le Heron J et al (2010) Opportunities to improve the global assessment of medical radiation exposures. In: International conference on radiation protection in medicine. National Centre of Radiobiology and Radiation Protection, Bulgarian Ministry of Health, Varna
6. European Commission (2008) Radiation Protection 154. European guidance on estimating population doses from medical X-ray procedures. EU Publications Office, Luxembourg
7. Hart D, Wall BF (2004) UK population dose from medical X-ray examinations. Eur J Radiol 50(3):285–291
8. International Atomic Energy Agency (2002) Measures to strengthen international cooperation in nuclear, radiation, transport and waste safety: international action plan for the radiological protection of patients. GOV/2002/36-GC(46)/12. IAEA, Vienna
9. International Atomic Energy Agency (2011) Radiation protection and safety of radiation sources: international basic safety standards – Interim edition, IAEA Safety standards series GSR Part 3 (Interim). IAEA, Vienna

10. International Commission on Radiological Protection (1991) Recommendations of the international commission on radiological protection. ICRP Publication 60. Annals of the ICRP. 21(1–3). Pergamon Press, Oxford
11. International Commission on Radiological Protection (2007) ICRP Publication 103. The 2007 recommendations of the international commission on radiological protection. Annals of the ICRP. 37(2–4). Elsevier, Oxford
12. Jacob P (2012) Uncertainties of cancer risks estimates for applications of ionizing radiation in medicine. In: International conference on radiation protection in medicine – setting the scene for the next decade. IAEA, Bonn
13. Mettler FA Jr, Davis M, Kelsey CA et al (1987) Analytical modeling of worldwide medical radiation use. Health Phys 52(2):133–141
14. Müller W (2012) Can we attribute health effects to medical radiation exposure? In: International conference on radiation protection in medicine – setting the scene for the next decade. IAEA, Bonn
15. National Council on Radiation Protection and Measurements (2009) Report No. 160. Ionizing radiation exposure of the population of the United States. NCRP, Bethesda
16. Shannoun F, Blettner M, Schmidberger H et al (2008) Radiation protection in diagnostic radiology. Dtsch Arztebl Int 105(3):41–46
17. Shannoun F (2012) A comparison of classification methodologies of health care levels used by UNSCEAR and WHO. In: Dose Datamed II project workshop on European population doses from medical exposure. Greek Atomic Energy Commission, European Commission, Athens
18. United Nations Scientific Committee on the Effects of Atomic Radiation (1958) Official Records of the General Assembly, Thirteenth Session, Supplement No. 17 (A/3838). UNSCEAR 1958 report. United Nations, New York
19. United Nations Scientific Committee on the Effects of Atomic Radiation (1962) Official Records of the General Assembly, Seventeenth Session, Supplement No. 16 (A/5216). UNSCEAR 1962 report. United Nations, New York
20. United Nations Scientific Committee on the Effects of Atomic Radiation (1972) Ionizing radiation: levels and effects. Volume I: levels. UNSCEAR 1972 report to the General Assembly, with annexes. United Nations, New York
21. United Nations Scientific Committee on the Effects of Atomic Radiation (1977) Sources and effects of ionizing radiation. UNSCEAR 1977 report to the General Assembly, with annexes. United Nations, New York
22. United Nations Scientific Committee on the Effects of Atomic Radiation (1982) Ionizing radiation: sources and biological effects. UNSCEAR 1982 report to the General Assembly, with annexes. United Nations, New York
23. United Nations Scientific Committee on the Effects of Atomic Radiation (1988) Sources, effects and risks of ionizing radiation. UNSCEAR 1988 report to the General Assembly, with annexes. United Nations, New York
24. United Nations Scientific Committee on the Effects of Atomic Radiation (1993) Sources and effects of ionizing radiation. UNSCEAR 1993 report to the General Assembly, with scientific annexes. United Nations, New York
25. United Nations Scientific Committee on the Effects of Atomic Radiation (1996) Sources and effects of ionizing radiation. UNSCEAR 1996 report to the General Assembly, with scientific annex. United Nations, New York
26. United Nations Scientific Committee on the Effects of Atomic Radiation (2000) Sources and effects of ionizing radiation. Volume I: sources. UNSCEAR 2000 report to the General Assembly, with scientific annexes. United Nations, New York
27. United Nations Scientific Committee on the Effects of Atomic Radiation (2010) Sources and effects of ionizing radiation. Volume I: sources. UNSCEAR 2008 report to the General Assembly with scientific annexes A and B. United Nations, New York
28. United Nations Scientific Committee on the Effects of Atomic Radiation (2013) UNSCEAR 2012 report to the General Assembly, with annexes. United Nations, New York (in press)
29. World Health Organization (2008) WHO global initiative on radiation safety in health care settings technical meeting report. WHO, Geneva

Part II
Procedure Justification, Protection Optimization, and Error Reduction

Chapter 6
National, Regional and Global Radiology and Medical Imaging Referral Guidelines: Issues and Opportunities

Lawrence Lau

Abstract This chapter focuses on the access to and use of referral guidelines in radiology and medical imaging worldwide. The rationale; national, regional and global actions; development and implementation processes; and issues and opportunities for referral guidelines are discussed. (The discussions are related to other chapters. Please refer to the following for more detail: (a) Use of referral guidelines as a justification tool: national perspective [Chap. 8]. (b) Use of imaging appropriateness criteria for decision support during radiology order entry: the MGH experience [Chap. 7]. (c) Improving quality through lean process improvement: the example of clinical decision support [Chap. 16]).

Radiology referral guidelines are tools to assist referrers and providers towards Good Medical Practice and procedure justification by using radiology and medical imaging appropriately, thus minimizing unnecessary radiation exposure and waste of resources. The development and provision of up-to-date evidence-based contents in an innovative user-friendly package lay the foundation in a multi-stage process involving many stakeholders. For system-wide implementation, the support from competent authorities is needed. Strong advocacy from organizations and agencies on the use of guidelines will facilitate their incorporation into policy. The active involvement and participation of the end-users along these steps will improve awareness, ownership, acceptance and use. Leadership from individual experts, organizations, and agencies by facilitating stakeholder engagement, collaboration and participation will improve guideline use leading to better patient care, safer and more appropriate use of procedures.

Keywords Computerized physician order entry • Imaging guidelines • Innovation • Leadership in radiology • Procedure justification • Quality radiology • Radiology referral guidelines • Unnecessary exposure

L. Lau (✉)
International Radiology Quality Network, 1891 Preston White Drive,
Reston, VA 20191-4326, USA
e-mail: lslau@bigpond.net.au

L. Lau and K.-H. Ng (eds.), *Radiological Safety and Quality: Paradigms in Leadership and Innovation*, DOI 10.1007/978-94-007-7256-4_6,
© Springer Science+Business Media Dordrecht 2014

105

1   Rationale

1.1   Emerging Challenges

Radiology and medical imaging (i.e. radiology in this chapter) play an indispensible role in and support Good Medical Practice [7]. Computed Tomography (CT) and Magnetic Resonance Imaging (MRI) are rated as top innovations in healthcare by leading physicians. Over three and a half billion radiological procedures are performed globally each year and this figure is rising [30].

In the United States, medical to total population exposure from ionizing radiation has increased from 15 % in 1980 to 50 % in 2006 [21] and health expenditure has increased from 13.9 % of GNP in 2001 to 15.3 % in 2003 [4]. The health sector is under-resourced in many countries, with considerable inequality. According to World Health Statistics, per capita health expenditure in 2007 ranged from US$7 to US$7,439 with a median cost of US$248 [34].

An increase in use does not necessarily imply overuse or misuse. Advances in techniques and technologies will naturally lead to new applications towards an earlier diagnosis and less invasive treatment. Aging populations is another reason for an increase in demand. However, inappropriate use will lead to more radiation exposure and health cost, and will threaten health system sustainability.

In a large health care plan, cross-sectional imaging nearly doubled from 1997 through 2006 [27]. While many procedures could be suboptimal and unnecessary, the exact figure is difficult to quantify [19]. There is a wide variation in the use of CT procedures within and between facilities. Up to 20 million adults and a million children in the US could have unnecessary CT each year [3]. A Finnish series in patients under 35 showed that 77 % of lumbar spine, 36 % of head, and 37 % of abdominal CT procedures were not justified [23]. Stakeholders will benefit from guidance on an appropriate use of radiological procedures and radiation safety [20].

The key stakeholders directly involved in the use of radiological procedures are patients, referrers and providers. Patients are more informed, but some incorrectly equate the use of investigative procedures with good medicine and quality care. The referrers and providers face similar challenges, i.e. workforce not keeping up with workload, which has increased significantly both in volume and complexity; insufficient time for consultation, communication and continuing professional development; rising practice cost and declining re-imbursement. There is limited training in radiation safety, radiation protection and an appropriate use of radiological procedures in undergraduate and postgraduate curricula. Failure to keep up with the current practice trend; attempt to meet patient expectation; self-referral; defensive medicine and wellness screening contribute to inappropriate use, unnecessary exposures and waste of resources [28].

1.2 Quality Radiology

The quality elements in radiology can be grouped under four categories: safety, appropriateness, effectiveness, and patient centricity [2]. Quality radiology may be described as "... a timely access to and delivery of integrated and appropriate radiological procedures in a safe and patient-centred facility and a prompt delivery of accurately interpreted reports by capable personnel in an efficient, effective, and sustainable manner... " [16].

1.3 Procedure Justification

Procedure justification and optimization of protection are the pillars of radiation protection [31]. Quality assurance and error minimization reduce unintended and accidental exposure. The justification of radiological procedures is considered at three levels. First, a radiological procedure will do more good than harm to society; radiation exposure, economic and social issues being considered. Second, the objective for a procedure is clearly defined and justified, i.e. it will improve the diagnosis or treatment, or provide information to assist care. Third, the procedure for a particular individual is justified, i.e. it will do more good than harm to that individual. Procedure justification applied at the beginning of the patient journey plays an important "gate-keeper" role by reducing unnecessary procedure and exposure (Fig. 6.1). The referrers and providers are jointly responsible for procedure justification, but in some settings this lies solely with the providers.

1.4 Role of Guidelines

In clinical radiology, the stakeholders' common goals are to deliver patient-centered care, to ensure *patient safety and an appropriate use of procedures by maximizing the available resources* and to work towards health care system sustainability.

Evidence-based radiology referral guidelines (i.e. guidelines in this chapter) support these goals and are used to promote an appropriate use of radiological procedures. Guidelines are useful clinical and justification tools, underpinning quality radiology and radiation safety respectively. Outcome data would confirm the usefulness and effectiveness of guidelines and should convince authorities of their role in patient care and health care system sustainability. This chapter provides an account of the rationale; national, regional and global actions; development and implementation processes; and issues and opportunities to improve the use of guidelines in practice.

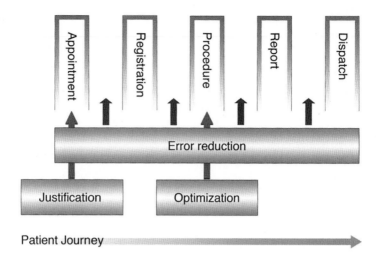

Fig. 6.1 Patient journey and procedure justification. The five key processes along the journey are: appointment, registration, procedure, report, and dispatch. Procedure justification and optimization of image quality and radiation protection (*long arrows*) are the pillars of radiation protection. Error reduction actions (*short arrows*) minimize human errors along this journey before, during and after a procedure. Procedure justification, applied at the beginning of this journey serves as a "gate-keeper" to minimize unnecessary procedure and exposure

2 Existing Actions

2.1 *National, Regional and Global Guidelines*

Over three decades ago, with the rapid advances in technology and the development of new diagnostic and interventional procedures, three leading professional organizations published guidelines to inform referrers of new applications and to promote an appropriate use of procedures. These proactive actions started well before the recent reports on increased utilization and population exposure.

Working independently, three formats were used. The American College of Radiology's Appropriateness Criteria is a comprehensive document outlining its recommendations in a tabulated format for different conditions and variants [1]. The Royal Australian and New Zealand College of Radiologists "Imaging Guidelines" [17] adopted a diagnostic algorithm approach while The Royal College of Radiologists used a concise pocket handbook format [29]. There are pros and cons for each format and they remain as the main approaches to the present time.

These publications were followed by a progressive release of guidelines by other organizations in Argentina, Austria, Canada, France, Germany and Hong Kong China. In Europe, there is a legal basis for the distribution of guidelines, i.e. the Euratom Treaty, BSS and Directive on Medical Exposures, specifically states that recommendations should be available to the prescribers of medical exposures. The European Commission published its guidelines as "Radiation Protection 118" [5].

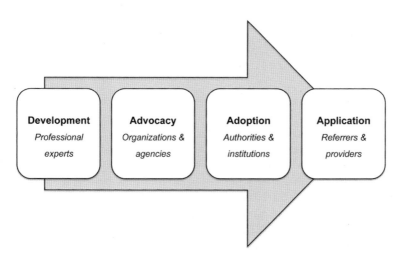

Fig. 6.2 Key actions to improve the use of referral guidelines. These include the presentation of evidence-based content in a user-friendly interface, strong advocacy, adoption of policy and participation of the end-users. The contribution and support from many stakeholders with different expertise are required

The Government of Western Australia publishes "Diagnostic Imaging Pathways" [9], which is endorsed by the RANZCR.

Together with the Pan American Health Organization, a regional project is underway to develop a set of guidelines for Latin America. The World Health Organization is collaborating with the International Radiology Quality Network and over 30 other organizations and agencies to develop, pilot and implement a set of global referral guidelines under the Global Initiative on Radiation Safety in Health Care Settings [14, 33, 35].

2.2   Lessons Learnt

Despite these actions, the use of guidelines in practice is low. To improve use, innovative actions focusing on two key areas are required, by providing *up-to-date evidence-based contents in a more user-friendly interface* and encouraging *local authority and end-users uptake* (Fig. 6.2). The former relies on regular, on-going, and pro bono contributions from radiology experts and presenting these in an interface appropriate for the end-users. The latter involves implementation actions by strengthening advocacy, adopting policy and providing implementation tools for different settings. Experience suggests providing guidelines per se does *not* guarantee use in practice; the support from national authorities and end-user improves this probability. With the range of referral guidelines and justification actions, there are many opportunities for stakeholders to work together and to avoid duplication.

3   Development and Implementation

3.1   Development Process

Initially, guidelines were essentially consensus-based, limited by the availability of evidence-based material. The proportion of evidence-based recommendations in modern guidelines is increasing as more and better data are available and organizations adopt an evidence-based approach. Guidelines are either developed *de novo* or adapted, adopted or translated from an existing publication. In either case, consultation with and involvement of external stakeholders are important to ensure ownership of the recommendations, their acceptance and use. However, the timing of engagement may vary. The steps involved in the development of new guidelines are summarized in Table 6.1.

Guidelines should recommend the most appropriate procedure with the lowest exposure, based on the best evidence. The hierarchy model used to describe efficacies in radiology is a useful concept [6]. To assist guideline developers and end-users, the grading of the quality of evidence in the recommendations is critical. The grading tools include the criteria used by the: US Preventive Services Task Force (USPSTF), National Health Service (NHS) and Grading of Recommendations Assessment, Development and Evaluation (GRADE) Working Group.

3.2   Content

Guidelines should be transparent, practical, population appropriate and endorsed by the stakeholders. Topics are discussed under clinical symptoms, signs or working diagnoses after taking disease prevalence into consideration. Some publications include a range of clinical variations. These recommendations are presented in a tabulated or algorithm (flow chart) format.

The appropriateness of a procedure is graded and compared. This is usually expressed as "usually not appropriate, may be appropriate and usually appropriate"

Table 6.1 Guideline development process	
	Define goal and project scope
	Convene expert panels
	Prepare a list of conditions/topics
	Conduct literature search
	Draft recommendations
	Assign appropriateness rating
	Include exposure comparison for procedures
	Review by expert panels
	Decide by Delphi process or modification
	Select format and media
	Prepare, edit and publish

or "indicated, specialized investigation, indicated in specific circumstances or not indicated". Radiation exposure for a procedure is expressed as: relative radiation levels (RRLs) in mSv, number of CXR equivalents, or multiples of natural background radiation.

3.3 Interface

Guidelines are published in print or electronic format. Each medium has its strengths and weaknesses. Print is widely accepted over a long period of time and is practical in many settings. However, once published it cannot be revised or edited. CDs and DVDs are durable but cannot be changed once produced and a computer is required, which may not be available in resource-poor settings.

When computers and Internet are available web-based solutions open more options. They are more convenient, allow access to more users, provide interactivity and enable timely updates. A 'printer-friendly' version enables referrers to choose the section they wish to print and provide to their patients. Mobile computing devices, i.e. personal digital assistant, Smartphone, tablet computer, ultra-mobile PC etc. are compact and elegant means for mobile web access, but are relatively expensive and many have short battery life.

The combined use of different media and interface improves use but costs more. Ultimately, the choice is based on: local setting and requirements; available resources; cost-effectiveness; guideline developer preference for flexibility in preparation and update; user preference and acceptance; durability and distribution strategy.

3.4 Implementation

Guideline access and implementation vary. Some organizations provide free Internet access to individuals, while a licence fee is needed for institutional use. In some cases, the professional organizations collaborate with the local authorities and distribute guidelines to end-users, including medical students and practitioners.

For more effective implementation, simply offering printed material or open on-line access is not enough. End-users need assistance and support to overcome inertia and initiate change. For example, when The Royal Australian and New Zealand College of Radiologists released the 4th edition of "Imaging Guidelines" in 2002, in addition to print, CD and on-line versions were released. Complimentary on-line help and phone support to assist and educate the end-users were available as part of the implementation strategy.

A range of tools is used to improve uptake. Examples include: information sheets, on-line or phone support, indication-specific request forms, referrer's checklist (Table 6.2), referrer feedback and Computerized Physician Ordered Entry

Table 6.2 A referrer's checklist (Adapted from "Making the Best Use of Clinical Radiology Services" 6th edition Royal College of Radiologists, London)

Has this procedure been done already?

Will the procedure and the result change patient care?

Is this the optimal time for the procedure, i.e. not too early before the disease could have progressed or resolved?

Is this the most appropriate procedure?

Has the relevant data and reasons for procedure been indicated on the request?

Are all the procedures necessary?

(CPOE). CPOE systems offer many advantages, i.e. integration of procedure appointment into the daily workflow for the referrers and providers; provision of interactive reminders and educational feedback; and evaluation of utilization.

3.5 Evaluation

The actions on guideline development and implementation are ongoing. The recommendations, format, media, implementation methodology, practical use and project framework should be regularly evaluated and updated to maintain currency, relevance and end-user acceptability. Some organizations include surveys as part of implementation to evaluate end-user knowledge, needs and feedback. Clinical audit is a useful quality improvement tool to encourage and monitor guideline use.

4 Issues and Solutions

The issues limiting and solutions improving guideline use are outlined in Table 6.3. They are related to the guidelines, end-users, and health care system and are discussed in the following sections.

4.1 Guidelines

Development and Update

Research and publication will improve the volume of evidence-based recommendations on procedure choice in radiology literature. For the guidelines to be relevant, regular update by incorporating evidence-based recommendations is required. This action relies heavily on the pro bono contributions from professional experts. Support from others is needed to share commitments and responsibilities.

Table 6.3 Guideline development and use: issues and solutions

Issues	Solutions
Guidelines	
Low evidence-based contents	Incorporate new evidence by regular update
Quickly out-dated contents	Update by sharing literature search
Contents too detailed	Prepare more concise and clearer advice
Varying exposure data	Standardize terminology
User-unfriendly, difficult to locate advice quickly	Improve format, search and interface Provide links to diagnoses, references etc.
Development and update rely on pro bono efforts	Stakeholders to provide support to experts
Turnover of leadership and key personnel	Work in teams and share corporate memory
Collaboration beyond radiology	Strengthen contents with other specialties
Language	Accurate translation into the local language
End-users	
Low awareness and acceptance	Strengthen advocacy and engagement to improve participation, ownership and use
Concerns for legal implication	Guidelines do not constraint practice but define and support Good Medical Practice
Knowledge and competency	Strengthen medical school curricula Improve ongoing professional development
Unnecessary procedures	Review medical records and confirm appropriateness before each request
Inertia to change, improvement transient	Provide change and implementation tools, e.g. education, audit and decision support
Health care system	
Varying needs, resources and procedure access	Adapt to suit, but the principle of Good Medical Practice is the same worldwide
Varying guideline access	Improve access, e.g. open on-line access
Poor infrastructure and weak radiation safety policies	Agencies to advocate and facilitate policy uptake and infrastructure strengthening
Limited funds for guideline development, implementation and evaluation	Pool resources and collective mobilization, UN agencies could be a neutral platform
Little incentive to encourage change	Consider pay for performance to reward excellence

The sharing of resources will minimize preparation time, e.g. by using coordinated literature search. Collaboration beyond the radiology community will strengthen guideline content by engagement with other clinical specialties. For such long-term projects, there is an inevitable turnover of leadership and key personnel. Working in teams and sharing corporate memory will ensure project continuity and direction.

Guidelines should be more user-friendly and concise as some of the existing is too detailed. If end-users have difficulty in locating the information quickly, it will limit acceptance and discourage use. More concise and clearer recommendations in a better format and interface are needed. Better programming will provide faster and more efficient search function. The provision of useful links to diagnosis, differential diagnosis and references will save end-user time. The provision of exposure data for procedures is important but presentation is variable. The stakeholders agree there is a need to standardize and improve data presentation. When adopting guidelines written in a foreign language, precise translation is paramount.

4.2 End-Users

Promote Awareness

Low awareness limits end-user acceptance and guideline use. The strengthening of advocacy, communication and engagement will improve end-user understanding of the rationale and ownership of the recommendations resulting in better participation and use. Public awareness on the use of radiation medicine has improved significantly in recent years with professional organizations and agencies leading this charge [10–12, 15, 25, 32].

Referrers support professionalism, good medicine and the primacy of patients. A consideration of the consequence of inappropriate use and the long-term sustainability of the health care system are good reasons to support guideline use. Further, children and young adults are particularly susceptible to the health effects of irradiation. An awareness of their roles and responsibilities will improve procedure justification by the end-users and the use of risk-benefit analysis prior to requesting procedures especially for the vulnerable groups. There were some concerns about the legal implication of guideline use. Guidelines do not constraint practice but define and support good medicine.

Facilitate Education

The knowledge and awareness on the appropriate use of radiology, radiation protection and radiation safety radiation among medical students and referrers is limited [8, 26]. A study of medical students and junior doctors showed 25 % and 11 % incorrectly believed that radiation is used in MRI and ultrasound respectively [36]. Education will improve knowledge and minimize unnecessary exposure.

It is useful for new graduates to appreciate the difference in practice between academic and community settings, i.e. the high-tech procedures in tertiary centers seldom apply because disease prevalence is different. The strengthening of undergraduate curricula and postgraduate continuing professional development will improve competency. The role of guidelines as an educational tool should be stressed. The relevant stakeholders: academic institutions, teaching departments, professional organizations and national authorities can contribute and collaborate toward this important action.

Appropriate Use

Prior to requesting, it is essential to verify if the procedure is necessary and appropriate at that moment, for the condition and the patient. Review of the medical records and clarification with the patient will confirm if the information seeking is not already available and minimize duplication (Table 6.2). Procedures should only be used if patient management hinges on the findings. In conditions where radiation and non-radiation techniques are both available and appropriate, non-radiation techniques such as ultrasound and MRI should be chosen. While a certain procedure is generally indicated for a given condition, it is important to ensure there is no contra-indication for a particular individual. If uncertain or in complex clinical scenarios, e.g. when more than one procedure is needed, direct or phone consultation with the providers is complementary to referral guideline.

Foster Change

Studies have shown that most actions used for dissemination and implementation brought about only transient improvements and the relapse rate was high. Tools are needed to foster change in practice by addressing time-poor, inertia, knowledge, and attitude issues.

It is possible to bring about attitude change and for the stakeholders to accept good advice, especially if they are better informed (awareness), committed to a common goal, convinced about the methodology employed and the recommendations provided (appropriateness). The change is more successful and sustainable if individuals are motivated and receive regular feedback (audit). This "triple-A" approach, i.e. awareness, appropriateness and audit, will encourage use and foster change [13]. Other strategies and tools supporting behavior change include education, publication, web-based information and clinical decision support integrated into electronic request.

4.3 Health Care System

Access to Resources

There are variations in resources, needs and access to radiology procedures and guidelines in different settings [22]. Improve access to trustworthy recommendations will facilitate use, such as open on-line access. Guidelines are adaptable to suit local needs, but the principle of Good Medical Practice is the same worldwide. Utilization issues differ between settings. In resource-rich settings, challenged by tight consultation schedule, expectation from specialists and patients and concerns of possible litigations, the referrers rely heavily on radiological procedures. In resource-poor settings, radiology is limited to basic radiography, even though other techniques could be more appropriate [22]. In any setting, guidelines assist the referrers by selecting appropriate and safe procedure within the available resources. In settings without an on-site radiologist, guidelines are important supporting tools for the radiographers when they play an extended role in justification and radiation protection.

Strengthen Infrastructure

Improvement in guideline use and radiation safety requires the strengthening of a health care system's infrastructure, e.g. distribution of guidelines; provision of decision-making tools; promotion of quality and safety culture; building workforce capacity; ensuring practitioner competency; supporting research towards more appropriate use of procedures; and application of supporting policies, e.g. the endorsement, monitoring and evaluation of referral guidelines use, the implementation and regulation of radiation safety standards. Beyond referral guidelines, other actions to strengthen system infrastructure and improve appropriate use could include undertaking reforms to address defensive medicine and self-referral issues.

The communication strengths and facilitating roles of WHO and IAEA with national authorities, academic institutions and professional organizations should be leveraged towards infrastructure strengthening.

Reward Excellence

Pay for performance (P4P) is a mechanism to financially recognize end-users for their commitment to quality care and excellence in performance through bonuses or penalties. It is a possible mechanism the stakeholders could use to encourage change and to improve a wider use of guidelines in practice.

5 Opportunities

5.1 Bridge the Gap

The challenges are to bridge the gap between knowledge and practice and to ensure guidelines are current by regular review and update (Fig. 6.3). To promote a wider use of guidelines require leadership and innovation. Leadership facilitates stakeholder engagement and the sharing of knowledge and resources. Collaboration is strength and saves time and cost. Innovative actions focusing on the improvements of these processes should be developed and applied.

Fig. 6.3 Bridging the gaps
between evidence and
practice. The work on
guidelines is ongoing. The
challenge is to bridge the
gap between knowledge and
practice and to ensure
guidelines are current by
regular review and update

5.2   Provide Leadership

Leadership could be described as "effectively influencing a group of people to
achieve a common goal". The initiation and maintenance of multi-faceted referral
guidelines actions require strong leadership from individual experts and leaders
from organizations and agencies. Leadership for guideline actions is particularly
important, as there is usually a long lead-time before the benefits of guidelines are
realized. Leaders provide a long-term vision and facilitate stakeholder engagement
and team collaboration towards a common goal. With potential population health
and radiation exposure impacts from an increasing use of radiological procedures
worldwide, UN and other inter-governmental agencies are well placed to lead and
advocate their appropriate use.

5.3   Engage Stakeholders

The stakeholders for radiology, justification of radiological procedures and the
development and use of referral guidelines are summarized in Table 6.4. The
end-users are the referrers and providers. Radiology referral guidelines are the
subject of much interest to organizations, agencies and authorities working in health
and radiation safety at national, regional and global levels. It is important to engage
the stakeholders and end-users to share ownership, promote collaboration, improve
acceptance, and ensure guidelines are relevant and used in practice.

5.4   Global Actions

Organizations and agencies are conducting or planning actions on the various aspects
of appropriate use of radiology, referral guidelines, procedure justification, radiation
protection, risk communication, change management, education methodology etc.,

Table 6.4 Stakeholders for radiology

Consumers: patients and general public
Referrers: general practitioners, specialists, other eligible referrers, and medical students
Providers: radiologists, radiation oncologists, nuclear medicine specialists, radiographers, nuclear medicine technologists, medical physicists, other eligible providers, and trainees
Payers: public authorities, private insurers, social services, individuals and others
Regulators: governments, health ministries, competent authorities, other related sectors, and policy makers
International organizations and UN agencies
Professional, academic and scientific organizations
Academic and research institutions
Medical defence organizations, malpractice insurers
Equipment manufacturers
Equipment vendors

which will benefit from collaboration and cross-fertilization. For example, conducting literature search in guideline development and update is fast becoming labor and resources intensive. This could soon be beyond the means of many organizations and would benefit from a joint approach and the sharing of resources. Other examples of actions include the adaptation of existing guidelines for trial and use in different settings; the preparation and dissemination of guidelines into local language and format; the evaluation of referral guidelines use; the organization of conference and workshops to promote awareness, advocacy and stakeholder engagement; and the studies to improve guideline implementation in facilities and health care systems.

5.5 Global Collaboration

The stakeholders have different perspectives and responsibilities, but play unique roles in the development, publication, distribution, implementation and evaluation of the use of referral guidelines. While some could have specific goals or particular interest in a certain aspect of guideline use, communication will identify value-adding opportunities and minimize working in isolation. Such multi-facetted projects require diversified expertise and adequate resources, which are beyond the means of a single organization. Professional organizations may not have the financial resources and international agencies may not have the professional expertise. Collaboration is a win-win solution.

Individual experts from professional organizations prepare evidence-based guidelines. UN agencies are excellent focal points to engage stakeholders and collate feedback, to liaise with their national counterparts, and to facilitate pilot and implementation. The WHO and IAEA have a well-developed network of national and regional contact points for health and radiation protection and safety. There is a higher likelihood of acceptance by national authorities if the material is endorsed and advocated by these agencies. Collaboration between the national authority, local professional organizations and other stakeholders paves the way for more effective pilot and implementation.

5.6   Global Guidelines

The stakeholders collaborating in the WHO/IRQN referral guidelines project recommended the preparation of a single set of guidelines rather than separate versions for different settings because the principle of Good Medical Practice is the same worldwide [14, 35]. The project scope includes guidelines development, trail, implementation and evaluation. This set of global guidelines represents best practice in ideal settings, but resources do vary between and within countries. For those settings with less resources, i.e. remote, rural or resource-poor countries, an adaptation of these guidelines in accordance to the local resources is required. Disease epidemiology varies and the gaps in the current guidelines will be filled over time, e.g. neglected tropical diseases.

5.7   Global Communication Platform

Guideline projects and actions require persistence and patience. Good communication overcomes barriers. Focusing on the different aspects, each action will improve use by addressing an individual element. However, these actions have the potential for wider impact by value adding and providing synergy to each other. Among the wide range of actions, there is likely overlap of project parameters and participating experts. Awareness of what others are doing will minimize duplication. However, this may not necessarily be the case and better communication between teams is desirable.

An informal and inclusive global communication platform for the stakeholders working in radiology referral guidelines will facilitate liaison, engagement, experience sharing and collaboration. The platform's possible roles are outlined in Table 6.5. Members of project teams can share knowledge and experience leading to cross-fertilization, better project design and outcome. However, the platform should not interfere with an individual project or infringe on a project's ownership.

5.8   Resources Mobilization

Funding is rather limited for guideline development, implementation and evaluation. As the guideline recommendations are applicable worldwide, it is desirable for the stakeholders to work together. Apart from collaboration between projects, collective resources mobilization to strengthen the funding for guideline actions could be considered. The UN agencies could serve as a neutral global platform.

Table 6.5 The possible roles of a global communication platform

Strengthen stakeholder communication and collaboration
Promote leadership and team building
Share experience and knowledge
Identify synergies and promote cross-fertilization
Minimize undertaking actions in isolation
Streamline development and update of guidelines
Collate and improve access to guidelines and related education tools
Trial and implement guidelines
Evaluate the use of guidelines
Facilitate innovations toward better end-user uptake
Be the registry for referral guidelines related projects
Identify gaps and proactively design future actions
Advocate and raise awareness as a coordinated group
Mobilize, share and maximize resources
Facilitate policy adoption and infrastructure strengthening
Advocate the reward to quality service providers

6 Conclusion

Radiology is indispensible in medicine and saves lives. The increasing number of studies showing an increase in cancer risk from the use of medical radiation [24] highlights the need to strengthen procedure justification and an appropriate use of radiological procedures. Unnecessary radiological procedures increase population exposure. Referral guidelines are effective procedure justification tools and promote Good Medical Practice. A major push for guidelines use is required.

This chapter outlines the rationale; current actions; development and implementation processes; and issues limiting and solutions improving the use of guidelines. Improvement strategies include actions to: advocate and promote awareness; conduct research and document evidence-based recommendations; provide education and training; strengthen system infrastructure; and apply effective policies [18].

Publication of guidelines alone is not enough to guarantee use and there is need to narrow the gap between knowledge and practice. It is unrealistic to expect user-friendly implementation tools will result in a dramatic increase in uptake and change. Change will be gradual but hopefully progressive with on-going support.

Each stakeholder plays a unique role. Experts from professional organizations draft guidelines with scientific vigor. With their experience, reputation, independent roles in health and radiation safety, records in successfully delivering complex global projects, and well-established connections with national authorities, the UN agencies are well placed to lead by providing a common platform for stakeholder engagement and participation. The links with the national counterparts facilitate implementation, monitoring and evaluation. There are opportunities yet to be explored to strengthen the existing arrangements.

Along this journey, leadership and collaboration between the stakeholders are crucial. Synergy is achieved by putting all the elements together into innovative packages, which is the responsibility of the leaders and users. The important role of the end-users play by accepting the recommendations, changing practice and using guidelines in daily decision-making must not be underestimated. Therefore, *evidence-based recommendations, strong advocacy, policy adoption and end-users support* are the key elements to improve guideline use (Fig. 6.2). An inclusive multi-sectorial global platform could be used as a forum for stakeholder engagement. These collective leadership and innovative actions will result in better, safer and more appropriate use of radiology.

References

1. American College of Radiology (2012) Appropriateness criteria. Available from http://www.acr.org/Quality-Safety/Appropriateness-Criteria Accessed 30 June 2012
2. Blackmore CC (2007) Defining quality in radiology. J Am Coll Radiol 4(4):217–223
3. Brenner DJ, Hall EJ (2007) Computed tomography: an increasing source of radiation exposure. N Engl J Med 357(22):2277–2284
4. Dunnick NR, Applegate KE, Arenson RL (2005) The inappropriate use of imaging studies: a report of the 2004 intersociety conference. J Am Coll Radiol 2(5):401–406
5. European Commission (2000) Radiation Protection 118. Referral guidelines for imaging. European Communities, Luxembourg
6. Fryback DG, Thornbury JR (1991) The efficacy of diagnostic imaging. Med Decis Making 11(2):88–94
7. General Medical Council, United Kingdom (2012) Good medical practice. Available from http://www.gmc-uk.org/guidance/good_medical_practice.asp. Accessed 14 July 2012
8. Georgen S (2010) They don't know what they don't know. J Med Imaging Radiat Oncol 54(1):1–2
9. Government of Western Australia (2012) Diagnostic imaging pathways. Available from http://www.imagingpathways.health.wa.gov.au/includes/index.html. Accessed 30 June 2012
10. Image Gently (2012) The alliance for radiation safety in pediatric imaging. http://www.pedrad.org/associations/5364/ig/. Accessed 30 June 2012
11. Image Wisely (2012) http://www.imagewisely.org/. Accessed 30 June 2012
12. International Atomic Energy Agency (IAEA1) (2012) Radiation protection of patients. http://rpop.iaea.org/RPOP/RPoP/Content/index.htm. Accessed 12 Mar 2012
13. International Atomic Energy Agency (IAEA2) (2012) Triple-A investment in patient health. Available from http://www.iaea.org/newscenter/news/2010/tripleinvestment.html. Accessed 30 June 2012
14. International Radiology Quality Network (2012) Referral guidelines project. Available from http://www.irqn.org/work/referral-guidelines.htm. Accessed 30 June 2012
15. InsideRadiology (2012) http://www.insideradiology.com.au/. Accessed 30 June 2012
16. Lau LSW (2006) A continuum of quality in radiology. J Am Coll Radiol 3(4):233–239
17. Lau LSW (ed) (2001) Imaging guidelines, 4th edn. The Royal Australian and New Zealand College of Radiologists, Sydney
18. Lau LSW, Perez MR, Applegate KE et al (2011) Global quality imaging: improvement actions. J Am Coll Radiol 8(5):330–334
19. Lehnert BE, Bree RL (2010) Analysis of appropriateness of outpatient CT and MRI referred from primary care clinics at an academic medical center: how critical is the need for improved decision support? J Am Coll Radiol 7(3):192–197

20. Malone JF (2008) New ethical issues for radiation protection in diagnostic radiology. Radiat Prot Dosimetry 129(1–3):6–12
21. National Council on Radiation Protection and Measurements (2009) Report No.160. Ionizing radiation exposure of the population of the United States. NCRP, Bethesda
22. Ng KH, Maclean ID (2011) Diagnostic radiology in the tropics: technical considerations. Semin Musculoskel Radiol 15(5):441–445
23. Oikarinen H, Meriläinen S, Pääkkö E et al (2009) Unjustified CT examinations in young patients. Eur Radiol 19(5):1161–1165
24. Pearce MS, Salotti JA, Little MP et al (2012) Radiation exposure from CT scans in childhood and subsequent risk of leukemia and brain tumors: a retrospective cohort study. Lancet 380(9840):499–505
25. RadiologyInfo (2012) http://www.radiologyinfo.org/. Accessed 30 June 2012
26. Smith-Bindman R (2010) Is computed tomography safe? N Engl J Med 363(1):1–4
27. Smith-Bindman R, Miglioretti DL, Larson EB (2008) Rising use of diagnostic medical imaging in a large integrated health system. Health Aff (Millwood) 27(6):1491–1502
28. Studdert DM, Mello MM, Sage WM et al (2005) Defensive medicine among high-risk specialist physicians in a volatile malpractice environment. JAMA 293(21):2609–2617
29. The Royal College of Radiologists (2012) RCR referral guidelines. Available from http://www.rcr.ac.uk/content.aspx?PageID=99. Accessed 30 June 2012
30. United Nations Scientific Committee on the Effects of Atomic Radiation (2010) Sources and effects of ionizing radiation. UNSCEAR 2008 report to the General Assembly with scientific annexes, United Nations, New York
31. Valentin J (ed) (2007) ICRP Publication 105. Radiological protection in medicine. Ann ICRP 37(6). Elsevier, Oxford
32. Virtual Departments (2012) http://www.goingfora.com/index.html. Accessed 30 June 2012
33. World Health Organization (2008) WHO global initiative on radiation safety in health care settings technical meeting report. WHO, Geneva
34. World Health Organization (2010) World health statistics 2010. WHO, Geneva
35. World Health Organization (2010) Medical imaging specialists call for global referral guidelines. Available from http://www.who.int/ionizing_radiation/medical_exposure/referral_guidelines.pdf. Assessed 30 June 2012
36. Zhou GZ, Wong DD, Nguyen LK et al (2010) Student and intern awareness of ionizing radiation exposure from common diagnostic imaging procedures. J Med Imaging Radiat Oncol 54(1):17–23

Chapter 7
Use of Imaging Appropriateness Criteria for Decision Support During Radiology Order Entry: The MGH Experience

Chris L. Sistrom, Jeffrey B. Weilburg, Daniel I. Rosenthal, Keith J. Dreyer, and James H. Thrall

Abstract This chapter will focus on the institutional perspective of The Massachusetts General Hospital (MGH) in Boston during the first decade of the twenty-first century. We will describe the MGH experience with design, implementation, adoption and current use of a computerized radiology order entry system that displays explicit normative "appropriateness scores" for outpatient CT, MR, and nuclear cardiology procedures and is embedded in routine clinical workflow. The radiology order entry (ROE) and decision support (DS) system has come to be known as ROE-DS within MGH and among a broader community of US radiologists, various industry stakeholders, and policy makers working in the field. One key point that will be emphasized is that a system like ROE-DS in wide use is a necessary, though by no means sufficient, first step in executing comprehensive institutional programs of imaging utilization management, quality improvement, and radiation dose mitigation.

The chapter will take the form of a historical narrative with emphasis on the following aspects over time: external stakeholders (payers, patients), internal stakeholders (physician's organization cost and quality management), service providers (radiology informatics, radiology leadership), and users (referring clinicians). Historical periods covered include time leading up to decision to create ROE-DS (up to 2003), planning and architecture (2003–2004), implementation and deployment (2004–2005), refinement and evaluation (2005-present), and future directions. Prior to telling the story of ROE-DS at MGH, it is useful to briefly

C.L. Sistrom (✉)
Department of Radiology, University of Florida, Gainesville, FL, USA
e-mail: csistrom@partners.org

J.B. Weilburg
Department of Psychiatry, Massachusetts General Hospital, Boston, MA, USA

D.I. Rosenthal • K.J. Dreyer • J.H. Thrall
Department of Radiology, Massachusetts General Hospital, Boston, MA, USA

L. Lau and K.-H. Ng (eds.), *Radiological Safety and Quality: Paradigms in Leadership and Innovation*, DOI 10.1007/978-94-007-7256-4_7,
© Springer Science+Business Media Dordrecht 2014

describe a broader historical context about the concepts of utilization variation and appropriateness, the RAND/UCLA Appropriateness Methodology, and the ACR Appropriateness Criteria for diagnostic imaging.

Keywords Appropriateness criteria • Decision support • Imaging utilization • Procedure justification • Quality improvement • Radiology order entry • Radiation exposure reduction • Utilization management

1 Healthcare Utilization, Variation, and Appropriateness

In 1973 Dr. Wennberg reported striking regional variations in the utilization of several common procedures [28, 45]. For example, tonsillectomy varied from 7 % in Middlebury to 70 % in Morrisville. Subsequently, Dr. Wennberg and colleagues expanded the survey and the Dartmouth Atlas of Healthcare (http://www. dartmouthatlas.org/) continues to document substantial variations in healthcare procedures including imaging.

Concerns were raised by these large unwarranted variations, which could not be explained by demographics or disease burden. There were no discernable differences in population health in the high and low utilization areas. These variations led to the concept of over-utilization, under-utilization and "right" levels of use and that higher use could possibly imply more inappropriateness.

The RAND/UCLA appropriateness method (RAM), which formed part of the RAND/UCLA healthcare utilization study (HCUS) was used to create and validate appropriateness criteria (AC) [6, 16]. As clinical trail data and technology assessment results did not exist at the time when the HCUS was conducted, a formal consensus process from a group of domain experts was used to increase reliability and validity [7, 12].

The RAM was initially applied to the procedures targeted for analysis in the HCUS, e.g. coronary angiography, coronary bypass surgery etc. The investigators collated a procedure's diagnostic or therapeutic efficacy and clinical outcome for different clinical scenario and invited the expert panel to assign appropriateness scores to each scenario. The experts assigned integer scores from 1 to 9 (1 = very inappropriate to 9 = most appropriate) to each procedure/clinical scenario pair. A determination of whether there was consensus was based on the dispersion of the scores and a summary score, asserted only if consensus was reached [6, 14]. In subsequent Delphi Rounds (numbering from 2 to 3) the experts discussed the scenarios for which they did not reach consensus and rescored them. Any items for which the panel failed to reach consensus were assigned a score of 0 while all others got the median score (1–9) derived from the first successful round.

Once AC scores were developed for the procedure/scenario pairs, hospital records and claims data were 'scored' for appropriateness. The age-/sex-adjusted population based rates of the same medical procedures were enumerated thus forming an analytic data set. Two landmark papers confirmed the high use regions had essentially the same distribution of appropriateness as the low use regions across all but one procedure. The only health service demonstrating any significant correlation between utilization and appropriateness was an imaging examination. Specifically, in regions with the highest rates of coronary angiography a significantly greater fraction were scored as inappropriate [11, 23].

The RAM has grown in popularity and has been studied extensively to determine its reliability, validity, and predictive power. The conclusion is that formally developed consensus of informed experts is an efficient way to estimate the appropriateness of specific types of medical care [3, 4, 20–22, 25–27, 29, 33, 39–42]. AC are probably best suited for a retrospective analysis of clinical records and claims data to measure appropriateness of various procedures [34]. An emerging use for AC is in clinical decision support especially together with computerized order entry [32].

2 ACR Appropriateness Criteria

The development and use of practice guidelines for medicine has been advocated by influential medical organizations for over 15 years [2, 18]. Printed guidelines are user-unfriendly and research is needed to better understand the barriers and improvements for the use of guidelines in radiology [17]. Dorfman stressed the importance of assessing utilization, developing guidelines, and performing outcomes-based research to determine efficacy of instituting guidelines in routine practice [13]. As the leading professional organization, the American College of Radiology (ACR) produces substantive and specific clinical guidelines that can be simply categorized as the why, when, and how of imaging. The ACR Practice Guidelines represent the response to *how* to perform imaging whereas the ACR Appropriateness Criteria (ACRAC) deal with the *why* and *when* questions.

The ACRAC were introduced in 1993 to define good patient care and to eliminate inappropriate use of radiologic procedures [9, 10]. The ACRAC embody the best current evidence and are the most comprehensive North American practice guidelines for diagnostic imaging, radiotherapy protocols, and image-guided interventional procedures for a large number of clinical scenarios. The ACR is committed to their on-going updates and has set aside the necessary resources and informatics infrastructure for this purpose.

When the MGH ROE system was updated to include appropriateness scores, the ACRAC were distributed in print and on CD. More recently, the criteria were available from the ACR website as downloadable pdf documents, which are organized by specialty/organ system and topics. This has increasingly become the dominant mode by which most users access the content.

The ACRAC is contained in one or more tables for each topic. For example, for "Neurological Imaging – Low Back Pain", there are six tables each representing a different "variant" (clinical scenario) within the broader rubric of low back pain (Fig. 7.1). The "Ratings" in column 2 are on the same scale as RAM and are grouped similarly but with slightly different wording. The MGH ROE-DS system

Clinical Condition:	Low Back Pain
Variant 1:	Uncomplicated acute low back pain and/or radiculopathy, nonsurgical presentation. No red flags (red flags defined in text).

Radiologic Procedure	Rating	Comments	RRL*
MRI lumbar spine without contrast	2		O
X-ray lumbar spine	2		☢ ☢ ☢
Myelography and postmyelography CT lumbar spine	2	In some cases postinjection CT imaging may be done without plain-film myelography.	☢ ☢ ☢ ☢
X-ray myelography lumbar spine	2		☢ ☢ ☢
Tc-99m bone scan with SPECT spine	2		☢ ☢ ☢
CT lumbar spine without contrast	2		☢ ☢ ☢
MRI lumbar spine without and with contrast	2		O
Rating Scale: 1,2,3 Usually not appropriate; 4,5,6 May be appropriate; 7,8,9 Usually appropriate			*Relative Radiation Level

Fig. 7.1 Appropriateness table from the 2011 version of the ACRAC Neurologic Imaging section and Low Back Pain topic. This is the first of six variant tables contained in the document

Fig. 7.2 Relative radiation level (*RRL*) symbol key included in every ACRAC topic document. It explains the symbols found in the far right column of each variant table

Relative Radiation Level Designations		
Relative Radiation Level*	Adult Effective Dose Estimate Range	Pediatric Effective Dose Estimate Range
O	0 mSv	0 mSv
☢	<0.1 mSv	<0.03 mSv
☢ ☢	0.1-1 mSv	0.03-0.3 mSv
☢ ☢ ☢	1-10 mSv	0.3-3 mSv
☢ ☢ ☢ ☢	10-30 mSv	3-10 mSv
☢ ☢ ☢ ☢ ☢	30-100 mSv	10-30 mSv
*RRL assignments for some of the examinations cannot be made, because the actual patient doses in these procedures vary as a function of a number of factors (eg, region of the body exposed to ionizing radiation, the imaging guidance that is used). The RRLs for these examinations are designated as NS (not specified).		

used the same scoring scale with the addition of colors (Red for 1–3, Yellow for 4–6, Green for 7–9). In the right hand column, "Relative Radiation Level" (RRL) complements the numeric rating scores. Figure 7.2 provides an explanation of the RRL designations. However, in theory, the RRL are already incorporated into the rating scores themselves [37].

3 History of ROE and DS at MGH

In the US, imaging interpretation (professional component) in all settings is reimbursed on a fee-for-service basis by the public payers (Medicare and Medicaid) and by almost all private insurance companies. The reimbursement for the production of the images (technical component) is mostly handled on a fee-for-service basis for outpatient imaging. All fee-for-service claims are coded with a "reason for procedure" by using the ICD-9-CM system. Accuracy and consistency of ICD coding became increasingly important as payers tried to reduce expenditure by creating increasingly complex and restrictive systems to limit coverage to a defined list of indications (ICD-9 code) and procedures (CPT/HCPCS code). Manual coding, claim submission and denial processing represent the largest component of billing expenses. There are two possible solutions to control coding costs: first, for the radiologists to attach ICD-9 codes to their reports, which could be problematic [36] or for the requesting clinicians to provide them at the time of ordering. This is precisely what the first MGH ROE system in 1999 was designed to do and the basic design and architecture reflects this legacy.

The ROE system is web-based and covers all but the most unusual outpatient imaging procedures. A master menu page (Fig. 7.3) provides a list of modalities, each with a drop-down submenu of procedures. After a procedure is selected, the system displays a procedure/scenario "landing page" (Fig. 7.4). At the same time, a scheduling system for outpatient imaging was incorporated into the ROE system. Procedure scheduling was and continues to be a major logistical and computational undertaking for MGH, which consumes as much resource as charge capture.

The ROE is a win-win system for both MGH and its referrers. For the Radiology Department, the administrative and revenue enhancing aspects were of great value. For the referrers, specialists and Primary Care Physicians (PCP), the system is time-efficient and user-friendly. It is integrated into their electronic medical record (EMR), is accessible by a single click, and covers all outpatient imaging procedures, including cardiac stress testing.

Further, ROE made the process of requesting imaging and health insurance reimbursement more streamlined. This was especially true with local providers such as the Massachusetts Blue Cross Blue Shield (BCBS) because the ROE system provided an "automatic preauthorization" number for these cases. This obviates any requirement to interact with radiology benefit management organizations (described below).

Health insurance providers were concerned by the increasing volume and use of "high cost" procedures and as a result their impact on profit and premium. Aiming to contain the rising reimbursement costs for imaging, insurers engaged Radiology Benefit Managers (RBMs) as authorization agents. To obtain prior approval, a referrer's staff was required to contact and discuss the procedure with a non-clinical RBM staff who would make a determination based on a list of questions. If the constellation of factors were consistent with those deemed by the RBM to be appropriate, the RBM would grant an "authorization number", which

Fig. 7.3 First level of outpatient radiology order entry (*ROE*) web interface. In addition to a dynamic menu of modalities and body regions, pending procedures for the currently active patient are displayed

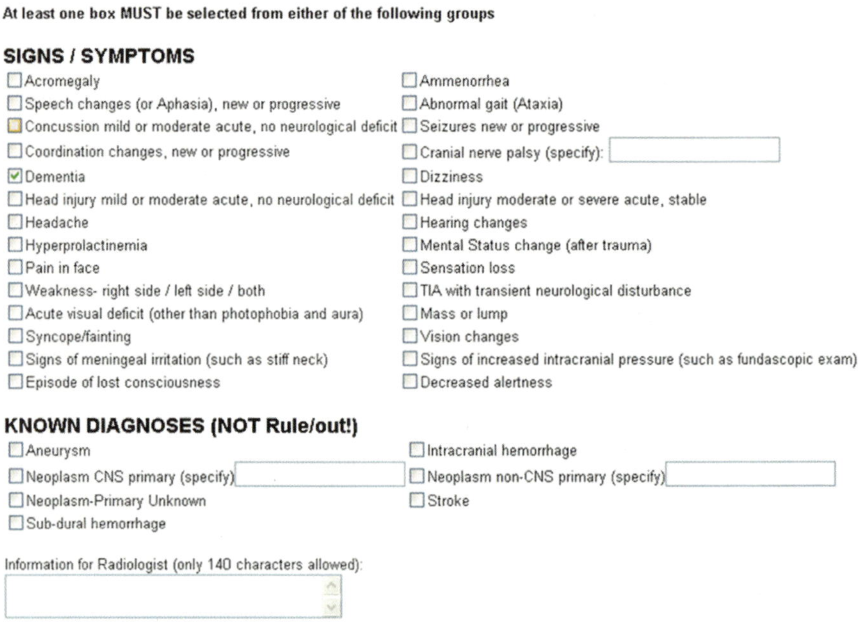

Fig. 7.4 Indication selection screen from the ROE-DS system interface. This one appears after CT of the head is selected on the first level menu and its contents are specific to CT of the head

was essential for a successful reimbursement claim. If approval were refused, the RBM would provide a nurse to review the material. If this review did not provide an approval, the requesting physician would be required to discuss the request with a RBM physician utilization manager. This process was both frustrating and time consuming and became the source of significant dissatisfaction for physicians and patients. The RBMs did not publish their approval criteria and the assessment process was neither transparent nor consistent. For example, different determinations might be reached with different RBM entities and even different case reviewers within the same RBM.

The MGH had poor experience with this RBM model, which was previously used in psychiatric and mental healthcare. The hospital network therefore negotiated with BCBS, leading to an agreed plan that was consistent with the "Pay-for-performance" schemes that were gaining traction in the early 2000s. Under this plan, the MGH network took 'risk' for imaging. A yearly utilization target, e.g. imaging procedures/1,000 members, was agreed and funds (typically 15 %) were set aside by the payer as a "withhold". These funds would be returned to MGH if imaging were equal to or lower than the target; otherwise the payer would retain it. This allowed the payer to mitigate the costs if imaging use continued to rise, and provided an incentive to the physicians and the hospital to control its use.

The MGH ROE-DS system provided a way for the physicians and hospital to identify and minimize imaging that was not appropriate. Inappropriate order could be abandoned, and this would at least in theory control the growth of costs in a medically appropriate manner. When the requested procedure was appropriate, an automatic authorization was granted. This innovative process saved time for the patients, referrers and provider and was clearly beneficial to all stakeholders.

The physicians and patients liked the system because scheduling was also streamlined. The physician's staff could show the patient a range of times and locations, and have the patient select the most convenient. A customized information sheet was printed and given to the patient outlining the confirmed appointment details, direction to the imaging facility, and procedure preparations. As a result most patients were very satisfied with this process.

To proceed from the initial version of ROE and add appropriateness feedback, it was necessary to add a new screen between the assertion of indications and scheduling of procedure. An example is shown in Fig. 7.5 from which the ROE-DS designers adopted the ACRAC method of scoring for procedure/indication groups (1–9 ordinal scale with appropriateness/utility increasing) and broke them up into three color-coded subgroups (Red for 1–3, Yellow for 4–6, Green for 7–9). The ACRAC did not and have not used or endorsed any such color scheme but preferred to stick with anchor phrases (see bottom of Fig. 7.1).

Populating the database and state tables was a substantial task, which rivals the scheduling logic in difficulty and complexity. Luckily, there were several years of data (through early 2004) to work with because the existing ROE system was already collecting information about the procedures and their indications. The first step was to map commonly occurring single indications or small groups of indications (clinical scenarios) to the ACRAC topic/variant descriptions. The

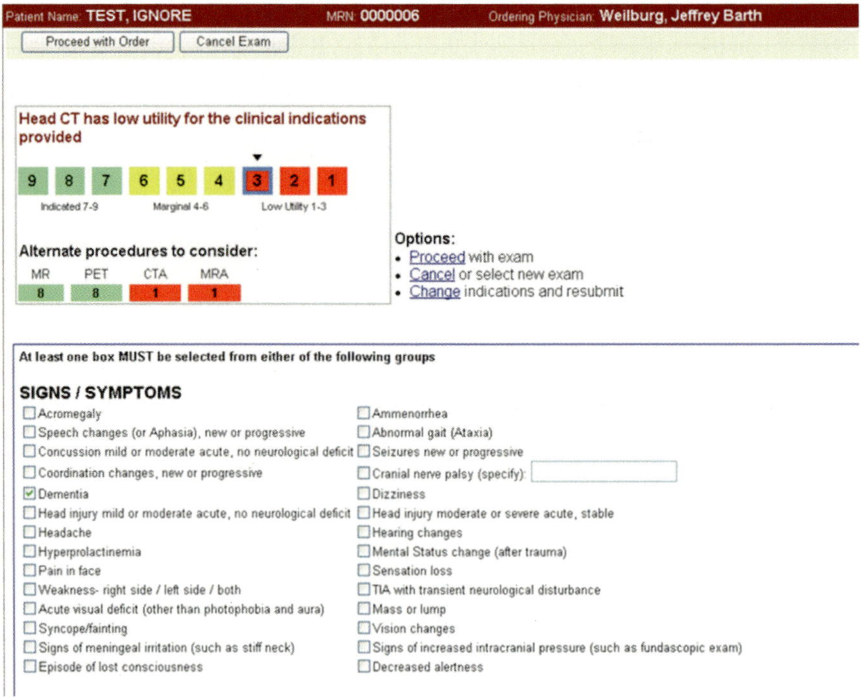

Fig. 7.5 Decision support feedback screen for CT of the head with only "Dementia" selected from the indications choices. In this case the score is 3, which is indicated as red/low utility. Note that "Alternate procedures to be considered" for the same clinical scenario (dementia not otherwise specified) and their scores are shown below the selected procedure. Clicking on one of the alternate modality links brings the user to the specific indication screen for the new modality and body area from which they can proceed to order the new procedure

current ACRAC contains ~2,000 scored scenario/procedure pairs for CT, MR, and nuclear medicine and the ones that were not accounted for in the empirical data were added to the ROE-DS system. Additionally, at least three times the number of scenario/procedure pairs had to be accounted for and scored such that 2006 version of the DS rules had 10,000 scenario/procedure pairs [31]. These "extra" rules (~8,000) were adjudicated by locally constituted expert panels of radiologists and physician colleagues from relevant disciplines and this process continues to the present through a standing committee made up of MGH Physician's Organization (MGH-PO) and radiology leadership that meets at least bi-weekly.

Subsequent to the introduction of the full ROE-DS system in late 2004, there have been two major re-designs of the web interfaces and underlying databases with the original functionality preserved and improved. One major milestone was to provide a ROE-DS "portal" that exposed both the DS rules and the scheduling engine to referrers from outside the MGH system. A "hard stop on red" policy and administrative change in 2007 is described more fully below.

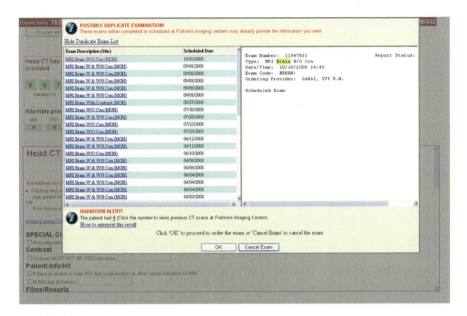

Fig. 7.6 Duplicate procedure pop-up screen which activates after indication selection only if the patient has had other procedures of the same body area in the past. Note that by clicking on any of the procedure instances to the *left*, the report contents are displayed on the *right*

Additionally, in late 2007, a "duplicate procedure" feature was added that displayed all prior procedures of the same body part of the procedure being ordered. A sophisticated internally developed system for "crawling" a patient's medical records with directed natural language processing throughout the whole Partners Healthcare System called the Queriable Patient Inference Dossier (QPID) was used for this enhancement [46]. QPID is triggered when a ROE-DS session is started after entering a patient's medical record number. By the time the ROE-DS user selects a procedure and asserts the indication(s), all pertinent imaging procedures from any setting at any Partners facility are ready for display along with the complete report texts. After completing the indication(s), but before the DS score is rendered, a screen like that shown in Fig. 7.6 is displayed. The user can click on any of the procedures and read the previous report(s). If the clinical question has already been addressed, the user can simply cancel out of the session by using the "cancel exam" button.

4 ROE-DS Evaluation and Results

Our group has published several papers that analyze and describe the results of the ROE-DS system for outpatient diagnostic radiology [31, 38, 43]. The first detailed the process of adapting and expanding the ACRAC to form the substrate of the

Table 7.1 Scored procedures through the first full year of ROE-DS (Nov 2004–2005)

Modality	Count	Score median (Mean)	Percent low utility (1–3)	% of total orders
CT	43,995	8 (8.1)	1 %	61 %
MRI	25,200	8 (7.7)	5 %	35 %
Nuclear cardiology	2,801	8 (7.0)	6 %	4 %

appropriateness (utility) ratings for CT, MR, and nuclear cardiology in ROE-DS [31]. Additionally, the paper gave some results from the time period between the introduction of DS scores (Nov 2004) through the first full year (2005) in terms of distribution of DS scores across modalities and body areas. One quantity of intense interest and speculation in terms of health system imaging costs and population radiation burden is the fraction of "inappropriate" or "unjustified" procedures. These data from the first 14 months of ROE-DS afford a unique opportunity for empirical estimation of this fraction in actual clinical practice rather than retrospective analysis of claims data and/or chart reviews. In ROE-DS, procedure clinical scenario combinations that receive scores of 1–3 are considered low utility and are displayed in red. Accordingly, the fraction of procedures with scores of 1–3 was called the "red rate" and has since then become a "term of art" in the field of imaging decision support. Perhaps the most interesting observation about the red rate is how it decreased during the first year of ROE-DS. During the first 2 months the red rate varied between 6 % and 7 % and then dropped to between 2 % and 3 % by the end of the first full year of operation. As we described in a subsequent paper, the red rate drifted back towards the 5–6 % level during 2006. A change in policy and interface functionality to force physician "unlocking" of low utility (score = 1–3) procedures brought the red rate down below 2 % where is has remained [43].

During that first full year of ROE-DS, patterns of red rate by modality and body area emerged which are also of interest as estimators of relative contribution of modalities and body areas to the burden of inappropriate (unjustified) procedures [31]. These are shown in Tables 7.1 and 7.2 respectively. Of note is that while CT scans comprise the majority of scored procedures, the red rate is quite low at 1 %. This should be of some reassurance and may prompt a shift in emphasis in CT from justification to optimization in terms of radiation dosage. On the other hand, it would seem that MRI might warrant more attention to the justification aspects of management strategies with red rate of 5 %. Even though MRI does not expose patients to ionizing radiation, it does have substantial cost for patients and/or the health finance system. It is hard to know how much effort to make towards improving justification of nuclear cardiology specifically (6 % red rate) because the field is currently in flux with likelihood of substantial shifts towards cardiac CT. That being said, there probably is considerable overlap among the common clinical scenarios associated with cardiac CT and nuclear cardiology. Therefore, justification of anatomic and functional imaging of the heart will become increasingly important. Fortunately, both the ACR and the American College of Cardiology have produced robust and comprehensive appropriateness criteria for cardiac

Table 7.2 Procedures with highest number of low utility scores through the first full year of ROE-DS function (Nov 2004–2005)

Procedure	Count	Score median (Mean)	Number low utility (1–3)	Percent low utility (1–3)	% of total low utility
Spine MRI	6,654	9 (7.8)	788	12 %	20 %
Extremity MRI	5,449	9 (8.0)	295	5 %	8 %
Spine CT	1,154	5 (5.6)	231	20 %	6 %
Head CT	2,575	5 (5.4)	181	7 %	5 %
Nuclear Cardiology	2,801	8 (7.0)	156	6 %	4 %
Face or sinus CT	869	8 (7.0)	91	10 %	2 %
Abdominal or pelvic MRI	1,215	6 (6.2)	52	4 %	1 %
Extremity CT	971	7 (6.3)	49	5 %	1 %
Chest MRI	202	5 (5.1)	37	18 %	1 %
Genitourinary MRI	318	6 (6.3)	27	8 %	1 %
Total	22,208	–	1,907	9 %	49 %

imaging exams and both organizations are committed to continued and expanded efforts.

The row entries in Table 7.2 represent specific studies that were highest in terms of absolute numbers of low utility (red) scores and in aggregate these ten account for roughly half of all of low utility procedures [31]. *Of interest is that spine imaging alone (CT and MRI) makes up over 20 % of the low utility procedures.* Upon review of these cases, a large fraction fit into the scenario described in Fig. 7.1 (uncomplicated acute low back pain without red flags). Given that all the modalities listed in that table have a score of 2, it seems logical that many low utility procedures would arise from that scenario. This should come as no surprise given that the issue of unnecessary/inappropriate (unjustified) advanced imaging of the spine (especially lumbar) has been the subject of a whole body of literature both in medical and health economics journals [1, 5, 8, 15, 19, 30, 35, 44]. The aggregate red rate of these, the most common low utility (inappropriate) procedures, was still only 9 %. This is quite low compared with more recent estimates made using retrospective review of studies that were ordered without a ROE-DS type system. For example, Lehnert looked at 459 MRI and CT procedures and determined that 26 % were inappropriate [24]. Other more informal estimates frequently range up to 30 %. As already noted, the overall red rate at MGH for outpatient MRI, CT, and nuclear cardiology has remained well below 10 % after introduction of ROE-DS and is currently hovering below 2 %. We assert that this is powerful evidence that clinicians are very responsive to comprehensive, normative, and immediate appropriateness feedback for outpatient imaging. We cannot make any such claims concerning inpatient or emergency department imaging and are working actively to expand the ROE-DS system into these settings.

In 2008, we retrospectively analyzed all MGH outpatient imaging utilization data from 2000 to 2008 [38]. Recall that the DS feedback for MR and CT was

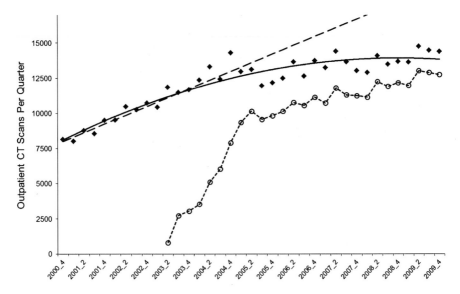

Fig. 7.7 Nine year time series of outpatient CT volumes at MGH. *Diamonds* indicate quarterly outpatient procedure volume. The *dashed line* is the linear trend calculated through 2004 and projected through 2009. The *solid line* is a second order polynomial trend calculated from 2000 through 2009. *Circles* and *dotted line* indicate the quarterly volume of outpatient procedures ordered through the ROE-DS system. Note, the published time series only went through 2007, additional quarterly volumes have been added for 2008 and 2009

turned on during the fourth quarter of 2004 and that the basic ROE system (without DS) had been operational since mid-2003. We also obtained quarterly volumes of outpatient visits from the MGH-PO for the same period. Individual piecewise linear regression models were estimated for CT and MR quarterly volumes respectively. These models had their "breakpoint" at the quarter 4 of 2004 and thus simultaneously estimated two slopes, one before ROE-DS and one after ROE-DS. We were also to estimate the magnitude and significance of the change in slope between the two periods. The models included quarterly outpatient visit volumes as adjustment variables to prevent confounding by external changes in overall practice activity.

For CT scanning the compound yearly growth rate in procedure volumes was 12 % before ROE-DS and 1 % afterwards with the difference in slopes being -11 % ($p < 0.001$). For MR scanning, the compound yearly growth rate in procedure volumes was 11.6 % before ROE-DS and 6.8 % after ROE-DS with the difference in slopes being -4.8 % ($p = 0.16$). These two time series are shown graphically in Figs. 7.7 (CT) and 7.8 (MR). Although the paper only reported on volumes through 2007, we have added quarterly outpatient volumes for each modality through 2009. It would seem fair to say that the volume of both CT and MR have not 'rebounded' and seem to be relatively stable although we have not extended the formal quantitative modeling including visit volume correction. Even if we were to extend the models into 2008 and beyond, the "great recession" undoubtedly has influenced

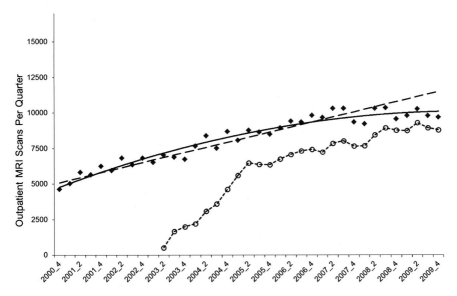

Fig. 7.8 Nine year time series of outpatient MRI volumes at MGH. *Diamonds* indicate quarterly outpatient procedure volume. The *dashed line* is the linear trend calculated through 2004 and projected through 2009. The *solid line* is an order 2 polynomial trend calculated from 2000 through 2009. *Circles* and *dotted line* indicate the quarterly volume of outpatient procedures ordered through the ROE-DS system. Note, the published time series only went through 2007, additional quarterly volumes have been added for 2008 and 2009

both overall practice activity (visits) and imaging volumes in ways that would likely confound any long run inferences about trends due to ROE-DS.

Another phenomenon documented in the time series was that the fraction of procedures ordered through the ROE system as opposed to phone and/or paper requests increased steadily over time to where well over 95 % of all outpatient diagnostic imaging procedures are now ordered and processed through the ROE-DS system. This includes all modalities though, to date, only for CT, MR, and nuclear cardiology are DS scores shown routinely. We would credit this to at least three factors. First, is that doctors, staff, and patients all find the ROE-DS system to be quite satisfactory in serving their respective needs in the ordering, scheduling, authorizing process. Secondly, the initial design choice for the ROE system to serve as a "one stop shop" for all outpatient imaging orders (even if DS is not offered) has been retained and reinforced. Lastly, intense and innovative medical management has the highest priority at upper levels of the MGH-PO leadership and universal ROE-DS use is emphasized strongly and consistently in these efforts.

Since inception, the ROE system was designed to allow order entry and scheduling of procedures by office staff. Each ROE user has a unique account and non-MD staffs are required to identify the clinician with whom they are working prior to ordering or scheduling procedures. After the DS rules were put into effect in late 2004, both clinicians and office staff users were permitted to over-ride a DS

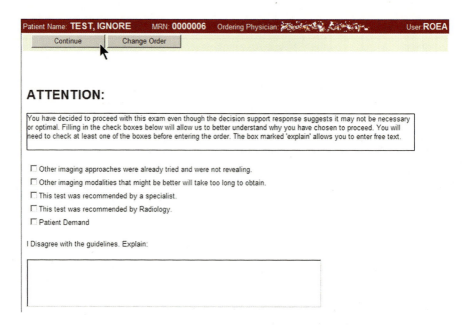

Fig. 7.9 Screen that is displayed when a clinician elects to "proceed on red". It provides for assertion of several common reasons for ordering an examaination despite a "red" DS score

score of 1–3 (red) for CT, MR, and nuclear cardiology procedures. Thus, if after selecting the procedure and asserting the indications a "red" DS score was returned, the user could proceed to order and schedule the procedure. Prior to doing so, an additional information screen (Fig. 7.9) would be presented where the user was requested to indicate why they wished to "proceed on red". In addition to a free text entry box the choices were:

- Other imaging choices were already tried and were not revealing
- Other imaging modalities that might be better will take too long to obtain
- The procedure was recommended by a specialist
- The procedure was recommended by a radiologist from a prior study
- Patient demand

In April of 2007, in an attempt to decrease the "red rate" a policy change was agreed upon by the MGH-PO and implemented in the ROE-DS system. Specifically, non-clinician (office staff) ROE users were no longer permitted to "proceed on red" by themselves. In such cases, the procedure request was "locked" and the user presented with a notification screen. The message informed the user that the clinician under whose proxy they were working would have to log on to ROE themselves and "unlock" the procedure. This became known around the institution as "hard stop on red" (HSOR). To estimate the effect of the HSOR intervention, we evaluated all ROE ordering events for CT, MR, and nuclear cardiology from two

Table 7.3 Results before and after implementation of "Hard Stop on Red" (HSOR)

	Pre HSOR (2006)		Post HSOR (2007)	
	Number	%	Number	%
Order initiated by				
MD	11,243	26.31 %	41,450	54.37 %
Non-MD	31,494	73.69 %	34,788	45.63 %
Total	42,737	100.00 %	76,238	100.00 %
Preliminary score				
1–3 (Red)	2,708	6.34 %	3,636	4.77 %
4–9 (Yellow, Green)	40,029	93.66 %	72,602	95.23 %
Total	42,737	100.00 %	76,238	100.00 %
Scheduled exam score				
1–3 (Red)	2,106	5.43 %	1,261	1.92 %
4–9 (Yellow, Green)	36,695	94.57 %	64,504	98.08 %
Total	38,801	100.00 %	65,765	100.00 %

periods: April-December 2006 (pre-HSOR) and April-December 2007 (post-HSOR). Ordering events included all sessions where any user logged on to the ROE system, selected a CT, MR, or nuclear cardiology procedure, asserted indications, and received a DS score. It is important to note that not all ordering events resulted in scheduled procedures and, in fact, the rate of ordering events that became scheduled procedures was what we specifically examined in the study [43].

Two main outcomes were analyzed stratified over period (pre-HSOR, post-HSOR) and DS score category (1–3 red, 4–9 yellow/green). These were the fraction of ordering events where the ROE user was a clinician (so called hands on keyboard rate) and the fraction of orders resulting in scheduled exams. Cochrane-Mantel-Haenszel methods were used for significance testing to account for the 2×3 table design structure. Table 7.3 gives a succinct summary of the most important results. The rate of clinicians logging on themselves for the initial ordering session instead of staff doubled from 26 % pre-HSOR to 54 % post-HSOR ($p < 0.0001$). The red rate of scheduled procedures fell from 5.4 % to 1.9 % ($p < 0.0001$). When we looked at the fraction of orders with initial yellow/green (DS score of 4–9) that resulted in a scheduled procedure it remained stable through both periods with pre-HSOR being 91 % (36,695/40,029) and post-HSOR being 88 % (64,504/72602) for a relative risk of NOT being scheduled (pre/post HSOR) of 0.988 (95 % CI = 0.984–0.992). This is in contrast to ordering events with initial red score with relative risk of NOT being scheduled (pre/post HSOR) of 3.54 (95 % CI = 3.20–3.92). We concluded that the HSOR intervention was quite successful in that the clinician "hands on keyboard" rate doubled, the red rate of scheduled procedures decreased by 65 % and yellow/green procedures were unaffected.

Finally, use of ROE allowed for the collection over several years of highly detailed data regarding which physicians ordered which images, for which patients, and for which reasons. This data was used to provide feedback to physicians regarding their performance. In addition to red rates, we were able to show the PCPs at MGH how their individual use of imaging overall, and then of CT, MR and nuclear cardiology, compared to the PCPs across the entire hospital and to the other PCPs within their practice. These Imaging Variation Reports are now provided to PCPs every 6 months, and have been very useful in engaging them in attempting to find ways to reduce imaging utilization variation. Variation analysis feedback was also provided for specialists in the group practice, starting with the neurologists. Analysis of other specialty areas, including oncology, is currently underway.

5 Future Directions

The implementation of a ROE-DS system is an innovative action aiming to improve quality care and reduce unnecessary exposure by a more appropriate use of imaging. Its on-going success and refinements hinge on strong leadership, stakeholder collaboration and end-user participation.

An obvious next step is to expand the ROE-DS system into inpatient and emergency room settings and this presents special challenges. Firstly, appropriateness scores for inpatient and emergency room clinical scenarios are not widely covered in the ACRAC and will have to be generated through local expert panel process as with many of the outpatient ROE-DS scores previously. Unlike the outpatient setting where convenient real-time scheduling of procedures was a substantial side-benefit for clinicians, office staff, and patients using the ROE system, the same advantage is harder to obtain for inpatient and emergency room imaging orders. Inpatient procedures are already done on a same day basis with triage rules for acuity (stat, urgent) already operating satisfactorily. Emergency room imaging orders are fulfilled in near real time on dedicated equipment. Further, neither inpatient nor emergency imaging studies are currently subject to external prior authorization/notification schemes by payers (e.g., radiology benefits management). Thus, the case for timesaving or convenience for clinicians and staff will be somewhat harder to make especially if a regime of collecting indications and giving DS score feedback is part of the package. Nonetheless, the institution as a whole has compelling motivation to understand, manage, and control inpatient and emergency room imaging utilization and this will only increase as payment schemes move towards greater shared financial risk by providers.

After a comprehensive ROE-DS system is in place, sophisticated and actionable medical management of utilization and appropriateness becomes possible. This entails comprehensive, granular, and complete capture and availability of all data generated during the ordering, scheduling, performing, reporting, and billing cycle. Any system for "profiling" providers in terms of resource utilization or "quality" metrics (i.e., appropriateness of imaging) requires very careful attention to methodically sound and transparent case-mix and risk-adjustment methods. This requires access to a large corpus of ancillary administrative and clinical data at the individual patient level in the institutional clinical data repositories. We are actively exploring many methods and models for analyzing, displaying, and distributing feedback to providers, divisions, and departments about utilization, variation, and appropriateness of imaging in outpatient, inpatient, and emergency room settings. Perhaps the most challenging is to expand these efforts to larger populations (e.g., insurance pools, regional Medicare/Medicaid beneficiaries) so as to understand and predict the mixture and costs of imaging that should be provided such that people get the all procedures that are necessary and few or none that are inappropriate.

Finally, it must be acknowledged that "appropriateness" in the sense of conformity with standard practice, is only an intermediate goal. The real target of utilization management must be patient outcome. Getting to that point will require some very difficult decisions about what percentage of procedures should lead to detectable changes in patient management in order for the procedure to qualify as "appropriate".

References

1. Ackerman SJ, Steinberg EP, Bryan RN et al (1997) Patient characteristics associated with diagnostic imaging evaluation of persistent low back problems. Spine 22(14):1634–1640
2. American Medical Association (1990) Attributes to guide the development of practice parameters. The Association, Chicago
3. Ayanian JZ, Landrum MB, Normand SL et al (1998) Rating the appropriateness of coronary angiography – do practicing physicians agree with an expert panel and with each other? N Engl J Med 338(26):1896–1904
4. Bernstein SJ, Hofer TP, Meijler AP et al (1997) Setting standards for effectiveness: a comparison of expert panels and decision analysis. Int J Qual Health Care 9(4):255–263
5. Boden SD, Swanson AL (1998) An assessment of the early management of spine problems and appropriateness of diagnostic imaging utilization. Phys Med Rehabil Clin N Am 9(2):411–417, viii
6. Brook RH (1994) The RAND/UCLA appropriateness method. In: McCormick KA, Moore SR, Siegel RA (eds) Methodology perspectives. Public Health Service, U.S. Department of Health and Human Services, Rockville
7. Brown BB (1968) The delphi process: a methodology used for the elicitation of opinion of experts. The RAND Corporation, Santa Monica
8. Carey TS, Garrett J (1996) Patterns of ordering diagnostic tests for patients with acute low back pain. The North Carolina Back Pain Project. Ann Intern Med 125(10):807–814
9. Cascade PN (1994) Setting appropriateness guidelines for radiology. Radiology 192 (1):50A–54A
10. Cascade PN (2000) The American College of Radiology. ACR appropriateness criteria project. Radiology 214(Suppl):3–46
11. Chassin MR, Kosecoff J, Park RE et al (1987) Does inappropriate use explain geographic variations in the use of health care services? A study of three procedures. JAMA 258 (18):2533–2537
12. Dalkey NC (1969) The Delphi method: an experimental study of group opinion. The RAND Corporation, Santa Monica
13. Dorfman GS (1999) Utilization of diagnostic tests: assessing appropriateness. Acad Radiol 6 (Suppl 1):S40–S51
14. Fitch KBS, Bernstein SJ, Aguilar MD et al (2001) The RAND/UCLA appropriateness method user's manual. The RAND Corporation, Santa Monica

15. Freeborn DK, Shye D, Mullooly JP et al (1997) Primary care physicians' use of lumbar spine imaging tests: effects of guidelines and practice pattern feedback. J Gen Intern Med 12 (10):619–625
16. Hicks NR (1994) Some observations on attempts to measure appropriateness of care. BMJ 309 (6956):730–733
17. Hillman BJ (1991) Practice policies. Rationale, methods of development, and implications for radiologic practice. Invest Radiol 26(7):689–693
18. Institute of Medicine (1990) Medicare: a strategy for quality assurance. National Academy Press, Washington, DC
19. Kendrick D, Fielding K, Bentley E et al (2001) The role of radiography in primary care patients with low back pain of at least 6 weeks duration: a randomised (unblinded) controlled trial. Health Technol Assess 5(30):1–69
20. Kravitz RL, Laouri M, Kahan JP et al (1995) Validity of criteria used for detecting underuse of coronary revascularization. JAMA 274(8):632–638
21. Kravitz RL, Park RE, Kahan JP (1997) Measuring the clinical consistency of panelists' appropriateness ratings: the case of coronary artery bypass surgery. Health Policy 42 (2):135–143
22. Kuntz KM, Tsevat J, Weinstein MC et al (1999) Expert panel vs decision-analysis recommendations for postdischarge coronary angiography after myocardial infarction. JAMA 282(23):2246–2251
23. Leape LL, Park RE, Solomon DH et al (1990) Does inappropriate use explain small-area variations in the use of health care services? JAMA 263(5):669–672
24. Lehnert BE, Bree RL (2010) Analysis of appropriateness of outpatient CT and MRI referred from primary care clinics at an academic medical center: how critical is the need for improved decision support? J Am Coll Radiol 7(3):192–197
25. McClellan M, Brook RH (1992) Appropriateness of care. A comparison of global and outcome methods to set standards. Med Care 30(7):565–586
26. McDonnell J, Meijler A, Kahan JP et al (1996) Panellist consistency in the assessment of medical appropriateness. Health Policy 37(3):139–152
27. Oddone EZ, Samsa G, Matchar DB (1994) Global judgments versus decision-model-facilitated judgments: are experts internally consistent? Med Decis Making 14(1):19–26
28. Paul-Shaheen P, Clark JD, Williams D (1987) Small area analysis: a review and analysis of the North American literature. J Health Polit Policy Law 12(4):741–809
29. Quintana JM, Escobar A, Arostegui I et al (2006) Health-related quality of life and appropriateness of knee or hip joint replacement. Arch Intern Med 166(2):220–226
30. Rao JK, Kroenke K, Mihaliak KA et al (2002) Can guidelines impact the ordering of magnetic resonance imaging studies by primary care providers for low back pain? Am J Manag Care 8(1):27–35
31. Rosenthal DI, Weilburg JB, Schultz T et al (2006) Radiology order entry with decision support: initial clinical experience. J Am Coll Radiol 3(10):799–806
32. Shekelle P (2004) The appropriateness method. Med Decis Making 24(2):228–231
33. Shekelle PG, Chassin MR, Park RE (1998) Assessing the predictive validity of the RAND/UCLA appropriateness method criteria for performing carotid endarterectomy. Int J Technol Assess Health Care 14(4):707–727
34. Shekelle PG, Kahan JP, Bernstein SJ et al (1998) The reproducibility of a method to identify the overuse and underuse of medical procedures. N Engl J Med 338(26):1888–1895
35. Shye D, Freeborn DK, Romeo J et al (1998) Understanding physicians' imaging test use in low back pain care: the role of focus groups. Int J Qual Health Care 10(2):83–91
36. Sistrom C, Drane W (2001) Networked ICD-9 coding system for a radiology department. AJR Am J Roentgenol 176(2):335–339
37. Sistrom CL (2009) The appropriateness of imaging: a comprehensive conceptual framework. Radiology 251(3):637–649

38. Sistrom CL, Dang PA, Weilburg JB et al (2009) Effect of computerized order entry with integrated decision support on the growth of outpatient procedure volumes: seven-year time series analysis. Radiology 251(1):147–155
39. Tobacman JK, Scott IU, Cyphert ST et al (2001) Comparison of appropriateness ratings for cataract surgery between convened and mail-only multidisciplinary panels. Med Decis Making 21(6):490–497
40. Tobacman JK, Zimmerman B, Lee P et al (2003) Visual acuity following cataract surgeries in relation to preoperative appropriateness ratings. Med Decis Making 23(2):122–130
41. Vader JP, Burnand B, Froehlich F et al (1997) Appropriateness of upper gastrointestinal endoscopy: comparison of American and Swiss criteria. Int J Qual Health Care 9(2):87–92
42. Vader JP, Porchet F, Larequi-Lauber T et al (2000) Appropriateness of surgery for sciatica: reliability of guidelines from expert panels. Spine 25(14):1831–1836
43. Vartanians VM, Sistrom CL, Weilburg JB et al (2010) Increasing the appropriateness of outpatient imaging: effects of a barrier to ordering low-yield examinations. Radiology 255 (3):842–849
44. Volinn E, Mayer J, Diehr P et al (1992) Small area analysis of surgery for low-back pain. Spine 17(5):575–581
45. Wennberg J, Gittelsohn A (1973) Small area variations in health care delivery. Science 182 (4117):1102–1108
46. Zalis M, Harris M (2010) Advanced search of the electronic medical record: augmenting safety and efficiency in radiology. J Am Coll Radiol 7(8):625–633

Chapter 8
Use of Referral Guidelines as a Justification Tool: National Perspective

Osnat Luxenburg, Esti Shelly, and Michal Margalit

Abstract The increasing volume of medical imaging procedures performed has raised the issue of radiation protection to be a matter of concern for the public, health professionals, regulatory authorities, as well as health facilities.

Justification and optimization are the cornerstones of radiation protection. Referral guidelines have been produced worldwide with the aim of promoting evidence-based good medical practice. It is an important tool for ensuring appropriate use and for reducing unnecessary exposure.

Implementation of referral guidelines on the national level is a complex issue requiring leadership, organizational change, education, infrastructure and others. Presented here are detailed and easy to follow steps for their successful implementation: (1) Advocate by *national leadership* is crucial for the success of the reform and achieving compliance. (2) Set up a *Task Force* of a group of stakeholders working in cooperation and coordination with the national leading authority. (3) *Select the appropriate referral guidelines* to suit the national policy, local needs and obligations. (4) *Identify the primary target groups* who will receive the greatest impact from the implementation of guidelines. (5) *Implement* by three synergistic pathways: *facilitating education and training* of the physicians, *creating organizational infrastructure*, and *developing information technology tools*. (6) *Monitor the use and evaluate the clinical impact*. (7) *Supervise and control*.

Successful implementation of such strategies and actions requires leadership, innovation and collaboration among the stakeholders. It is essential to sustain the implementation by maintaining a *cycle* of on-going updates and managing based on continuous feedback from the medical community, health institutions and the public, as well as from the data and information collected.

O. Luxenburg (✉) • E. Shelly • M. Margalit
Medical Technology Administration, Ministry of Health, 39 Yirmiyahu St.,
Jerusalem 9446724, Israel
e-mail: Osnat.Luxenburg@moh.health.gov.il

L. Lau and K.-H. Ng (eds.), *Radiological Safety and Quality: Paradigms in Leadership and Innovation*, DOI 10.1007/978-94-007-7256-4_8,
© Springer Science+Business Media Dordrecht 2014

143

Keywords Appropriate use • Implementation • Information technology • Justification • Medical education • Medical exposure • National leadership • National perspective • Organizational infrastructure • Policy making • Radiation protection • Referral guidelines • Task force

1   Introduction and Background

Every day around the world, ionizing radiation is used for the imaging of patients in more than 10 million diagnostic radiology procedures, and 100,000 diagnostic nuclear medicine procedures [23]. The increasing use of medical imaging procedures in the last decades, particularly CT and nuclear medicine, has raised the issue of radiation protection to be a matter of concern for the public, health professionals, regulatory authorities, as well as health organizations.

Physicians are referring their patients for many imaging procedures in order to get a quick and accurate diagnosis. However, it is a mixed blessing, since up to 2 % of cancers may be attributable to the radiation exposure itself during CT scanning [9].

Related issues such as: appropriate use, radiation exposure, quality, safety and cost, are raised and discussed in order to formulate quality improvement measures, strategies, and actions to maximize the benefits and minimize the risks. Successful implementation of such strategies and actions requires leadership, innovation and collaboration among the stakeholders both in the country and on an international level [30, 31].

Justification and *optimization* are the cornerstones of radiation protection. Good medical practice includes a justified and adequate referral, an optimized procedure and a correct diagnosis corresponding to the clinical request. In the process of carrying out an imaging procedure, justification serves as a gatekeeper while optimization applies to the way the procedure is performed [2, 32].

Once a procedure has been justified, the *optimization* of image quality and radiation protection should be obtained, while applying the ALARA principle (As Low As Reasonably Achievable) [38, 46]. For each imaging procedure performed, the radiation dose to the patient should be documented and monitored under quality control measurements and local or international guidelines [41].

The term *justification* relates to the appropriateness of the procedure. A medical exposure is justified when the benefit for the patient is larger than the expected harm [23, 25–27]. Prior to referring patients for procedures with radiation exposure, the physician should consider alternative imaging modalities with less, or no, radiation exposure, if the diagnostic yield is expected to be sufficient [2].

Regarding radiation safety and radiation protection, there is an increase in interest and actions in recent years from international agencies, professional organizations and competent authorities to promote appropriate use and to reduce unnecessary exposure [1, 14, 24–27]. For example, The IR(ME)R 2000 regulations require justification of every medical exposure and the establishment of recommendations concerning referral criteria for medical exposures [48].

Justification is applied at three levels: society, procedure, and individual. *On the society level*, the imaging tool must prove to be beneficial to society in terms of public health and economy. Potential benefits include high sensitivity in diagnosis, reduced morbidity and mortality, improved efficiency and cost containment. *Justification of the procedure* includes suitability between the clinical question and the imaging modality that was chosen. Referral guidelines can assist medical professionals to improve the quality of diagnosis and treatment. Lastly, the procedure needs to be justified on an *individual level* when the benefit for the patient is larger than the expected harm and after all other diagnosis alternatives were considered [23, 26, 27, 32].

For justified and optimized procedure, it is necessary for physicians to be aware; they must know the magnitude of the radiation dose given and weigh its potential risk and harm against its benefits.

Moreover, the global trend of the patient becoming an active participant in decision making regarding their disease management, forces the doctors and health care professionals to be more knowledgeable and updated. They must assist the patient in balancing the immediate risks of their presenting condition against the long-term risk of the radiation dose involved in the procedure they are being advised to undergo [22, 36].

1.1   Impact from Innovations

Several countries conducted surveys in order to assess the appropriate use of medical imaging of referring physicians. Among these countries are Sweden and Israel. One of the goals of the Swedish Radiation Protection Authority (SSI) research project of 2006 was to assess the degree of appropriateness for radiological procedures (focusing on CT). A sample of 4,714 referrals was evaluated by both clinicians and radiologists and the justification of these procedures was determined. This study showed a need for improvement concerning the justification of CT procedures. The study highlighted the importance of promoting the use of referral criteria for both prescribers and radiologists [2].

In Israel, A national survey performed to evaluate the appropriateness of CT and MRI procedures included a sample of 3,215 patients, who were interviewed upon arrival for CT and MRI procedures. Appropriateness levels were determined to be 65–90 %, (varies according to the body system that was being examined and the medical indication for which it was ordered). Following this survey, the second largest health care provider in Israel introduced a CT and MRI preauthorization program based on the American College of Radiology's (ACR) "Appropriateness Criteria" and the Royal College of Radiologists' (RCR) "Making the Best Use of a Department of Clinical Radiology". Preauthorization of CT according to these referral guidelines resulted in a significant decrease (33 %) in the number of referrals given. As a result, there was a reduction in unnecessary patient radiation exposure and imaging costs [7, 33].

Some organizations go a step further and focus on different ways to minimize radiation exposure from medical procedures, especially in vulnerable populations such as young children. In light of this, the Alliance for Radiation Safety in Pediatric Imaging developed the "*Image Gently*" campaign. Its goal is to change practice by increasing awareness of the opportunities to lower the radiation dose in the imaging of children [1, 18].

The uniqueness of this project is its approach to the parents as well as to the medical community, by using a media advertising campaign. The concept is to create higher awareness among parents as the health consumers so that the service providers will be required to meet the radiation protection standards [39]. A recent study performed in the United States demonstrated a decrease in the number of CT procedures performed in children's hospitals [50]. The Image Gently Alliance Campaign is one of the major contributors to this important reduction [19].

As mentioned above, several studies have shown that the degree of appropriateness of radiological procedures varies between countries and medical organizations [13, 40]. Unjustified imaging procedures may result from the referrers' knowledge gap, inadequate clinical assessment, self-referral or excessive practice of defensive medicine. Use of clinical guidelines can be a good solution to this problem.

2   Referral Guidelines

Referral guidelines may offer the clinician a simple, useful tool for referring patients to the most appropriate diagnostic, interventional or screening procedure for their condition. Referral guidelines cover a wide variety of clinical or diagnostic problems that physicians face daily and define the relevant diagnostic tools to use for each scenario. Clinicians, policy makers, and health facilities utilize referral guidelines as a pathway to lead clinical decision-making that creates consistent, efficient and cost-effective medicine [17, 20, 54], thus contributing to a more sustainable health care system.

The guidelines created by the Royal College of Radiologists in the UK [43], the ACR [3, 4], the EU (outlined by the Medical Exposures Directive) [15], and others, are examples that reflect the broad interest all over the world in improving quality of care as well as controlling the expenditure on expensive imaging technologies.

The greatest benefit for *patient care* that could be achieved by referral guidelines is to spare unnecessary procedures while improving health outcomes.

Various *health professionals* could benefit from referral guidelines. In hospitals, the guidelines may be useful for young physicians. In the community, the guidelines can serve as an updated tool for informed and effective use of the imaging procedures available. Clinicians may also turn to guidelines for medico-legal protection or to reinforce their position in dealing with administrators who disagree with their practice policies. Medical students should be trained to use referral guidelines as part of their routine decision-making process [22].

The potential benefit for *health care systems*, clinical departments, regulatory bodies and insurers is the reduction of the overload on health care organizations, which are already stressed by a global shortage of health care practitioners, while optimizing value for money [36].

However, referral guidelines may also have their limitations. Guidelines may act as a starting point for the clinicians but should not prevent them from tailoring care to the individual needs of the specific patient [53]. Unless guidelines are frequently reviewed and updated, they may result in sub-optimal, ineffective or harmful practice. This is a long term and on-going process that requires strong professional contribution along with investment of time and resources for its maintenance [16]. A fundamental limitation of the availability of practice guidelines per se is that they often do not change practice behavior. Most studies indicate that passive dissemination of guidelines, such as publishing them in a medical journal, is ineffective in changing behavior [11]. There are a variety of implementation strategies that can be used to increase the chances of effectively changing practice patterns. Some of these strategies include decision-support tools, reminder systems, audit, and feedback [22, 45].

3 National Perspective

There are various strategies to promote radiation protection of patients. One of them is justification of imaging procedures, which serves as a gate-keeper with an impact on population health status, medical organizations workflow and national expenditure on health.

This chapter will expand the discussion to address the various aspects of the design and implementation on a national level. Policy makers who are setting public health strategies have an incentive to implement referral guidelines (as a justification tool) since they ensure good medical practice. Additionally, they reduce medical expenditures and produce effective management of resources in the health system. Although referral guidelines may have local influence on specific health organizations, regional authorities, or sub-groups of physicians, the overall significant impact would only be manifested when it is implemented on a national level. For those countries who wish to implement guidelines widely, a detailed and structured model will be presented (Fig. 8.1).

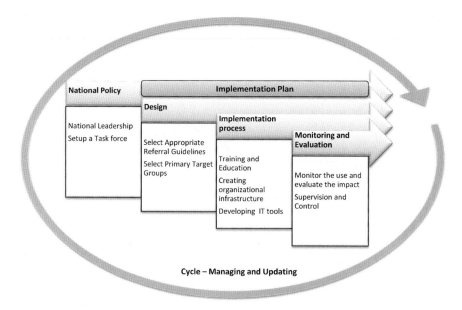

Fig. 8.1 Referral guidelines on a national level – implementation model

4 Steps for Successful Implementation of Referral Guidelines on a National Level

4.1 Constructing National Policy

Radiation protection is a broad issue, has a clear implication on public health and is relevant to many patients with a variety of diseases. In response, the policymakers lead by investing resources to construct and implement appropriate national policies. A wide view and comprehensive solutions are required. Clear national policy should be constructed and guide health facilities and professionals to promote public health. Relevant legislation should accompany the key actions to provide the required infrastructure for the implementation of the defined national policy.

Countries may vary in their health care infrastructure, legal framework and culture, which together influence the key elements of national policy. However, each country recognizes certain basic principles that are articulated in its health policy. In Israel, policy making on medical technologies and infrastructure has common factors, since they are all influenced by the health system, health care facilities, reimbursement mechanisms and delivery pathways to the community. The model presented in this chapter is a generic step-by-step guide, covering the key elements necessary for successful implementation of referral guidelines at a national level, based on the experience gained in recent years from the various Israeli programs in the medical technology field, including radiation protection.

National Leadership

National leadership is crucial for creating successful reforms, changing processes and achieving compliance among the different stakeholders. Once a decision is made regarding the importance of a certain national policy, the path to implementation must be constructed. Although countries may vary in their governing and regulatory bodies, the first step towards the implementation of national policy should be appointing a leading senior official authority to be accountable for achieving the goals.

The importance of national leadership is based on several points:

- It enables the actions to be executed and promoted on the national level. For example, the allocation of resources and the creation of the specific legislation or enterprise infrastructure at a national level.
- A comprehensive point of view is needed to integrate and align the main concept of the initiative with other national policies or actions.
- "Rome wasn't built in a day". Major processes may have long timelines, therefore it is important that once national regulatory bodies are held accountable, this sense of obligation will ensure the continuity of processes, duties and responsibilities.
- The implementation is more efficient due to process synchronization and the creation of national uniform methodology.
- There is a greater likelihood of achieving high compliance among stakeholders due to the leadership of a senior official.
- Only senior official regulatory bodies can interact with other international bodies in order to harmonize major policies and to discuss important issues with their counterparts, such as IAEA, WHO etc.

Gathering of all governmental agencies involved in radiation protection is an essential part of the appointed leader's responsibilities. For example, the initiatives of various agencies, which focus on different perspectives of radiation protection, should be integrated (such as protection of the worker, patient and the environment). This can be done through the Radiation Protection Board that meets periodically and consists of all senior representatives from the relevant government agencies.

In order to attain a high level of health care, it is necessary to interweave actions in a crisscross pattern – one at the Ministry level with other governmental bodies (horizontal axis), and the other with operating health organizations (vertical axis) (Fig. 8.2).

Once a decision is made that it is important to include referral guidelines within the core principles of a country's national radiological protection policy, this policy should be promulgated in the formal professional channels, as well as in the popular media, in order to facilitate professional and public dialogue and produce broad legitimacy. Each country should use the appropriate official tool required to disseminate its policy.

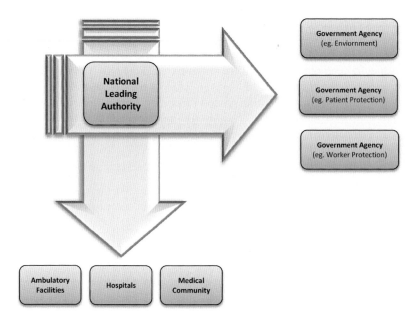

Fig. 8.2 National leadership – criss-cross pattern

Set up a Task Force

Due to the vast human resources and the long period of time required for the project, a task force should work alongside the national leading authority. The task force responsibilities include: defining the goals of the project, designing and approving implementation plan, evaluating and monitoring the success of the program and reporting to the national leading authority.

At the outset, it is of great importance to *identify* the relevant major stakeholders within the health system – including in-house stakeholders within the agency, governmental agencies, regulatory bodies, medical centers and professional societies [30, 31, 51]. From these, potential opposition organizations should be taken into consideration in order to anticipate possible points of contention that could be a stumbling block to implementing national guidelines.

The aim is to *collaborate* with as many different groups as possible – the government sector, public sector and private sector – including referrers, providers and consumers. Each of these groups can bring important contributions and insights to the process. In other words, the task force should include both policy makers as well as people who can implement the guidelines in "the field". This is expected to significantly increase the applicability of the decisions.

Of course it is not always possible to include all of the relevant players in this process. Therefore, in order to have a successful outcome, it is necessary to find other ways to include them in the process and harness their abilities. These measures may include multiplayer sessions, public hearings or periodic open meetings, allowing them to make comments and participate throughout the process. It is also important to hear the voices of practicing physicians in "the field", as well as the patients they treat. Since they are not a part of the decision-making process, the web can provide a tool for dialogue and raising suggestions, which encourages public participation and makes the process more transparent.

4.2   Design and Implementation Plan

Implementing a national program is a step-wise process; with many aspects that have far-reaching implications. It is important to recognize the "big picture" and identify the right pathway to accomplish these goals. Time and efforts should be invested in defining the goals, designing the project as well as recruiting the appropriate team. The plan may include incentives to promote the use of referral guidelines on an institutional level and on an individual physician level. For example, by offering economic incentives for cost savings to health organizations due to using referral guidelines, or institutional reward for physicians achieving the most justified referrals. The implementation process should be accompanied by appropriate publicity covering both the medical community as well as the general public.

4.3   Design

Select the Appropriate Referral Guidelines

Creating detailed referral guidelines requires intensive work and resources. As mentioned earlier, there are many referral guidelines from different professional groups that are evidence-based and resemble one another. Among others, referral guidelines have been produced by the Royal College of Radiologists in the UK [43], the American College of Radiology [3, 4], the European Commission [15], Canada [10] and Australasia [12, 42].

An example is the American College of Radiology's (ACR) appropriateness criteria, which are evidence-based guidelines developed by expert panels in diagnostic imaging, interventional radiology, and radiation oncology. The criteria currently include more than 160 topics with over 700 variants (ACR Website).

While many guidelines may validate each other, some may sometimes contradict. This could result in confusion and frustration for practitioners searching for appropriate guidelines to suit their needs. A compendium of national guidelines for imaging of pediatric patients, published in 2011 [52], tried to address this issue. This compendium generated a comprehensive list of 155 guidelines both for the use of specific imaging modalities, as well as guidelines to evaluate and manage specific diseases and clinical presentations.

Moreover, since there are different international guidelines, it is important to harmonize them with the national guidelines. Currently, a WHO core group was formed to spearhead the production of a set of global international referral guidelines for the appropriate use of radiation imaging under the Global Initiative on Radiation Safety in Health Care Settings. In this project, the experts from the International Radiology Quality Network (IRQN), International Atomic Energy Agency, the European Commission, as well as other experts from international, regional and national professional societies, agreed upon a roadmap to develop an international set of evidence-based referral guidelines and facilitate their implementation [28, 55]. It can be assumed that the WHO/IRQN guidelines will probably become the "gold standard" for professional societies, governmental agencies, health organizations and other stakeholders.

It is recommended that a country would select and adopt existing guidelines instead of "reinventing the wheel". Adjustments to the country's existing medical infrastructure, as well as to the current medical reimbursement policies, would increase the feasibility of adopting these guidelines. The national referral guidelines should be mandatory on one hand, but flexible on the other hand. This will enable appropriate adaptation on the institutional level according to local equipment availability, medical practice and population perception.

Select Primary Target Groups

It is essential to prioritize the most important and sensitive populations where the impact of the intervention will be the greatest. This can be accomplished by considering several factors, including the risk level, the population size and the compliance of the population. For example special populations like children and women of child bearing potential could be among the first populations to be addressed (Image Gently Website, [19]).

Different countries choose different methods to achieve the goal of minimizing radiation exposure to populations at risk such as young patients. It is recommended to start with a pilot project focusing on one of the groups selected, and only after gaining experience with this population, extend the project to other groups. Down the road, the experience and insight gained in the pilot project will guide future program design and implementation in a variety of medical situations.

4.4 Implementation Process

As mentioned in the previous sections, the implementation of a national program is complex. Sometimes national programs can encounter some initial resistance. Based on the experience gained in Israel, the following measures can be taken to reduce this resistance: increase the transparency of the process; outline the goals and outcomes clearly; appoint opinion leaders as part of the program, as spokespersons; and consult and coordinate with the different stakeholders before publishing the mandatory directives. Ideally, the initiative will receive full cooperation from these opinion leaders and stakeholders but when consensus cannot be reached, the leading authority must forge ahead in a thoughtful but assertive way in order to make progress in the long run.

It is crucial to synchronize and progress forward in parallel on three pathways: facilitating education and training of the physicians, creating organizational infrastructure, and developing information technology tools.

Education and Training

The current curriculum at medical schools does not sufficiently address the issue of radiation protection. Most referring physicians (who are not radiologists) do not receive enough formal education and training regarding radiation exposure and its potential risk. More emphasis is given to the introduction of new medical technologies and their advantages in diagnosis and treatment than to the accompanying disadvantage and potential hazards. Several studies have revealed that referring physicians possess limited knowledge about ionizing radiation and its effects. Less than 50 % of them were aware of referral guidelines and used them for many common procedures [5, 8, 29, 34, 35, 37, 44, 47].

Physicians need to be educated and trained in order to utilize referral guidelines. Bautista et al. examined in 2009 the use of ACR criteria among physicians when making decisions regarding radiologic imaging. Only 2.4 % of clinicians reported using ACR guidelines when making decisions. More popular and widely accessible resources, such as Up To Date, MD Consult, PubMed, and the physician's specialty journal, are preferred. Interestingly, 27.8 % of the physicians selected Google as one of their top three resources [6].

There are other studies confirming physicians are not using referral guidelines. Therefore, in order to succeed and reach a higher level of compliance, efforts should be made to promote *education and training* of physicians, starting with medical students, both on strengthening the knowledge and on practicing using the guidelines. This can be done by using conferences and continuing medical education setups, and internet-based education programs [31]. Other ways to strengthen education and training, especially in rural areas where the physician's population is scattered, are by providing long-distance learning and designing training-packages to be delivered on-site.

While the upgrade of the physician's knowledge-base and training is occurring, parallel steps should be taken to educate the general public. Since the average patient does not read scientific literature, but instead receives most of his/her information from other less scientific channels. More effective information channels should be used, such as: websites of health organizations, social networks and media. A balance between increasing awareness of the risks associated with ionizing radiation, while also emphasizing the clinical importance of receiving the appropriate procedures, should be achieved.

Creating Organizational Infrastructure

As part of the national strategy, it is crucial to construct the "vertical-axis" of the project – direct connection between the task force and the specific medical organizations (Fig. 8.2). In each health organization that takes part in the national project, a focal project manager or team should be appointed. They will have an ongoing liaison with the task force and will lead the implementation of the project within its organization. Their responsibilities will be to act as middle-men between the task force, the institution and the physicians. On one hand they will report and transfer data to the task force, as well as point out problems in the implementation process, and on the other hand they will actively collaborate with colleagues to develop, advise and help implement the referral guidelines.

The task force will instruct the medical organizations to follow these steps:

• Assess the local infrastructure and its ability to cope with the selected referral guidelines.
• Formulate adequate organizational practices to fit the national policy.
• Create incentives to use referral guidelines on an institution level and on an individual physician level.
• Promote consultation between referral physicians and radiologists within health organizations or through various interactive communications. In rural settings, where the availability of radiologists is limited, referring physicians "hotline" and teleradiology may serve as a solution.
• Monitor the use of referral guidelines and evaluate the clinical impact.

The institution team will follow the instructions of the task force within the boundaries of the institution's infrastructure and economical abilities.

Develop Information Technology Tools

In practice, building an implementation plan must include information technology (IT) development to ensure friendly and available decision support tools. Studies have shown that creating a user-friendly path to access referral guidelines leads to better dissemination [21, 49]. It is recommended to design easily available tools which (if possible) will integrate into the institution's existing clinical decision support systems. The principle of "KISSS" – Keep It Simple, Smart and Sustainable – should be applied.

Physicians often struggle to balance their time and, when searching for information, may use more easily accessible, and popular, sources such as PubMed or a simple Google search. Developing an effective IT system to house referral guidelines is crucial for their successful implementation and may decrease the use of other sources of information.

The decision on the appropriate level of IT development is influenced by the differences between countries in terms of population's size, level of IT, available financial resources for the project, as well as the characteristics of their health system. Some countries will choose to develop IT in a centralized way, while others will empower the medical organizations to develop individual systems. Some countries may also benefit from navigating (through IT systems) the referral physicians to consult with radiologists within their health organizations, or in pre-authorization centers [22].

4.5 Monitoring and Evaluating

Monitor the Use and Evaluate the Impact

An integral part of managing the national program is to monitor the use of referral guidelines and evaluate their clinical impact. In order to achieve these aims, the task force should map the initial status by using a survey. The data generated from this survey will define the baseline to evaluate the success of the project. The most important parameters used in the survey should reflect the utilization of referral guidelines and the existing infrastructural support, in terms of IT. When choosing the parameters that will be collected, it is necessary to consider the availability and the ease of accessing the data in the existing information system. Sometimes it is better to compromise and measure indirect data rather than trying to collect new data that is not supported in the existing system. Where possible, collecting data in a centralized way is preferable to combining data from individual sources.

Upon completion of the first phase of the project (pilot groups), there is a need to assess the change among physicians in using referral guidelines. Each country should define locally its standard for success, considering its national baseline. According to the results of the survey, presented to the national leading authority by the task force, alterations should be made before broadening the project to other populations.

Supervision and Control

Supervision entails inspections and audits in the health organizations, as well as data collection on national level. The use of existing control systems is recommended, although they may have been created for other purposes, in order not to over-burden the existing system by adding additional mechanisms. Therefore, the task force should operate with regulatory bodies that are periodically inspecting health organizations. It is useful to identify the existing supervision and control systems and integrate them as much as possible in the new mechanism. For example, databases of large reimbursement bodies may contain the required data on the use of certain imaging modalities, which can indicate a change in utilization patterns.

Each health organization that takes part in the national project, should report annually to the task force, by using a dedicated electronic report system. The task force will publish periodic reports and individual feedback will be given to the reporting institution. The importance of the feedback and the follow-up is to create incentives for continuous improvement and raise awareness in the medical community.

4.6   Cycle of Managing and Updating

Effective implementation on a national level depends on the ability to update and adapt to changing situations over time. At the last quarter of each fiscal year, a summary and conclusion regarding the current year will be concluded by the task force and the national leading authority. According to this assessment, an annual update process must take place and should review the following items:

1. Possible weaknesses in terms of cooperation.
2. Special populations that may need more attention and effort.
3. The degree of change observed in relation to the baseline.
4. New scientific evidence that may need to be integrated into the guidelines, or has already influenced other international guidelines.

Continuous feedback from the medical community, health institutions and the public, along with data and information collected by the task force, will contribute to the design of the next year's work plan, including allocation of resources. In the initial part of the project, the budget has to be funded by the leading national authority to facilitate the changes, to establish the infrastructure and to update the

Fig. 8.3 Cooperation and feedback – key for success

guidelines. The resources saved from appropriate referrals and prevention of unnecessary procedures in the following years will provide on-going funding for the updates of the guidelines and surveys on implementation.

5   Summary

The development and implementation steps outlined in this chapter are the practical steps that can be used by policy makers worldwide. The key for a successful process relies on national leadership, an effective task force and cooperation from the medical community and the public (Fig. 8.3). It is crucial to progress in three parallel pathways: facilitating education and training of the physicians, creating organizational infrastructure, and developing information technology tools. Updating and managing by national leadership will ensure the continuity of the process, promote good medical practice and facilitate effective utilization of the health system resources.

References

1. Alliance for Radiation Safety in Pediatric Imaging (Image Gently). http://www.pedrad.org/associations/5364/ig/. Accessed 22 Jul 2013
2. Almén A, Leitz W, Richter S (2009) National survey on justification of CT-examinations in Sweden. Swedish Radiation Safety Authority. Report number: 2009:03, ISSN: 2000-0456
3. American College of Radiology (ACR) (2011) ACR practice guideline for performing and interpreting diagnostic computed tomography (CT). Revised 2011 (Resolution 35)
4. American College of Radiology (ACR) (2012) ACR Appropriateness Criteria Website. Available at: http://www.acr.org. Accessed 30 Jan 2012
5. Amis ES Jr, Butler PF, Applegate KE et al (2007) American College of Radiology white paper on radiation dose in medicine. J Am Coll Radiol 4:272–284
6. Bautista AB, Burgos A, Nickel BJ et al (2009) Do clinicians use the American College of Radiology Appropriateness criteria in the management of their patients? AJR Am J Roentgenol 192(6):1581–1585
7. Blachar A, Tal S, Mandel A et al (2006) Preauthorization of CT and MRI examinations: assessment of a managed care preauthorization program based on the ACR Appropriateness Criteria and the Royal College of Radiology guidelines. J Am Coll Radiol 3(11):851–859
8. Borgen L, Stranden E, Espeland A (2010) Clinicians' justification of imaging: do radiation issues play a role? Insights Imaging 1:193–200
9. Brenner DJ, Hall EJ (2007) Computed tomography-an increasing source of radiation exposure. N Engl J Med 357(22):2277–2284, Review
10. Canadian Association of Radiologists (CAR) (2005) Diagnostic imaging referral guidelines. A Guide for Physicians, Quebec
11. Davis DA, Thomson MA, Oxman AD et al (1992) Evidence for the effectiveness of CME: a review of 50 randomized controlled trials. JAMA 268:1111–1117
12. Diagnostic Imaging Pathways. Website Available at: http://www.imagingpathways.health.wa.gov.au/includes/index.html. Accessed 22 Jul 2013

13. Dunnick NR, Applegate KE, Arenson RL (2005) The inappropriate use of imaging studies: a report of the 2004 intersociety conference. J Am Coll Radiol 2:401–406
14. European Commission (EC) (1997) Council Directive 97/43/Euratom, on health protection of individuals against the dangers of ionizing radiation in relation to medical exposure, and repealing Directive 84/466/Euratom. Off J Eur Commun No L180/22-27, 9.7
15. European Commission (EC) (2000) Referral guidelines for imaging. http://ec.europa.eu/energy/nuclear/radioprotection/publication/doc/118_en.pdf. Accessed 30 Jan 2012
16. Feder G (1994) Management of mild hypertension: which guidelines to follow? BMJ 308:470–471
17. Field MJ, Lohr KN (1990) Clinical practice guidelines: directions for a new program. National Academy Press, Washington, DC
18. Goske MJ, Applegate KE, Boylan J et al (2008) The image gently campaign: working together to change practice. AJR 190(2):273–274
19. Goske MJ, Applegate KE, Bulas D et al (2011) Approaches to promotion and implementation of action on radiation protection for children. Radiat Prot Dosimetry 147(1–2):137–141
20. Grimshaw JM, Russell IT (1993) Effect of clinical guidelines on medical practice: a systematic review of rigorous evaluations. Lancet 342:131722
21. Grol R, Cluzeau FA, Burgers JS (2003) Clinical practice guidelines: towards better quality guidelines and increased international collaboration. Br J Cancer 89(Suppl 1):S4–S8
22. Hendee WR, Becker GJ, Borgstede JP et al (2010) Addressing overutilization in medical imaging. Radiology 257:240–245
23. Holmberg O, Maloneb J, Rehani M et al (2010) Current issues and actions in radiation protection of patients. Eur J Radiol 76(1):15–19
24. International Atomic Energy Agency, Radiation Protection of Patients (IAEA-RPOP). http://rpop.iaea.org. Accessed 22 Jul 2013
25. International Atomic Energy Agency (IAEA), Food and Agriculture Organization of the United Nations (FAO), International Labour Organization (ILO), Organization for Economic Co-operation and Development–Nuclear Energy Agency (NEA), Pan American Health Organization (PAHO), World Health Organization (WHO) (1996) International basic safety standards (BSS) for protection against ionizing radiation and for the safety of radiation sources, vol 115, Safety series. IAEA, Vienna
26. International Commission on Radiological Protection (ICRP) (2007) The 2007 recommendations of the international commission on radiological protection. ICRP Publication 103, Ann ICRP, 37 (2–4). Pergamon Press, Oxford
27. International Commission on Radiological Protection (ICRP) (2007) Radiation protection in medicine. ICRP Publication 105, Ann ICRP 37(6). Pergamon Press, Oxford/New York
28. International Radiology Quality Network (IRQN) (2012) Global initiative on radiation safety. http://www.irqn.org. Accessed 22 Jul 2013
29. Kumar S, Mankad K, Bhartia B (2007) Awareness of making the best use of a department of clinical radiology among physicians in Leeds teaching hospitals. UK Br J Radiol 80:140–141
30. Lau LS (2007) Leadership and management in quality radiology. Biomed Imaging Interv J 3 (3):e21
31. Lau LS, Pérez MR, Applegate KE et al (2011) Global quality imaging: improvement actions. J Am Coll Radiol 8(5):330–334
32. Lau LS, Pérez MR, Applegate KE et al (2011) Global quality imaging: emerging issues. J Am Coll Radiol 8(7):508–512
33. Luxenburg O, Vaknin S, Polak G et al (2004) Annual research report. The Israel National Institute for Health Policy and Health Service Research, Hebrew
34. Malone JF (2008) Invited paper: new ethical issues for radiation protection in diagnostic radiology. Radiat Prot Dosimetry 129:6–12
35. Malone JF (2009) Invited paper: radiation protection in medicine: ethical framework revisited. Radiat Prot Dosimetry 135:71–78

36. Malone J, Guleria R, Craven C et al (2012) Justification of diagnostic medical exposures, some practical issues: report of an International Atomic Energy Agency Consultation. Br J Radiol 85:523–538
37. Mankad K, Bull M (2005) Awareness of "making the best use of a department of clinical radiology" among physicians. Clin Radiol 60:618–619
38. McCollough CH, Primak AN, James Kofler NB et al (2009) Strategies for reducing radiation dose in CT. Radiol Clin North Am 47(1):27–40
39. Nickoloff EL, Alderson PO (2001) Radiation exposures to patients from CT: reality, public perception, and policy. AJR 177(2):285–287
40. Oikarinen H, Meriläinen S, Pääkkö E et al (2009) Unjustified CT examinations in young patients. Eur Radiol 19(5):1161–1165
41. Rehani MM, Frush DP (2011) Patient exposure tracking: the IAEA smart card project. Radiat Prot Dosimetry 147(1–2):314–316
42. Royal Australian and New Zealand College of Radiologists (RANZCR) (2001) Imaging guidelines. Western Australian Department of Health, Perth. Available at: http://www.imagingpathways.health.wa.gov.au/includes/index.html. Accessed 30 Jan 2012
43. Royal College of Radiologists (2007) Making the best use of clinical radiology services, Referral guideline, 6th edn. The Royal College of Radiologists, London
44. Shiralkar S, Rennie A, Snow M et al (2003) Doctors' knowledge of radiation exposure: questionnaire study. BMJ 327:371–372
45. Sistrom CL, Dang PA, Weilburg JB et al (2009) Effect of computerized order entry with integrated decision support on the growth of outpatient procedure volumes: seven-year time series analysis. Radiology 251:147–155
46. Slovis TL, Berdon WE (2002) The ALARA (as low as reasonably achievable) concept in pediatric CT: intelligent dose reduction. Pediatr Radiol 32:217–317
47. Soye JA, Paterson A (2008) A survey of awareness of radiation dose among health professionals in Northern Ireland. Br J Radiol 81:725–729
48. The Stationery Office (2000) Ionizing radiation (medical exposure) regulations. The Stationery Office, London
49. Thomson R, Lavendar M, Madhok R (1995) How to ensure that guidelines are effective. Br Med J 311:237–241
50. Townsend BA, Callahan MJ, Zurakowski D et al (2010) Has pediatric CT at children's hospitals reached its peak? Am J Roentgenol 194:1194–1196
51. Vano E (2011) Global view on radiation protection in medicine. Radiat Prot Dosimetry 147 (1–2):3–7
52. Williams CH, Frush DP (2012) Compendium of national guidelines for imaging of the pediatric patient. Pediatr Radiol 42(1):82–94
53. Woolf SH (1997) Shared decision-making: the case for letting patients decide which choice is best. J Fam Pract 45:205–208
54. Woolf SH, Grol R, Hutchinson A et al (1999) Potential benefits, limitations, and harms of clinical guidelines. BMJ 318:527–530
55. World Health Organization (WHO) (2012) Global referral guidelines. http://www.who.int/ionizing_radiation/about/med_exposure/en/index2.html. Accessed 22 Jul 2013

Chapter 9
Team Approach to Optimize Radiology Techniques

Robert George, Lawrence Lau, and Kwan-Hoong Ng

Abstract In radiology, a team approach is the key to successfully optimize image quality and radiation protection, thus contributing to good quality radiology and patient-centered care. This is achieved by a combination of leadership, collaboration and participation. This chapter outlines the actions used in radiology facilities and health care systems towards this common goal.

Keywords Leadership • Optimization of protection • Radiology technique • Radiation protection • Radiation safety • Teamwork

1 Common Goal

Radiation safety is one of the key elements of quality in radiology [3, 18]. The focus for radiology facilities, whether sited in a major teaching institution, a small town, a suburban private clinic or a remote rural practice is to provide the best possible patient-centered care. For health care systems, the stakeholders work together to undertake research, promote awareness, provide education and training, strengthen

R. George (✉)
International Society of Radiographers and Radiological Technologists,
Sturt, SA, Australia
e-mail: robgeo@bigpond.net.au

L. Lau
International Radiology Quality Network, 1891 Preston White drive, Reston,
VA 20191-4326, USA
e-mail: lslau@bigpond.net.au

K.-H. Ng
Department of Biomedical Imaging and Medical Physics Unit,
University of Malaya, Kuala Lumpur, Malaysia

University of Malaya Medical Centre, Kuala Lumpur, Malaysia
e-mail: ngkh@ummc.edu.my

L. Lau and K.-H. Ng (eds.), *Radiological Safety and Quality: Paradigms
in Leadership and Innovation*, DOI 10.1007/978-94-007-7256-4_9,
© Springer Science+Business Media Dordrecht 2014

infrastructure and apply effective policies to ensure practice safety and the radiation protection of patients and staff [20, 21].

Radiology and medical imaging (to be referred to as radiology for this chapter), by their very nature, are based on teamwork. Regardless of the facility, the critical issue is management leadership and team participation towards patient-centered care by implementing innovative actions along the patient journey from appointment, registration, procedure, and report finalization to result dispatch. For example, a team approach is essential to achieve and maintain a robust and effective quality assurance program. In radiation protection, the emphasis is to minimize the radiation dose in keeping with the ALARA principle, i.e. the exposure should be consistent with the clinical needs and relevant to economic and social factors [35]. The teams work together to maximize the benefits and minimize the risks of these radiation procedures, thus supporting one of the important tenets of the Hippocratic Oath ". . . to do no harm . . ." [31].

Amongst the many actions used to improve radiation safety and quality in radiology, the optimization of image quality and radiation protection in at the workplace is perhaps one of the most tangible. The outcome judged by the image quality and the exposures is immediate, which provides instant feedback and job satisfaction to the teams. Given the dramatic increase use of radiology worldwide, especially with CT procedures where a recent study showed a 330 % increase use in an emergency department [15], it is important to ensure that the ALARA principle is correctly applied. There is no doubt about the important role radiology plays in patient care, but it is the teams' responsibility to ensure that exposures are appropriately managed.

This chapter focuses on the teamwork in optimizing technique and radiation protection in radiology facilities and in health care systems. The same elements of leadership, communication, collaboration, participation and implementation of innovations are equally applicable. The outcome as a result of these actions supports Good Medical Practice, professionalism, quality care, radiation safety and appropriate use of radiology. Underpinning these actions is an understanding of the issues, leveraging of the opportunities and application of the solutions. In essence, it is about teamwork towards the optimization of technique and radiation protection together with justification and error reduction measures in every procedure.

2 Leadership and Teamwork

Regardless of the setting, i.e. well- or under-resourced states; high or low income nations; remote, rural or suburban sites; large or solo facilities; academic or private institutions; agencies or professional organizations, a strong and committed leadership is essential to champion and bring innovative ideas into tangible actions.

In many situations, collective team action is needed to address the different facets and to achieve better outcome. For example, the development, maintenance and improvement of an effective quality assurance program in a radiology facility depend on the combined efforts from radiologists, radiographers, medical physicists and other team members. For health care systems, professional organizations and regulatory authorities could jointly host training workshops to improve image quality and radiation protection for multi-detector computed tomography (MDCT) by applying clinical audit, justification and technique optimization [30]. To bring such knowledge into everyday practice, practitioner participation in these workshops and the subsequent application of the lessons learnt in the facilities are required.

2.1   Individual Leadership

The directors of radiology facilities, institutions, professional organizations and agencies provide the leadership to underpin the team approach by: setting direction; improving communication; raising awareness in the effects of radiation, radiation protection and safety standards; providing resources and support towards the education and training of the team members and the sharing of experience; implementing best practice elements of teamwork; using quality maps; and applying change management measures.

2.2   Team Efforts

Working effectively as a team requires the collaboration and active participation of the members. The improvement actions in radiology facilities require a collective approach to: enable team learning; implement quality control, assurance and improvement actions; conduct clinical audits; apply quality tools to monitor and minimize dose and improve image quality; learn and apply error reduction and risk minimization measures; develop a collaborative, quality and safety culture; participate in practice accreditation; and implement safety standards. For health care systems, the creation and maintenance of good communication channels and networks and the sharing of experience and resources between institutions, organizations and agencies will minimize duplication, facilitate collaboration and provide synergy.

2.3   Roles and Responsibilities

In radiology facilities, the best outcome will be achieved if the stakeholders are committed, play their roles and fulfill their responsibilities by: doing the right procedure, using the right technique, for the right patient, at the right time, for the right reason, and at the right cost [19]. In everyday practice, the stakeholders include the referring clinicians, radiologists, radiographers (medical imaging technologists), medical physicists, clerical and nursing staff.

Their roles and responsibilities are well documented. For example, the referrers and radiology providers shall: recognize and respect the primacy of the patients; avoid unnecessary duplication of procedures; ensure appropriate and effective procedures are selected and performed; use alternative procedures without ionizing radiation whenever possible and appropriate, especially in children and pregnancy; and beware and minimize the potential procedural risks including radiation exposure.

The referring clinicians shall: request the procedures only if the results will alter patient management; provide relevant clinical details; ensure the procedures are not a substitution for adequate clinical examination; and consult the radiology team if and when required. The radiology providers shall: perform procedures based on skill, competence and knowledge; work with other team members and ensure minimal radiation exposures are used to achieve a diagnosis, e.g. by using shielding, appropriate technical factors etc.; minimize repeats by applying quality control, quality assurance and quality improvement measures; and provide and deliver timely and accurate results.

Governments and regulatory authorities protect the public and ensure quality, safety and appropriate use of radiology by regulations and policies, e.g. the requirements for radiation safety are detailed in the International Basic Safety Standards [10]. The responsibilities of the competent authorities are to: promote appropriate access to radiology, irrespective of income, location and cultural background; ensure the services and patient care are integrated and coordinated across facilities; regulate and maintain a quality and safe health care framework and promote quality and safety culture; ensure provider competency by registration; encourage a cost effective use of resources; and aspire to a sustainable health care system supporting quality radiology services.

3 Team Approach in Radiology Facilities

The following discussions refer to the actions used by team members to optimize radiology technique in radiology facilities.

3.1 Team Members

The front-line team members in primary contact with the patients include radiographers, radiologists and radiology nurses. They work in close collaboration with medical physicists, engineers, administrators and other supporting staff. While organizational structures may vary, the members usually work in specific areas or rooms under an overall management. The efforts to improve radiation safety and quality in radiology cut across all areas and require a collective approach. Many actions would benefit from

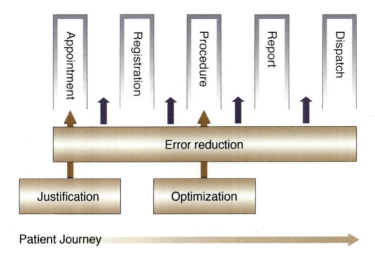

Fig. 9.1 Patient journey and protection measures. The five key processes along the journey are: appointment, registration, procedure, report, and dispatch. Procedure justification and optimization of image quality and radiation protection (*long arrows*) are the pillars of radiation protection. Procedure justification, sited at the beginning of this journey serves a 'gate-keeper' to minimize unnecessary and unintended exposures. Error reduction measures (*short arrows*) minimize human errors along this journey before, during and after a procedure

collaborations to achieve facility-wide improvement. For an individual patient, effective communication between team members will ensure better tailoring of the procedure and more useful diagnostic data by using optimal exposure.

3.2   *Patient Journey*

The patient journey describes the steps before, during and after a procedure. An analysis of these steps provides a better understanding of the roles of the team members and the possible improvement actions. Procedure justification and optimization of image quality and radiation protection are the pillars of radiation protection [35]. Error reduction minimizes human errors along this journey.

There are five key processes along the patent journey (Fig. 9.1), namely:

- Appointment;
- Registration;
- Procedure;
- Report; and
- Dispatch.

It is immensely valuable to regularly review the potential risks and improvement opportunities in these key processes and communicate the findings and actions to the members responsible. Mentoring of new members by informing them of their roles and responsibilities and the importance of a team approach is essential.

3.3 Appointment

Referrers usually request radiological procedures after careful consideration of the relevant criteria, and the choice is based on many factors including geographic access and available modality. It is important for referrers to have access to up-to-date referral guidelines. In complex cases, vulnerable groups or whenever there is uncertainty about choice of procedure, consultation with the radiology team is most valuable. An important component of radiological training is radiation safety and radiation protection. Radiologists provide useful guidance to the referrers on alternate procedures not utilizing ionizing radiation, if and when appropriate. There are many facilities offering only basic radiography, particularly in rural or remote areas. In such situations, it is not possible to consider ultrasound or other non-invasive modalities.

Regardless of the setting, the chosen procedures should be appropriately scheduled and the relevant patient preparations provided. Patient information for procedures could be available from the referrers as fact sheets or a web link. If not offered, it can almost be assumed that the patients will attempt to access these with potential misleading consequences. With interventional procedures, it is important for the patients and their families to be informed of any specific post-procedure protocols they must follow. Patients are generally stressed by their medical concerns. Therefore, it is immensely useful if facility location and other access details are provided to them when the appointments are made.

3.4 Registration

Registration is the first step in risk management and the optimization of the radiology processes. This step sets the impression that the patient will have of the facility. In management terms, this is referred as the "moment of truth" [4], i.e. "anytime a customer comes into contact with any aspect of a business, however remote, is an opportunity to form an impression."

The reception staff must ensure that: the procedure scheduled is indeed the one that was requested; the patient preparation conforms to the instructions, e.g. fasting, cessation of anti-coagulants etc.; and all the relevant personal information is accurately recorded. Sometimes the patients or the referrers' staff may have asked for a Chest X-ray when in fact a CT Scan of the chest is required. Each procedure must be assessed and verified by a team member before initiation.

Patient registration usually involves a series of processes and checks: patient name, procedure booked, referrer name, previous procedures, allergies, pregnancy, medical history, billing details, result delivery and follow up appointment, if applicable. Despite the use of sophisticated Radiology Information Systems (RIS), human errors could easily occur and result in incorrect entries to any one of these fields, leading to other errors along the journey. Relevant team members should be informed of the conditions or disabilities, which could be a barrier to good patient care, e.g. hearing or visual impairment or frailty. Patients should be given an accurate estimation of the waiting time during registration.

3.5 Procedure

Prior any procedure, the first step is the confirmation of the patient's identity and the procedure to be performed. The same rigor should be applied as for surgical procedures because unnecessary and unintended irradiation or life-threatening contrast reaction is totally unacceptable.

The procedure is the most critical of these five processes because its appropriateness and quality and the subsequent report will most likely influence the patient's ongoing management. Depending on the procedure, this is likely the area where teamwork is most important. For example, a team approach is most obvious for an interventional procedure as there may be several other related processes, e.-g. preparation, sedation and aftercare, when close collaboration is needed to achieve patient-centered care and minimize potential risks.

3.6 Optimization of Image Quality and Radiation Protection

There are concerns about an increasing exposure in radiology procedures despite technological advances, which should reduce patient dose. Most of this is due to poor quality assurance of the equipment or inappropriate radiographic technique, particularly with digital radiography in both Direct Digital (DDR) and Computed Radiography (CR) modes. Similarly, with MDCT, there have been many recent reports of severe unintended exposures and wide dose variations for similar studies. For example, a study in 19 developing countries showed that the paediatric CT doses varied by a factor of up to 55 times [23]. The technique variations for identical indications, between and within facilities suggest that the working procedures could be optimized [28], e.g. by using 'child-size' protocol, single series, appropriate Z coverage, and automatic exposure control.

The optimization of protection is applicable to the design, selection, construction and installation of appropriate equipment, and to the daily working procedures. The aim is to have measures in place to ensure good quality images and sufficient data to enable a diagnosis to be made by using the lowest dose possible, i.e. the dose commensurate with the medical purpose. The use and selection of measures will depend on resources and will have an impact on the exposure and risk to the patient, the staff, and sometimes the public as well as financial implications.

Equipment commissioning is a joint effort of the equipment vendor, facility and regulatory authority. The principles of good practice, radiation safety, radiation protection and dose management will be applied in daily working procedures.

In radiography, the minimum number of exposures required to achieve a satisfactory procedure should be obtained using well-maintained equipment with a fast film screen combination, adequate collimation and gonadal shielding.

CT protocols should be chosen to answer the clinical question and team members should communicate closely and select the minimal number of series and exposure factors to obtain this information. High mA settings and thin slices will result in increased radiation exposures. The dose is compounded if multiple series are performed. Teams must appreciate: (1) the difference in patient exposures between good quality diagnostic images and best quality images; and (2) the use of too many series to cover all possible scenarios in every patient is usually not necessary. For example, when using CT to follow-up a lung nodule, a low-dose technique will reduce the dose by 90 %. In children, indication-based and weight-based protocols should be employed. When choosing the exposure factors, the team must always balance the radiation exposure against the need to obtain adequate diagnostic information.

The optimization of protection includes the radiation protection of patients and staff. An experienced member whose duty it is to oversee radiation protection actions and their ongoing education should supervise this important task. Staff training in radiation safety and protection should cover all the members involved including other ancillary staff, e.g. anesthetists and anesthetic nurses. Regular monitoring of interventional procedures in particular by dosimetry should be mandatory and the results should be widely available as members could easily become complacent with certain work practices.

To audit performance and benchmark, meaningful, reliable and measurable metrics are used. For examples, diagnostic reference levels (DRLs) are used to improve awareness, evaluate a facility's performance, benchmark and identify improvement opportunities. A paediatric CT dose registry is being developed in the US to promote optimization by sharing 'optimal' technique and dose indices. Clinical audits followed by corrective actions will improve image quality and reduce exposure [24]. A dose reduction of 50 % could be achieved without reducing diagnostic accuracy by providing training in effective optimization [5].

The use of individual monitoring devices to record and monitor individual exposure is gaining momentum in better-resourced settings. These monitoring devices are useful tools to facilitate research in technique improvement. The development and integration of a collaborative, quality and safety culture requires leadership and participation of the team.

3.7 Quality Processes

There are three related quality processes: quality control, quality assurance and quality improvement [18]. The taking of a chest radiograph is used as an example to highlight the differences between these processes (Fig. 9.2):

- Quality control is the rejection and repeat of a poorly exposed film to ensure the final image for reporting is of diagnostic quality;

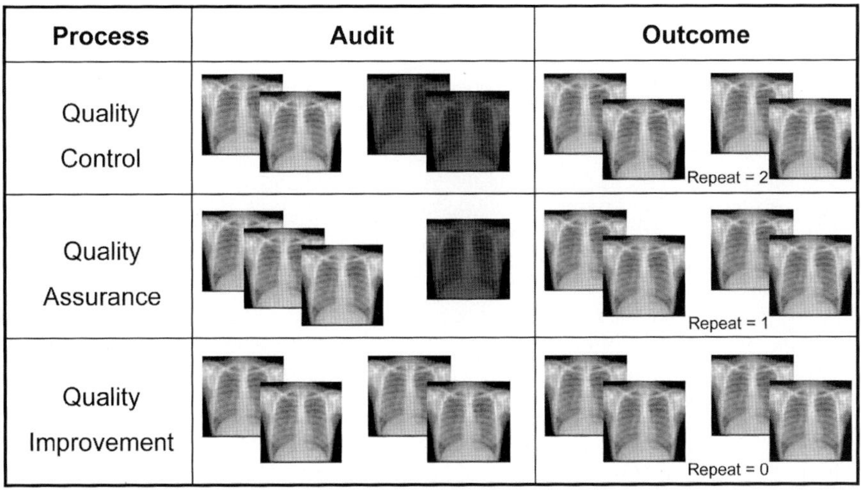

Process	Audit	Outcome
Quality Control		Repeat = 2
Quality Assurance		Repeat = 1
Quality Improvement		Repeat = 0

Fig. 9.2 Quality processes illustrated by the taking of chest radiographs. Quality control is the rejection and repeat of a poorly exposed film to ensure the final film is of diagnostic quality. Quality assurance employs equipment maintenance, equipment check and appropriate daily working procedures to reduce the number of un-diagnostic images. Quality improvement involves the analysis of faults, identification of improvement opportunities and development of solutions for each step of the process. This is followed by staff training, action implementation and impact evaluation, which ultimately will lead to better diagnostic images with less exposure

- Quality assurance employs regular equipment maintenance, equipment check and appropriate daily working procedures supported by well-documented procedure manual, exposure chart etc. to reduce the percentage of un-diagnostic images; and
- Quality improvement is more proactive. Audits are used to analysis faults, identify improvement opportunities and develop solutions for each step of the process. This is followed by staff training, action implementation and impact evaluation, which will ultimately lead to better diagnostic images with less exposure.

A team approach is important to develop and maintain a successful and effective quality assurance program. For mammography, the ultimate performance indicator is the detection of occult non-palpable cancer. Team members must ensure their efforts are focused on high quality performance. In mammography, similar to other quality assurance program in radiology, the medical physicists play important roles due to their in-depth knowledge in medical physics, radiation safety and technology. Both acceptance and annual testing are essential for digital mammography equipment. These units may be configured and operated in a wide range of exposure settings and thus they should be tested in accordance to their intended use. It is important for the teams to determine the clinical application and configuration setting prior to testing. The teams must have a sense of cohesiveness [16]. Members should understand their individual roles, the roles of others and the interdependency of the contribution from each team member.

3.8   Report

Qualified practitioners report the procedures. The aim at this stage is to carefully evaluate all the information available and prepare an accurate report. All data must be reviewed and an unambiguous report provided, with particular reference to the questions raised in the request. The use of structured and standardized reports will provide a more comprehensive document, minimize errors and reduce ambiguity. Good reports indirectly contribute to a reduction in patient exposure by minimizing the need for repeats or further procedures requiring ionizing radiation.

3.9   Dispatch

The importance of this last step could be overlooked by some facilities. Patient outcome will be suboptimal unless the results are appropriately communicated to the referrers. In cases of urgent or unexpected findings, e.g. a newly diagnosed case of breast cancer, urgent discussion with the referrer is desirable, to arrange a possible additional procedure, e.g. bone scan, an urgent biopsy or referral to a breast surgeon, while the patient is still in the facility. This is particularly helpful if the patient comes from a remote area and may have difficulties in arranging subsequent travel. The same applies to trauma when urgent surgical intervention is required. These situations emphasize the need to a team approach to radiology. The registration process should have paved the way for efficient management of such eventualities. The team members should have these examples explained to them in training or during induction to ensure an understanding of the whole process and how they can contribute to better patient-centered care.

4   Team Approach in Health Care Systems

The following discussions are some examples and by no means comprehensive. However, they illustrate the leadership, innovations and team approach used to optimize radiology techniques and improve radiation protection and radiology quality within health care systems.

4.1   Equipment Providers

The design, construction, selection and installation of equipment are important elements of optimization of protection. The increase in utilization of radiology resulting from advances in technique, technology and applications is a testament to the range of innovations that the equipment manufacturers have developed in recent years. Public concern with radiation exposure has encouraged manufacturers to focus on dose reduction, e.g. there is a 50–80 % dose reduction for the same image quality in the state-of-the-art CT scanners through hardware and software improvements. It is most desirable to extend these dose reduction innovations to older scanners, hardware and software permitting. There will be further innovations in the design and construction of equipment and software programming which will lead to better images with lower exposure.

In some situations, the personnel responsible for equipment tendering and selection may not be well informed of the current trend, latest development or the local needs. Better communication between the equipment vendor, the purchaser and the end-users will ensure the acquisition of a piece of equipment that is most suitable for the local needs.

The proper commissioning and use of equipment are fundamental to good practice. However, there could be gaps in this process resulting in improper or suboptimal use. As a result, the full benefits of these devices are not realized. Innovative and collective actions between the equipment vendor, facility and regulatory authority are desirable to ensure satisfactory commissioning. Before acceptance and annual testing, the team members should define their needs and be guided by the medical physicist to ensure the settings and configurations are in accordance to the intended applications. Adequate education and training should be provided by the vendor to the end-users, not only on image quality but radiation safety and protection. The teams would work closely with the vendors to ensure on-going proper use, appropriate quality assurance measures and maintenance. Following hardware or software upgrades, appropriate training and detailed instruction must be given.

4.2 Regulatory Authorities

Regulatory authorities implement and monitor: radiation safety and radiation protection regulations for patients and workers; operator registration, including qualification and on-going development requirements; and equipment registration, including commissioning and maintenance. By raising awareness and providing supporting tools, the authorities could collaborate with professional organizations to advance best practice in radiation safety and more effective implementation of the BBS.

Despite the United Nations Scientific Committee on the Effects of Atomic Radiation's [34] awareness raising efforts, more participation in population exposure surveys is needed. The regulatory authorities are the national leads for this initiative. The authorities should advocate the use of DRLs and clinical audits as useful quality assurance tools by conducting awareness campaigns to inform the practitioners. If resources are available, formal accreditation program for radiology facilities could be considered.

4.3   Institutions

Research and academic institutions conduct studies into the health effects of medical radiation. An innovative approach to radiation risk research is by promoting collaborations between institutions and programs to minimize the duplication of efforts. Research efforts focusing on the optimization of image quality and radiation protection are encouraged.

Academic institutions and professional organizations provide undergraduate and postgraduate education and training. Studies have shown that medical students, junior doctors and referrers are poorly informed on radiation safety, radiation protection, appropriate use of and cancer risk from radiology [7, 27, 29, 39]. Possible innovative approaches could be by: conducting awareness campaigns on basic radiation protection to school students, along similar approach as the Sun Smart and UV protection campaign [36, 38]; strengthening of teaching in radiation safety and appropriate use of radiology in undergraduate curriculum; and reviewing and applying innovative post graduate teaching methodology, which is appropriate for adult continuing professional development to meet future needs.

4.4   Professional Organizations

To tackle the global workforce shortage and the increase in workload, more innovations are needed. In some regions, e.g. the United Kingdom, some experienced technologists have extended their roles after undergoing additional formal training in image interpretation for radiography, and in some cases for CT, MR and some interventional procedures.

The training of radiographers and radiological technologists varies considerably from country to country. The International Society of Radiographers and Radiological Technologists has established guidelines for basic training, by identifying the essential and core components [14]. The teaching of radiation protection in this education program is critical and must be the subject of on-going, continual education and adherence. The radiographers deliver the radiation dose to the patients and must accept this responsibility.

Professional organizations such as the ISRRT collaborate with experts from other organizations and agencies, e.g. IAEA, WHO, International Organization for Medical Physics [11], International Radiology Quality Network [12, 17] and Image Gently [8] etc. to increase capacity, improve capability, optimize protection and standards of practice at the grassroots level by focusing on: awareness raising; education and training; developing material and training tools; providing books, materials and translations to developing countries; and developing and conducting training missions.

In resource-poor or remote facilities where only basic radiography is available, an innovative approach should be considered to ensure basic radiation safety.

In these settings, radiography procedures could be provided by a solo radiographer with or without the remote support of a radiologist and or medical physicist. Amongst the local health care team, the radiographer may be the sole person with any knowledge of the potentially harmful radiation effects and so must act as the patients' advocates to ensure procedure justification, optimization of radiation protection and the ALARA principles are applied. For this to be effective, recognition of this role from and support by the local health care team is important. This innovative approach to be championed by professional organizations and supported by agencies, regulatory authorities and local health care teams will lead to a safer and more appropriate use of radiology in such settings. Similar radiation protection and radiation safety model could be applied to the remote facilities employing teleradiology without the local support of a radiologist and or medical physicist.

4.5   UN Agencies

Radiation medicine is used by other medical specialties, e.g. interventional cardiologists, gastroenterologist, urologists, orthopedic surgeons, gynecologists and neurosurgeons. The International Atomic Energy Agency leads international efforts to promote awareness and educate these users by developing and distributing training material and conducting workshops on radiation safety, radiation protection and dose management [26].

There is a global shortage of radiology practitioners. Many radiological procedures are not supervised or reported by qualified radiologists. The World Health Assembly adopted Resolution WHA59.23 (2006) to encourage member states to train more health care workers to overcome this global shortage. The World Health Organization assists in capacity building by supporting the development of workforce planning teams; providing technical support to revitalize education institutions; seeking support from global health partners, donors and development agencies; promoting partnerships; and applying innovative technologies to teaching.

4.6   Working Together

Despite such modern advances, many millions of radiology procedures are still performed on basic non-digital equipment including wet film processing. In these settings, regular basic quality assurance and equipment maintenance are essential. An estimated two thirds of the world's population is currently without access to radiology. Most live in resource-poor tropical regions with harsh environments. In these settings, the challenges include poor infrastructure, non-functioning equipment, suboptimal maintenance, workforce shortage, inadequate education and training and absence of a quality culture [25].

International, regional and national agencies and professional organizations, collaborate and offer help and support in practical and educational ways by addressing management, infrastructure, personnel and equipment issues to promote safer and more appropriate use of radiology in the resource-poor settings. Possible approaches include the: provision of financial and administrative training to ensure economic sustainability of services; design, testing, and deployment of clinical strategies suitable for settings with limited resources; improvement of the role of volunteers; and implementation of information technology models to support digital imaging in the developing world [22].

Effective communication to other stakeholders, especially to the public, is a very important action to promote awareness and provide reputable and trustworthy information on radiology procedures, quality radiology, radiation protection and radiation safety. To achieve this, innovative approaches could be adopted by leveraging on the advances in information and communication technology. In addition to the provision of user-friendly websites, many organizations are exploring social media as an emerging means to strengthen this coverage.

Various organizations and agencies provide evidence-based messages and education material on the benefits and risks of radiology to the public and profession, in different languages, formats and media. An example is 'Image Gently', a very effective awareness campaign championed by the Alliance for Radiation Safety in Pediatric Imaging. Its goal is to change practice by lowering radiation dose in the imaging of children [8]. Its' CT protocols are based on patient size and are independent of manufacturer or machine. Other examples of web-based resources include: Radiation Protection of Patients [9], RadiologyInfo [1], InsideRadiology [32, 36] and Virtual Departments [33]. For practitioners, some organizations provide open access to high quality educational material in radiology and radiation protection, e.g. GO-RAD [13], Biomedical Imaging and Intervention Journal [2], ECR Digital Preview System [6], IAEA Radiation Protections of Patients [9] etc.

To underpin their support and commitment to quality patient care and radiation safety, the payers could consider a pay for performance re-imbursement model and provide funding for quality and safety initiatives [37].

5   Conclusion

Radiology facilities are potentially high-risk environments and it is the teams' responsibility to minimize the risks to patients and staff. The best way to optimize techniques is to recognize that every procedure is a team event. The teams include those within the facility and others who are intimately involved, i.e. the referrer, the anesthetist or the nurse assisting the procedure. Team members need to be aware of radiation and other safety issues and help to ensure that exposures are minimized for the benefit for all.

The five processes along the patient journey highlighted the potential errors, which unfortunately could occur in any facility, e.g. wrong patient, wrong procedure etc.

Paying detailed attentions to these potential risks by good management techniques and active team participation will help to achieve better quality and patient-centered care.

Good ideas and innovations come from individuals who are the experts in their fields. Leaders of facilities, institutions, organizations, agencies, or authorities are the champions in facilitating and bringing these ideas into practice. Communication and the sharing of these innovations with other like-minded stakeholders will facilitate collaborations and establish teams. Collaboration is strength, minimizes duplication and waste. The vital link to realize these leadership efforts and innovations is the engagement of the end-users. Their participation by translating these ideas into practice towards better optimization of radiology techniques, image quality, radiation protection and patient-centered care must be vigorously encouraged.

Teamwork is equally applicable to radiology facilities as well as health care systems. Strong leadership, trustworthy communication, unreserved collaboration and active participation underpin good teamwork and support Good Medical Practice, professionalism, quality patient care, radiation safety and appropriate use of radiology.

References

1. American College of Radiology and Radiological Society of North America (ACR and RSNA) RadiologyInfo. http://www.radiologyinfo.org/. Accessed 1 Oct 2013
2. Biomedical Imaging and Intervention Journal (BIIJ) http://biij.org/. Accessed 1 Oct 2013
3. Blackmore CC (2007) Defining quality in radiology. J Am Coll Radiol 4(4):217–223
4. Calzon J (1987) Moments of truth. Ballinger Books, Cambridge
5. Catalano C, Francone M, Ascarelli A et al (2007) Optimizing radiation dose and image quality. Eur Radiol 17(Suppl 6):F26–F32
6. European Society of Radiology (ESR) ESR Digital Preview System. Available at: http://edips-download.myesr.org/. Accessed 1 Oct 2013
7. Georgen S (2010) They don't know what they don't know. J Med Imaging Radiat Oncol 54 (1):1–2
8. Image Gently – The Alliance for Radiation Safety in Pediatric Imaging http://www.pedrad. org/associations/5364/ig/. Accessed 1 Oct 2013
9. International Atomic Energy Agency (IAEA) Radiation protection of patients. http://rpop.iaea. org/RPOP/RPoP/Content/index.htm. Accessed 1 Oct 2013
10. International Atomic Energy Agency (2011) Radiation protection and safety of radiation sources: international basic safety standards – Interim edition, IAEA safety standards series GSR Part 3 (Interim). IAEA, Vienna
11. International Organization for Medical Physics (IOMP) http://www.iomp.org. Accessed 1 Oct 2013
12. International Radiology Quality Network (IRQN) http://www.irqn.org. Accessed 1 Oct 2013
13. International Society of Radiology (ISR) http://www.isradiology.org. Accessed 1 Oct 2013
14. International Society of Radiographers and Radiological Technologists (ISRRT) http://www. isrrt.org/isrrt/default.asp. Accessed 1 Oct 2013
15. Kocher KE, Meurer WJ, Fazel R et al (2011) National trends in use of computed tomography in the emergency department. Ann Emerg Med 58(5):452–462
16. Kopans DB (2007) Breast imaging. Lippincott Williams and Wilkins, Philadelphia

17. Lau LSW (2004) International radiology quality network. J Am Coll Radiol 1(11):867–870
18. Lau LSW (2006) A continuum of quality improvement in radiology. J Am Coll Radiol 3 (4):233–239
19. Lau LSW (2007) The design and implementation of the RANZCR/NATA accreditation program for Australian radiology practices. J Am Coll Radiol 4(10):730–738
20. Lau LSW, Perez MR, Applegate KE et al (2011) Global quality imaging: emerging issues. J Am Coll Radiol 8(7):508–512
21. Lau LSW, Perez MR, Applegate KE et al (2011) Global quality imaging: improvement actions. J Am Coll Radiol 8(5):330–334
22. Mollura DJ, Azene EM, Starikovsky A et al (2010) White paper report of the RAD-AID conference on international radiology for developing countries: identifying challenges, opportunities, and strategies for imaging services in the developing world. J Am Coll Radiol 7(7):495–500
23. Muhogora WE, Ahmed NA, Alsuwaidi JS et al (2010) Paediatric CT examinations in 19 developing countries: frequency and radiation dose. Radiat Prot Dosimetry 140(1):49–58
24. Muhogora WE, Ahmed NA, Almosabihi A et al (2008) Patient doses in radiographic examinations in 12 countries in Asia, Africa, and Eastern Europe: initial results from IAEA projects. Am J Roentgenol 190(6):1453–1461
25. Ng KH, McLean ID (2011) Diagnostic radiology in the tropics: technical considerations. Semin Musculoskelet Radiol 15(5):441–445
26. Rehani MM (2007) Training of interventional cardiologists in radiation protection – the IAEA's initiatives. Int J Cardiol 114(2):256–260
27. Smith-Bindman R (2010) Is computed tomography safe? N Engl J Med 363(1):1–4
28. Smith-Bindman R, Lipson J, Marcus R et al (2009) Radiation dose associated with common computed tomography examinations and the associated lifetime attributable risk of cancer. Arch Intern Med 169(22):2078–2086
29. Soye JA, Paterson A (2008) A survey of awareness of radiation dose among health professionals in Northern Ireland. Br J Radiol 81(969):725–729
30. Sun Z, Aziz YF, Ng KH (2012) Coronary CT angiography: how should physicians use it wisely and when do physicians request it appropriately? Eur J Radiol 81(4):e684–e687
31. The Hippocratic Oath http://www.nlm.nih.gov/hmd/greek/greek_oath.html. Accessed 1 Oct 2013
32. The Royal Australian and New Zealand College of Radiologists (RANZCR) Inside radiology. http://www.insideradiology.com.au/. Accessed 1 Oct 2013
33. The Royal College of Radiologists (RCR) Virtual Departments. http://www.goingfora.com/index.html. Accessed 1 Oct 2013
34. United Nations Scientific Committee on the Effects of Atomic Radiation (UNSCEAR) (2010) Sources and effects of ionizing radiation. UNSCEAR 2008 report to the General Assembly with scientific annexes. United Nations, New York
35. Valentin J (ed) (2007) ICRP Publication 105. Radiological protection in medicine. Ann ICRP 37(6). Elsevier, Oxford
36. World Health Organization (WHO) (2003) Sun protection – a primary teaching resource. WHO, Geneva
37. World Health Organization (WHO) (2008) WHO global initiative on radiation safety in health care settings technical meeting report. WHO, Geneva
38. X-Rays: The Inside Story http://www.eric.ed.gov. Accessed 1 Oct 2013
39. Zhou GZ, Wong DD, Nguyen LK et al (2010) Student and intern awareness of ionizing radiation exposure from common diagnostic imaging procedures. J Med Imaging Radiat Oncol 54(1):17–23

Chapter 10
Advances in Technological Design to Optimize Exposure and Improve Image Quality

Daniel F. Gutierrez and Habib Zaidi

Abstract Multimodality imaging is playing a key role in the clinical management of patients in routine diagnosis, staging, restaging and assessment of response to treatment, surgery and radiation therapy planning of malignant diseases. The complementarity between anatomical (CT and MRI) and functional/molecular (SPECT and PET) imaging modalities is now well recognized and the role of fusion imaging is widely used as a central piece of the general tree of clinical decision making. Moreover, dual-modality imaging technologies including SPECT/CT, PET/CT and nowadays PET/MR represent the leading component of a modern healthcare facility. There have been significant advances in data acquisition along with innovative approaches to image reconstruction and processing with the aim to improve image quality and diagnostic information. However, CT procedures involve relatively high doses to the patient, which triggered many initiatives to reduce the delivered dose particularly in paediatric practice.

This chapter discusses the state-of-the-art developments and challenges of multimodality medical imaging technologies and dose reduction strategies. Future opportunities and the challenges limiting their adoption in clinical and research settings will also be addressed.

D.F. Gutierrez
Division of Nuclear Medicine and Molecular Imaging, Geneva University Hospital, Geneva CH-1211, Switzerland

H. Zaidi (✉)
Division of Nuclear Medicine and Molecular Imaging, Geneva University Hospital, Geneva CH-1211, Switzerland

Geneva Neuroscience Center, Geneva University, Geneva CH-1211, Switzerland

Department of Nuclear Medicine and Molecular Imaging, University of Groningen, University Medical Center Groningen, 9700 RB Groningen, Netherlands
e-mail: habib.zaidi@hcuge.ch

L. Lau and K.-H. Ng (eds.), *Radiological Safety and Quality: Paradigms in Leadership and Innovation*, DOI 10.1007/978-94-007-7256-4_10, © Springer Science+Business Media Dordrecht 2014

Keywords Hybrid imaging • Image quality • Multimodality imaging • Patient dose • Radiation exposure

1   Introduction

The fields of diagnostic radiology and medical imaging cover a plethora of techniques that are vital for evaluating and managing patients who require medical care. Conventional imaging techniques such as plain film radiography and more recent techniques such as x-ray computed tomography (CT) and magnetic resonance imaging (MRI) are used to evaluate a patient's anatomy with sub-millimeter spatial resolution to discern structural abnormalities and to evaluate the location and extent of disease. These methods offer relatively fast scan times, precise statistical characteristics, and good tissue contrast especially when contrast media are administered to the patient. In addition, x-ray fluoroscopy and angiography are used to evaluate the patency of blood vessels [7], the mechanical performance of the cardiovascular system, and structural abnormalities in the gastrointestinal or genitourinary systems. Similarly, CT and MRI can be performed with cardiac gating to the heart at different phases of the cardiac cycle. Computed tomography recently has experienced a significant increase in utilization with the advent of multislice helical scanning techniques that cover a large region of the patient's anatomy with a single breath-hold and with scan speeds that can capture both the arterial and venous phases of the contrast bolus. These increased scan speeds enhance patient comfort, and contribute to patient throughput and cost effectiveness [31].

In contrast to the anatomical imaging techniques described above, functional imaging methods including planar scintigraphy, single-photon emission computed tomography (SPECT), positron emission tomography (PET), and magnetic resonance spectroscopy (MRS), assess the regional differences in the biochemical status of tissues. In nuclear medicine, including SPECT and PET, this is done by administering the patient with a biologically active molecule or pharmaceutical, which is radiolabelled and accumulated in response to its biochemical attributes. Nuclear imaging relies on the tracer principle in which a minute amount of a radiopharmaceutical is administered to assess physiological function or the biomolecular status of a tissue, tumour, or organ within the patient. The volume of the radiopharmaceutical used is sufficiently small such that its administration does not perturb normal organ function. However, the radiopharmaceutical produces a radioactive signal that can be measured, and ideally imaged, by using an external array of radiation detectors. By design, the radiopharmaceutical has a targeted action, allowing it to be imaged to evaluate the specific physiological processes in the body. There are now many radiotracers which are suitable for clinical diagnosis, with additional radiotracers available for in vivo as well as in vitro biological experimentation. Extended descriptions of these techniques are well documented [14, 17, 78, 82].

2   Overview of Medical Imaging Techniques

2.1   Projection Imaging (2D) Techniques

Projection imaging was the first non-invasive imaging technique developed by Wilhelm C. Röentgen following the discovery of x-rays in 1895. It was quickly applied as a clinical tool, not only to demonstrate fractures or other anomalies of hands and arms, but also for surgical purposes. This is a transmission procedure involving the acquisition of a two-dimensional image of a body region. The x-rays produced by the x-ray tube on one side of the patient pass through the body. The image formation is based on the differential penetration of the x-rays, with the contrast of the image being the consequence of the different paths of the radiation going through different tissues. For example, bones cause significantly higher attenuation than soft tissues, resulting in a smaller exposure of the regions where x-rays encounter bone and thus the resulting image is whiter than soft tissue.

Mammography is a procedure specially designed to detect abnormalities in the breast, which can be very difficult owing to two reasons: lesions can be small and the density difference between them and breast tissues is also very small.

Fluoroscopy is a technique where x-ray images are acquired in a dynamic or real-time mode. The name of this technique comes from the fluorescence that Thomas Edison saw while looking at glowing plate screens of calcium tungstate bombarded with x-rays. The first detectors used on these systems were made of image intensifiers coupled to television screens, while the actual systems are composed of flat panel detectors based on thin film transistors (TFT) that are replacing image intensifiers.

2.2   Tomographic Imaging (3D) Techniques

The drawback of conventional radiography is that it does not provide any depth information when 3D objects are projected onto 2D detector geometries. Tomographic imaging presents the advantage of circumventing this limitation through 3D image reconstruction of the region examined to provide series of 2D image slices of the body. For this reason, it was predicted that tomographic imaging techniques will likely replace planar imaging in the future [2]. This remains a controversial topic, which is still being debated [7, 50].

The first tomographic imaging system was introduced by David Kuhl and Roy Edwards in 1965 [11] representing the departure of tomographic imaging in nuclear medicine (SPECT) as well as transmission imaging with radionuclides (^{131}I and ^{241}Am). Later, Sir Godfrey Hounsfield introduced the concept of computed tomography (CT) based on transmission of tube generated x-ray beam [34].

Computed Tomography (CT)

Although the first CT device was introduced by Hounsfield in 1973 [34], the mathematical principles of the filtered backprojection are much older and stem from radioastronomy-related research by Johan Radon in 1917 [62]. He demonstrated that a 3D image of an unknown object could be reconstructed from an infinite number of projections of the object. In the case of CT, the acquisition is realized by rotating both the x-ray tube on one side and the detector on the opposite side of the patient in a synchronized way. In this case, the line integrals of the transmission of the x-ray beam through the object represent the projections that will serve to reconstruct the object. The reconstruction of images from the acquired projections is performed using the filtered backprojection algorithm.

One of the significant advantages of CT is that the reconstructed image scale, referred to as the Hounsfield unit (HU) scale, is linearly correlated with the linear attenuation coefficient of the corresponding biological tissue, thus giving direct physical information of the chemical composition of the tissues within the patient.

Magnetic Resonance Imaging (MRI)

This modality is sometimes also called Nuclear Magnetic Resonance Imaging (NMRI) because it uses the property of nuclear magnetic resonance of atoms inside the body. This is achieved by positioning the patient inside a strong magnetic field (B_0) used to align the magnetization of the protons of some atoms inside the body (related to the proton density). The application of a short radiofrequency (RF) pulse then produces an electromagnetic field with the resonance frequency just needed to tilt the global magnetization away from the magnetic field (B_0). When the pulse goes off, the magnetization re-aligns back to the magnetic field, producing a radio frequency signal that can be measured with receiver coils. The mechanisms of this relaxation, i.e. the longitudinal (T_1) and the transverse (T_2) relaxation times, are at the origin of this signal measurement. These relaxation times are the characteristics of different tissues and can be measured by applying suitable sequences of RF pulses to the region of interest and recording the radiofrequency NMR signals with a receiver coil.

Single-Photon Emission Computed Tomography (SPECT)

SPECT is a tomographic nuclear imaging technique enabling the generation of images representing the bio-distribution of gamma-ray emitting radiopharma-ceuticals in the body acquired from standard planar emission projections around the patient. These images are usually obtained using one or more large scintillator detectors at the same time turning around the patient (180° or 360°), although new solid-state detector technologies dedicated for cardiac [26] and breast imaging [64] have recently been introduced in academic and corporate settings.

The reconstruction of images from the acquired projection data is usually performed by using either analytic algorithms (e.g. filtered backprojection, as in CT), or by the more popular iterative reconstruction methods which are now implemented on all commercial systems and widely used in practice [63]. Direct analytical techniques have the advantage of being straightforward and relatively fast but produce images with limited quality owing to the oversimplified line-integral model of the acquired projection data used. In contrast, iterative techniques are computationally more intensive but the resulting images are much improved (principally due to more accurate and complex modelling of the acquired projection data), which have enabled them to replace the analytic techniques.

Iterative algorithms involve a feedback process that allows a sequential tuning of the reconstruction to achieve a better match with the measured projection data. The iterative process starts with an initial estimate of the tracer distribution and calculates the corresponding projections using a forward projection operator incorporating an appropriate model of the emission and detection physics. Then, the difference between estimated and measured projections are be used to modify the original estimate of the tracer distribution through suitable additive or multiplicative corrections for each voxel using the backprojection operator. The adjusted tracer distribution then becomes the starting point for the next iteration and the process continues for a predetermined number of iterations using the continuous feedback loop until a final solution is reached.

Positron Emission Tomography (PET)

Similar to SPECT, PET is a nuclear medical imaging technique where a positron-emitting radiotracer is injected to the patient. ^{18}F-fluorodeoxyglucose (^{18}F-FDG), which is a glucose analogue, is the most widely used radiotracer in practice. Once a positron is emitted, it travels for a very short distance depending on its energy before annihilating with an electron to produce two anti-parallel annihilation photons each having an energy of 511 keV. The imaging procedure is based on the annihilation coincidence detection (ACD) of these two emitted photons. To achieve this, the patient is surrounded by a large number of detectors consisting of small scintillation crystals assembled in a cylindrical geometry. When both photons with energies of 511 keV are detected at the same time (with a time difference typically less than 500 ps, depending on the characteristics of the detectors), the event is accepted and the straight line joining these detectors defines a line of response (LoR).

The development of optimized detection geometries combined with high performance detector technologies and compact scanner design has become the goal of active research groups in both academic and corporate settings. Significant progress was achieved in the design of commercial PET instrumentation during the last decade allowing to reach a spatial resolution of about 4–6 mm for whole-body imaging, ~2.3 mm in PET cameras dedicated for brain imaging, and sub-millimeter resolution for female breast, prostate and small-animal imaging [16].

3 Basic Concepts of Image Quality

Assessing the quality of medical images is a very complex task. To understand the reasons for this complexity, one must keep in mind that the goal of medical imaging is not to produce pretty images but rather to provide non-invasive and informative diagnostic information that can help clinicians to make a diagnosis or to plan surgical or therapeutic procedures. The common problem is to determine the guidelines for an objective assessment of image quality by using established metrics [8].

The complexity of the whole process is shown in Fig. 10.1. The patient is usually referred for a medical imaging procedure to be carried out on a specific device depending on the clinical question. A plethora of structural and functional imaging techniques is available today. The collected information will usually be processed before it is displayed by a physical support, typically by a display monitor in a digitized facility or by conventional films in other modalities.

3.1 *Image Contrast*

The diagnostic accuracy and performance of the radiologist when reading these images was initially related to the concept of image quality [66]. The main factors influencing image quality are contrast, spatial resolution and noise. The contrast of an image can be defined as the difference (or in some cases the relative difference) between two neighbour signals coming from the object that is being imaged. For example:

- In projection radiography, this is the difference in x-ray transmission between two different structures, e.g. bone and soft tissue;
- In CT, this is related to the difference between the linear attenuation coefficients of two different tissues, e.g. soft tissue and fat;
- In PET, the difference is in the activity concentrations in two organs or tissues, i.e. normal and pathological tissues; and
- In MRI, the difference between the magnetic susceptibility of two tissues, e.g. white and gray matter in the brain.

3.2 *Image Resolution*

The spatial resolution is a measure of how close two adjacent objects can be distinguished. Usually, we express spatial resolution in units of spatial frequency, which is the inverse of the closest discrimination distance, or the number of sinusoidal cycles that can be fitted in unit space (cycles/mm). The spatial resolution

Fig. 10.1 Schema of
diagnostic decision based
on medical imaging
acquisition and
interpretation

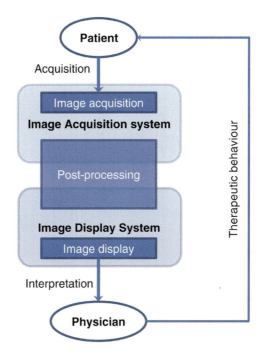

is usually characterized by the modulation transfer function (MTF), which reports the amount of sinusoidal signals transferred by the imaging system (Fig. 10.2). This particular metric can be measured in analogue systems (e.g. screen-film) by sinusoidal test patterns using Coltman's method [18] or by using sharp-edge test phantoms on digital systems [68]. Alternatively, the analysis of the noise response can be used for this purpose [47].

3.3 *Image Noise*

Image noise is defined as the random variation of the signals. Since the imaging process is inherently statistical because of the nature of quantum physics, the noise is well described by Poisson/Gaussian functions. For this reason, the noise can simply be characterized as the standard deviation (σ) of the signal in a uniform region (Fig. 10.3). Nevertheless, this measure doesn't reflect the frequencial content of the noise which can be derived from the noise power spectrum (NPS) (Fig. 10.4), which in turn can be measured according to the International Electrotechnical Commission (IEC) standards (International Electrotechnical [39]).

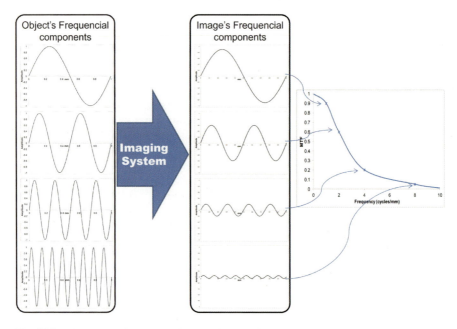

Fig. 10.2 Schema explaining the principle of modulation transfer function (*MTF*) measurement and its meaning. Four different frequential input signals (*left*), the signal output produced by the imaging system (*middle*) and the proportional restitution plotted as a function of spatial frequency.

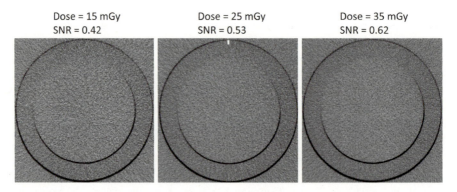

Fig. 10.3 CT images showing the effect of dose augmentation to noise and image quality. Uniform CT images with acquisition doses of 15, 25 and 35 mGy and their Signal to Noise Ratio (*SNR*). The signal is constant between these images because they are obtained from one unique object. However, the SNR improves by dose augmentation (from 15 to 35 mGy) due to a reduction of quantum noise

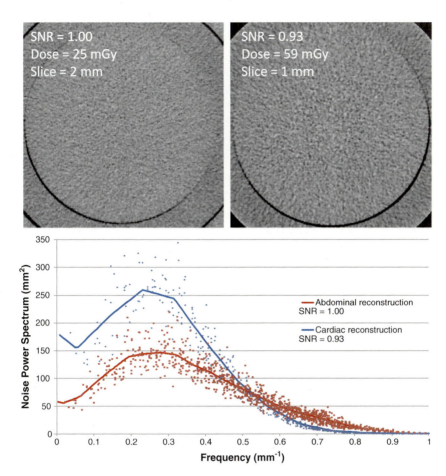

Fig. 10.4 CT images showing the effect of Weiner spectrum to noise and image quality. Uniform CT images with similar SNR meaning that both realisations contain the same magnitude of noise but different Weiner spectrum (also known as noise power spectrum), showing that the frequencial content of this noise is quite different

3.4   Signal Versus Noise

These image quality metrics described are closely linked and as such the interaction between them and the influence one has on others cannot be completely dissociated. For the task of image reading and interpretation by clinicians, other methods were proposed to better account for their imbrications. The limitations in terms of signal perception in noisy images was studied early in the Rose model [65] where the concept was extended to the so-called signal-to-noise ratio (SNR) (Fig. 10.3), which can be related to the probability of signal detection. The SNR can be defined as the difference in the mean output with (S_S) and without the signal (S_B) divided by the square root of the average variance of the noise, i.e. the standard deviation (σ):

$$SNR = \frac{S_s - S_B}{\sigma} = \frac{\Delta S}{\sigma}$$

This formalism was extended to new image quality assessment concepts that were introduced in the context of photographic imaging, such as detective quantum efficiency (*DQE*), which is a measure of the collective effects of the signal (associated with image contrast) and noise performance of an imaging system, usually expressed as a function of spatial frequency ω [21, 71]:

$$DQE(\omega) = \frac{SNR_{out}^2(\omega)}{SNR_{in}^2(\omega)}$$

where the SNR_{in} is the input *SNR* that originating from the object, or in other words, the *SNR* that will be achieved by an ideal detector, while SNR_{out} is the *SNR* that is obtained at the output of the imaging system. Obviously, all information should be obtained at the output of a perfect system, thus producing a *DQE* of 1, whereas a blind system will produce a *DQE* of 0.

One must keep in mind that the purpose of an image is to assist the observer (imaging specialist) in reaching a (clinical) decision. Therefore, image quality must be defined in terms of the targeted task [76]. Early contributions were made by Wagner for the task of discriminating between two different signals, i.e. present or absent, in the case of stationary Gaussian noise [75]. This task is commonly referred as the signal known exactly (*SKE*) and is used to introduce the concept of noise-equivalent quanta (*NEQ*), which depicts the minimum number of x-ray quanta necessary to generate a particular SNR by measuring change in the mean amplitude and in the variation in the amplitude of sine waves:

$$NEQ(\omega) = Q \cdot DQE(\omega)$$

where *Q* is the number of quanta incident on the detector that can be roughly equivalent to SNR_{in}^2, then:

$$NEQ(\omega) = \frac{|MTF(\omega)|^2}{NPS(\omega)}$$

The better the system performance, the higher the *NEQ* will be for all frequencies, while an *NEQ* of 0 means a blind system. As one can see, all the previous developments are closely related to statistical decision theory (SDT) that defines the framework where the *NEQ* was identified (*SKE*). This theory is based on the principle that any observer (human or mathematical) aiming at classifying an image (g) into one of two different categories, i.e. positive or negative, will perform this task through a statistical calculation (λ) to be compared with a threshold λ_0 that defines if the image belongs to any of the previously defined categories (Fig. 10.5).

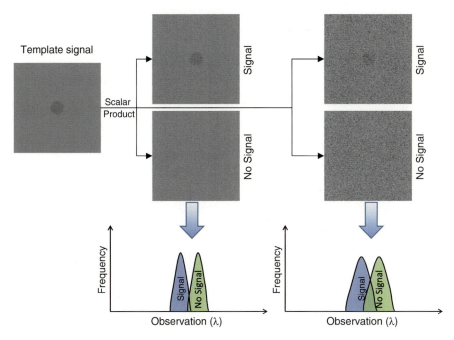

Fig. 10.5 Simple schema of the linear observer model principle. A template signal is compared (scalar product) with different noise realizations of signal-present and signal-absent images to produce task-dependent figures of merit that allow the discrimination between signal-present and signal-absent situations

If the decision is positive, the hope is that it is correct (True Positive: TP), but there are chances that the decision is wrong (False Positive: FP). The same could happen if the decision is negative (TN and FN, respectively). The compromise between the TP rate (TPR: Sensitivity) and the FP rate (FPR: 1 – Specificity) is defined by the decision threshold λ_0, and the variation of this threshold leads to the construction of the receiver operating curve (ROC) [54, 55], whose area under the curve (AUC) can be adopted as a image quality criteria (Fig. 10.6). For particular conditions, this metric can be related to a discrimination task [27] and as a consequence to the SNR. The AUC can change from 0.5 (random choice) to 1 (perfect performance) but its shape can be the sign of a highly sensitive or highly specific imaging system.

As stated earlier, the observer in the SDT can be mathematical, i.e. observer models. The most widely used are the linear observer models because the statistical calculation (λ) is a vectorial product of the image (g) and a discriminant function (u). Different variations of this model and optimal choice of the discriminant function are described elsewhere [9].

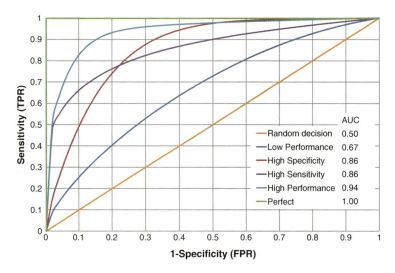

Fig. 10.6 Receiver Operating Curve (*ROC*) analysis as a metric in the statistical decision theory

4 Dose Reduction Strategies

To understand the magnitude of radiation exposure arising from medical imaging procedure, national surveys were carried out by different bodies in various countries worldwide. For instance, the National Radiological Protection Board (NRPB) performed periodic surveys of medical imaging practice in the UK to gauge individual and collective radiation doses. It has been shown in the NRPB-W4 report [30] that conventional radiography accounts for about 90 % of the total procedures and represents about 38 % of the collective dose. While mammography accounts for approximately 4 % of the total x-ray procedures, it represents roughly 2.4 % of the collective dose. Fluoroscopy represents just about 1.5 % of the total procedures but accounts for 16 % of the collective dose. For the latter, radiation protection is an important issue for the medical staff because this technique is widely used in interventional radiology.

CT has opened new opportunities in medical imaging in the last 40 years, but it has become a major source of medical exposure as it represents almost 40 % of the collective dose [30] despite accounting for only 3.3 % of the total procedures using ionizing radiation. The issue is that the frequency of CT procedures is still growing and recently became a public health concern for different governments [72], thus indicating that despite much worthwhile efforts targeting dose optimisation in CT, the topic still requires further research and development.

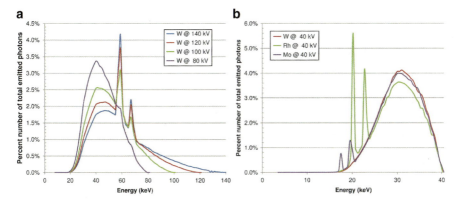

Fig. 10.7 Emitted x-ray spectra for different voltages and anodes. (**a**) Spectrum of Tungsten (*W*) anode x-ray tube for voltages of 80, 100, 120 and 140 kV. (**b**) Spectrum of x-ray tube voltages of 40 kV with anodes of Tungsten (*W*), Rhodium (*Rh*) and Molybdenum (*Mo*)

4.1 Hardware Innovations

Since projection radiography was the first introduced radiographic imaging technique, related technological advances are enormous. The first advance in technology was made by adapting the x-ray beam spectrum to energy changes (Fig. 10.7a). This allows the user to choose an optimal spectrum depending on the required procedure. For example, in chest radiography, higher energies (100–120 kV) are used because low contrast imaging between lungs and bones is desirable. On the opposite, low energy beams are used, for example, in orthopaedic radiography where high contrast between bones and soft tissue is needed. A big advance in x-ray tube design was achieved following the introduction of special dedicated mammographic anodes (molybdenum anodes) in 1965 (Fig. 10.7b). This was followed by rhodium anodes and filtrations made with the same material. These specific anodes present a high emission probability of two particular energies (characteristic radiation) at 17.9 and 19.5 keV for molybdenum, which are optimum to obtain very good contrast with smaller breast; while rhodium anodes have characteristic radiation at 20.3 and 22.7 keV, which are optimal for larger breasts, making it possible to save up to 40 % of the dose specially for larger breasts [45].

Detector technologies for projection imaging started with photographic film-based detectors that were subsequently improved following the introduction of fluorescent screens (analogue). The contrast capability of these detectors was defined by the characteristic curve [36], which measures how the film opacity (most precisely the log of the opacity) changes as a function of the log of the x-ray exposure. The higher the difference in the opacity for a smaller exposure, the higher is the contrast capability of the film. An example of such a curve is shown in Fig. 10.8, demonstrating the associated gradient of the curve calculated at each measurement point, which is representative of the film contrast.

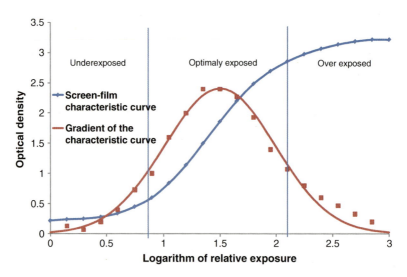

Fig. 10.8 Screen-film exposure characteristic curve

One can see that this curve has an "S" shape on the left and right sides (respectively underexposed and overexposed regions). Both regions are flat owing to the small contrast capability also put into evidence by the lower values on the gradient curve, while the central part (optimal exposition) shows a better contrast capability. For this reason, patient exposure is directly related to the type of screen-film employed since the exposure must be very close to the central part of this curve.

As a matter of fact, optimal exposure of films proved to be a difficult task to achieve given that the operator has to be consistent for different procedures to avoid errors. As a consequence, the concept of automatic exposure control (AEC) was developed with the aim to optimize the exposure of films under different settings [58]. At the present time, the technology changed to digital detectors where the contrast is no more dependent on the exposure (almost linear) as shown on the characteristic curve of two different detectors (Fig. 10.9). However, this remains a challenging issue when setting-up AEC devices [42]. Nevertheless, when these detectors were introduced, their performance was not optimal and as such their limited spatial resolution reduced their *NEQ* and *DQE* [57]. Recent advances in detector design overcome this drawback and result in better performance [56].

CT is a high dose procedure. Paradoxically the technical advances that allowed significant improvement in image quality were responsible for the increasing use of this technology by opening new avenues and offering novel diagnostic applications. The first CT scanner was built in 1972. It was first used to scan a brain inserted in a water-box at a speed of 4 min per rotation and required overnight reconstruction of the 8 mm thick images. The first waterless whole body scanner was developed in 1974 [48]. Given the slow acquisition time, motion blurring was an important limitation in terms of spatial resolution degradation. This was significantly reduced

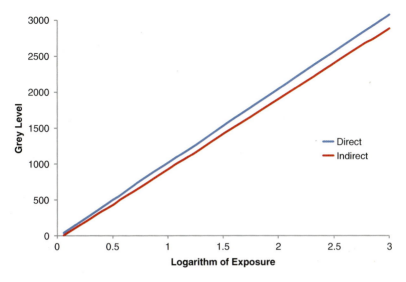

Fig. 10.9 Digital detectors exposure characteristic curve

by the introduction of the second-generation CT scanners through the use of narrow beams (contrary to its single-beam progenitor) and multiple detectors (3 in the beginning) that reduced motion blur by increasing rotation speed to 20 s per rotation. This idea was further exploited in the third generation CT scanners by introducing fan beams that cover the patient's width and by applying an array of detectors (250 to 750) to capture the entire beam, thus increasing the rotation speed. This was made possible by the introduction of high performance proportional detectors (e.g. Xenon detectors) that do not require recalibration prior to scanning [12] to enable axial translation (1st and 2nd generation scanners). This last advancement was crucial for the design of third generation scanners without axial translation, defining the basic geometry of current CT scanners (Fig. 10.10).

The slip ring technology was the next major advance in CT, allowing a continuous rotation of the x-ray tube/detector gantry, and was subsequently consolidated with helical scanning to reduce the acquisition time to about 1 s [41]. The helical pitch concept, defined as the table movement per rotation divided by the slice thickness, was also introduced with this technology. This allowed dose reduction when the pitch is higher than 1 by avoiding the irradiation of a smaller surface of the patient, but compromising image quality to some extent since the reconstruction is performed with interpolation of the missing data.

Another major leap in CT technology was the introduction of dual-slice scanning that opened the door to multi-slice CT scanners, thus enabling larger axial coverage per tube rotation. The consequence is faster acquisition without noticeable impact on individual patient dose [74]. Current technology permits acquisitions of up to 320-slices [70], with sub-millimetric slice thickness and rotation time of less than half a second.

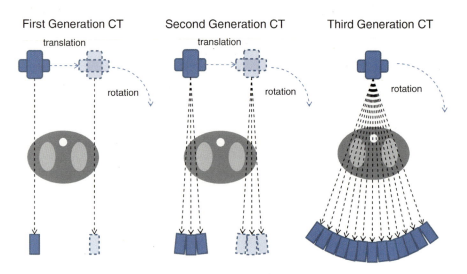

Fig. 10.10 Evolution of CT scanner acquisition techniques

With respect to dose reduction strategies, beam hardening filters have long been employed in x-ray CT to preferentially absorb soft and low-energy x-rays, which have little or no contribution to image formation, thus allowing the reduction of patient dose and beam hardening artefacts [5]. The introduction of AEC devices proved to be clinically relevant and had a significant impact in clinical practice (Fig. 10.11). The principle of AEC is to modulate the tube current as a function of patient attenuation during table movement (z-modulation) and also during the x-ray tube rotation (xy-modulation). The user can lever this modulation by choosing either a noise index or a reference tube current that will produce a preset noise level. The results have demonstrated that dose saving can be effective provided the scanner is used by a well-trained operator. This last observation, even if obvious and fundamental, is not so easy to implement in practice, since CT protocols must be adapted to new AEC devices. The motivations behind improvement in image quality are legitimate; however, the radiologist must keep in mind that the goal of imaging is to achieve clinical diagnosis and not simply to produce pretty images. Part of the responsibility can be attributed to scanner manufacturers who must train the end users to optimize the use of these devices. To avoid potential problems, the choice of noise reference level must be realistic and a maximum current exposure threshold must be set to avoid excessive exposure to patients, particularly in large and obese subjects [29, 52].

New research directions were explored during the last decade including dual-source CT (DSCT) [22–24] to achieve much better contrast discrimination and to improve spatial and temporal resolution. Increasing radiation dose remains the major concern. Even if an individual dose is not significantly increased, the increasing demand and utilization generated by improved image quality and increase applications will lead to a significant increase in the collective dose.

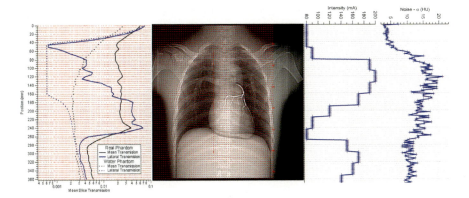

Fig. 10.11 Schema explaining the principle of automatic exposure control (*AEC*). *Left*: mean slice and lateral transmisions obtained with the experimental set-up (*continuous line*) and calculated with the hypothesis of a phantom having the same shape but filled with water (*dashed line*) showing the shape-only transmission dependency. *Middle left*: Topogram of the phantom. *Middle right*: Current intensity modulation obtained with AEC, very similar to the shape-only transmission curve. *Right*: The resulted mean noise with AEC, which should be as constant as possible

On the other hand, nuclear medicine forms a special category of medical imaging techniques in the sense that the information provided is functional rather than anatomical. The contrast and image quality reflecting tracer biodistribution in different tissues and organs are subjected to high statistical uncertainty given the small activity administered to the patient. For this reason, the performance of an imaging device is highly dependent on its sensitivity, which is defined as the number of counts detected per unit time per unit of activity present in the body region being studied [13]. The sensitivity depends on many parameters including geometric efficiency, crystal detection efficiency ... etc. In the early days of PET imaging, detectors were made of thallium-doped sodium iodide (NaI[Tl]) crystals similar to Anger cameras used for SPECT imaging. This technology was first replaced by Bismuth Germinate (BGO) crystals ($Bi_4Ge_3O_{12}$) introduced in 1973 [79], chosen mainly because of its higher efficiency at 511 keV. Gadolinium Oxyorthosilicate (GSO) (Gd_2SiO_5) is another popular scintillator used on commercial PET scanners [33], but the current materials of choice are Lutetium Oxyorthosilicate (LSO) crystals (Lu_2SiO_5) [53] and Cerium-doped Lutetium Yttrium Orthosilicate (LYSO) crytals ($Lu_{2(1-x)}Y_{2x}SiO_5$:Ce).

The availability of faster scintillation crystals and electronics renewed an interest in old technologies such as time-of-flight (ToF) PET and made this approach feasible on commercial clinical systems. First attempts relied on crystals such as Barium Fluoride (BaF_2) because of its very good time resolution of 156 picoseconds (ps) [4, 49]. Time-of-flight exploits the difference in the arrival time between the two crystals in coincidence to reduce the uncertainty in the annihilation point of the coincidence event to a limited region along the line of response, thus substantially diminish the noise in the reconstructed images. A timing resolution of 500 ps results in a reduction of a factor of five in the variance of the reconstructed image

Uncorrected PET   Attenuation corrected PET   PET/CT fusion

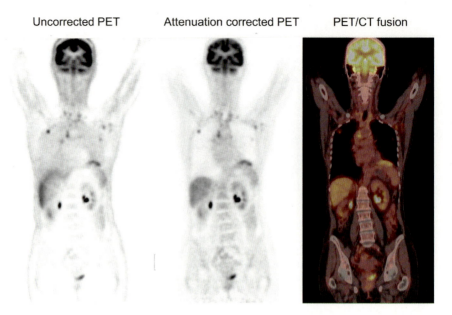

Fig. 10.12 Coronal PET images with and without CT-based attenuation correction and a fused PET/CT image. *Left*: Uncorrected PET image showing an underestimation of tracer uptake in the central part of the image. *Middle*: attenuation corrected PET image with the map derived from the corresponding CT image. *Right*: Fused PET and CT image putting into perspective complementary functional (*PET*) and anatomical (*CT*) information

compared to conventional PET [59]. It has been demonstrated that ToF-PET enables a much better compromise between contrast and noise, a property that is most invaluable in challenging situations such as: low statistic studies, scans requiring short examination time, large patients, low uptake ...etc. [19]. This technical innovation increases throughput, improves patient comfort, and reduces the patient dose by lowering the administered dose without deteriorating image quality. The administered activity of PET tracers could be tailored to an individual through the use of appropriate metrics reflecting data quality, such as noise-equivalent counting rate (NECR). It has been shown that the clinical NECR response matching individual patient scans can be accurately modelled to optimize the activity to be administered to an individual, thus reducing patient dose [77].

The introduction of dual-modality PET/CT imaging systems in clinical environments has revolutionized the practice of medical imaging [73]. The complementarity between the detailed anatomical (CT) and functional or metabolic (PET) information provided in a "one-stop shop" and the possibility of using CT images for attenuation correction of the PET data has been the driving force behind the success of this technology. Photon attenuation in the body during data acquisition is an important source of image degradation in PET. Despite the non-negligible increase in radiation dose [35, 84], the use of CT-based attenuation compensation has many advantages compared to conventional transmission-based scanning and has received a great deal of attention in the literature [46, 86]. Various dose reduction strategies including the use of low-dose CT protocols are becoming available for clinic use [80]. Attenuation correction is now recognized as an important image correction procedure not only for accurate quantification of PET images but also for lesion detection and accurate localization [6] (Fig. 10.12). Dedicated paediatric PET/CT protocols were developed to shorten acquisition time and reduce dose [1, 3].

Following the trend of hybrid PET/CT, PET/MR was introduced into clinical practice and is progressively becoming recognized by the medical community [25, 69]. MRI and PET offer complementary sensitivity and functional data respectively. When these technologies are combined together into a system that is capable of simultaneous acquisition will capitalize the strengths of each other [61]. The most important advantages of hybrid PET/MR technology in that anatomical MRI provides better soft tissue contrast compared to CT and this contrast can be adapted through the use of various sequences [85]. MR can also provide significant functional data of its own, such as diffusion-weighted MR for assessing tissue ischemia, contrast-based perfusion measurements in the brain and heart, fMRI and angiography. In addition, in contrast to CT, MRI does not use ionizing radiation and as such, it has major advantages when dose reduction is a concern, e.g. in paediatric application. Nevertheless, some general guidelines highlighting the major concerns for this imaging modality were developed for the users [37, 38]. These mostly emanate from the high-gradient fields that can stimulate some nerves and produce a sensation of discomfort and pain, particularly for those individuals taking drugs or afflicted with epilepsy. RF pulses can also induce heating of biological tissues.

There are several ways to combine PET and MRI data of the same patient [85]. The most straightforward approach is to adopt the configuration of PET/CT, in the so-called "tandem" arrangement, where the two procedures are performed sequentially in space and time. However, a very attractive solution is to acquire MRI and PET data simultaneously in space and probably in time. This innovative approach has resulted firstly in the design of an "insert" concept, where a small axial size MR-compatible PET insert fits inside a standard MRI scanner, and secondly to a completely integrated version, where a dedicated whole body PET scanner is built within a dedicated MRI scanner. This latter conceptual design is the most attractive but also the most challenging to achieve. However, it is certainly the way forward to optimize the use of combined PET/MRI systems in clinical diagnosis, therapy and follow-up.

Similar to PET/CT, the anatomical information from MRI can be useful for many other tasks including attenuation compensation, transmission-based scatter modelling, motion detection and correction, partial volume correction, and introducing *a priori* anatomical information into the reconstruction of the PET emission data. Given its clinical relevance and the challenges, the development of robust strategies for MRI-guided attenuation correction remains a major drawback and will probably be solved by strengthening research and developing innovative solutions in this area [32, 83].

With the advent of full digital systems, image display must be also ensure a high quality display of medical images by respecting well established standards of reliability. Different standards were proposed but the most widely used are those detailed in the German DIN 6868–57 regulation [20] and its US counterpart published in the AAPM TG18 report [67]. While most of the requirements of these standards are fulfilled by all commercial systems, the most important one is the greyscale display conformance with the DICOM 3.14 standard, which is the most difficult to comply since not all monitors (especially common desktop display systems) conform to this requirement [28]. Special attention should therefore be paid to this issue since it might compromise the whole medical imaging chain.

4.2 Software Innovations

In parallel to hardware solutions to reduce patient dose without sacrificing image quality, software approaches were developed to achieve this goal mostly through the use of advanced image processing to enhance image quality or innovative image reconstruction strategies in emission and transmission tomography to reduce the high noise typically present in low-dose scanning protocols. The reduction of image noise and streak artefact in low dose CT scans using specially designed noise reduction filters successfully achieved this goal at the expense of image sharpness and contrast as well as slightly altered modulation transfer function at higher spatial frequencies. Various filters have been suggested in the literature with varying degrees of success. Filtering can be performed either in image space [43] or projection space [40]. The latter has some advantages in terms of spatial resolution loss, which is important in clinical diagnosis.

Alternative techniques were suggested to deal with the challenging task of streak artefacts removal given the difficulty of discriminating them from the attenuation information of biologic tissues. One such technique, referred to as artefact suppressed large-scale nonlocal means, consists of a two-step processing scheme for suppressing both noise and artefacts in thoracic LDCT images. The idea is to exploit the specific scale and direction properties to discriminate noise and artefacts from the imaged structures [15].

On the other hand, the introduction and widespread commercial adoption of iterative reconstruction techniques in emission tomography spurred renewed interest in their use in other imaging modalities including x-ray CT. Basically, two major classes of image reconstruction algorithms are used in CT: direct analytical methods and iterative methods. Until recently, the most widely used methods for image reconstruction were direct analytical techniques because they are relatively quick and their derivation straightforward. However, the resulting image quality is limited by the over simplified line-integral model of the acquired projection data. In contrast, iterative techniques are computationally more intensive but the resulting images demonstrate improvements (principally arising from more accurate modelling of the acquired projection data) which have enabled them to replace analytic techniques not only in research but also in clinical settings [60].

Iterative reconstruction techniques came forward in x-ray CT after having been abandoned owing to the large amount of data produced by the latest generation of CT scanners and the associated computational burden. The widespread availability of high performance computing even on desktop computers (including GPU and cloud computing), and the continuing endeavours towards dose reduction in diagnostic procedures led to renewed interest in iterative techniques and made them a hot topic for the leading manufacturers of clinical CT scanners.

Iterative techniques are commonly used in problems that involve optimization. The reconstruction problem can be considered a particular case where one is trying to determine the 'best' estimate of the distribution of attenuation coefficients in CT based on the measured projections (sinograms). Iterative reconstruction has a number of impending advantages that make it attractive in comparison with conventional analytical reconstruction methods. Foremost is the limiting assumption in analytical techniques that the measured projection data are perfectly consistent with the object to be reconstructed, a requirement that is certainly not exact in practice given the presence of noise and other physical factors (e.g. beam hardening, scatter, detector spatial response, truncation ... etc.). Iterative algorithms are well suited to handling complex physical models of the transmission and detection processes. This ability to directly model the system response, including some consideration of the noise characteristics, provides substantial flexibility in the type of data that can be reconstructed.

The popularity of iterative image reconstruction stems from substantial improvement of image quality, particularly for low-dose protocols [10]. Most manufacturers have recently implemented iterative reconstruction techniques into their platforms, thus enabling the use of ultra-low dose protocols without sacrificing image quality [44, 51, 81]. As such, substantial reduction in patient dose (in the range 40–80 %) could be achieved using this class of reconstruction algorithms. It is expected that iterative CT reconstruction implemented on commercial cloud computing environments will be available in the near future.

Computer-aided diagnosis (CAD) has been one of the major research topics in medical imaging and diagnostic radiology. As such, it has become an integral part of routine clinical workflow for the detection of breast cancer by mammography at many screening facilities. However, the technique is still in its infancy given the enormous potential expected for its application to the detection of other diseases based on analysis performed on various single and hybrid imaging modalities.

5 Summary

Over the past three decades, we have witnessed significant advances in medical imaging, which have revolutionized the assessment of a multitude of disorders with high accuracy and precise spatial resolution. Recent advances in medical imaging system design have resulted in significant improvements in anatomical, functional, and dynamic imaging procedures taking advantage of innovative approaches in image reconstruction and analysis strategies. With these developments, computer-aided diagnosis is becoming a reality. These computer-based tools allow physicians to understand and diagnose disease through virtual interaction. The role of medical imaging is not limited to the visualization and evaluation of structure and function, but goes beyond that to diagnosis, surgical planning, simulation, and radiotherapy planning.

Better communication and collaboration between the end-users and equipment vendors in research (by providing feedback) and better clinical utilization (through better commissioning, quality assurance, education and training, developing standards and implementing policies) improves daily practice and the optimization of radiation protection. Senior leadership teams can inspire and encourage other individuals, team members and health care organizations to change, adapt, grow, and prepare for the future challenges. This can be achieved through creative dialogue between multidisciplinary teams, thus facilitating a culture of best practice, and providing the necessary skills to better-prepare future medical imaging leaders. With the ever-increasing costs and challenges in health care, there is a strong need for innovative leaders to develop new models for the healthcare systems where medical imaging plays a pivotal role.

On reflection, it is gratifying to witness the vast progress that multimodality imaging has made in the last two decades through leadership and innovation in the various disciplines. It is expected that with the availability of high-resolution multimodality imaging systems in the near future, molecular imaging-based and personalized medicine will become clinically feasible.

Acknowledgements This work was supported by the Swiss National Science Foundation under grant SNSF 31003A-135576, Geneva Cancer League, the Indo-Swiss Joint Research Programme ISJRP 138866, and Geneva University Hospital under grant PRD 11-II-1.

References

1. Accorsi R, Karp JS, Surti S (2010) Improved dose regimen in pediatric PET. J Nucl Med 51(2):293–300
2. Alavi A, Basu S, Torigian D et al (2008) Is planar imaging in radiology and nuclear medicine a viable option for the 21st century? Q J Nucl Med Mol Imaging 52(4):319–322
3. Alessio AM, Kinahan PE, Manchanda V et al (2009) Weight-based, low-dose pediatric whole-body PET/CT protocols. J Nucl Med 50(10):1570–1577
4. Allemand R, Gresset C, Vacher J (1980) Potential advantages of a cesium fluoride scintillator for a time-of-flight positron camera. J Nucl Med 21(2):153–155
5. Ay MR, Mehranian A, Maleki A, Ghadiri H, Ghafarian P, Zaidi H (2013) Experimental assessment of the influence of beam hardening filters on image quality and patient dose in volumetric 64-slice x-ray CT scanners. Phys Medica 29(3):249–260
6. Bai C, Kinahan PE, Brasse D et al (2003) An analytic study of the effects of attenuation on tumor detection in whole-body PET oncology imaging. J Nucl Med 44(11):1855–1861
7. Ballinger J (2008) Re: planar and SPECT imaging in the era of PET and PET-CT: can it survive the test of time? Eur J Nucl Med Mol Imaging 35(12):2340

8. Barrett HH, Myers KJ (2003) Foundations of image science. Wiley, New Jersey
9. Barrett HH, Yao J, Rolland JP et al (1993) Model observers for assessment of image quality. Proc Natl Acad Sci U S A 90(21):9758–9765
10. Beister M, Kolditz D, Kalender WA (2012) Iterative reconstruction methods in X-ray CT. Phys Med 28(2):94–108
11. Bonte FJ (1976) Nuclear Medicine Pioneer Citation, 1976: David E Kuhl MD. J Nucl Med 17(6):518–519
12. Boyd D, Coonrod J, Dehnert J et al (1974) A high pressure xenon proportional chamber for x-ray laminographic reconstruction using fan beam geometry. IEEE Trans Nucl Sci 21:184–187
13. Budinger TF (1998) PET instrumentation: what are the limits? Semin Nucl Med 28(3):247–267
14. Bushberg JT, Seibert JA, Leidholdt EM (2002) The essential physics of medical imaging, 2nd edn. Lippincott Williams & Wilkins, Philadelphia
15. Chen Y, Yang Z, Hu Y et al (2012) Thoracic low-dose CT image processing using an artifact suppressed large-scale nonlocal means. Phys Med Biol 57(9):2667–2688
16. Cherry SR (2006) The 2006 Henry N. Wagner lecture: of mice and men (and positrons) -advances in PET imaging technology. J Nucl Med 47(11):1735–1745
17. Cherry SR, Sorenson JA, Phelps ME (2004) Physics in nuclear medicine, 3rd edn. Elsevier Health Sciences, Philadelphia
18. Coltman JW (1954) The specification of imaging properties by response to a sine wave. J Opt Soc Am 44:468–469
19. Conti M (2011) Focus on time-of-flight PET: the benefits of improved time resolution. Eur J Nucl Med Mol Imaging 38(6):1147–1157
20. Deutsches Institut für Normung (2012) Publication DIN 6868-157 Image quality assurance in diagnostic X-ray departments – Part 157: RöV acceptance and constancy test of image display systems in theirs environment. Deutsches Institut für Normung, Berlin
21. Fellgett BP (1958) Equivalent quantum-efficiencies of photographic emulsions. The Observatories, Cambridge
22. Flohr TG, Bruder H, Stierstorfer K et al (2008) Image reconstruction and image quality evaluation for a dual source CT scanner. Med Phys 35(12):5882–5897
23. Flohr TG, Leng S, Yu L et al (2009) Dual-source spiral CT with pitch up to 3.2 and 75 ms temporal resolution: image reconstruction and assessment of image quality. Med Phys 36 (12):5641–5653
24. Flohr TG, McCollough CH, Bruder H et al (2006) First performance evaluation of a dual-source CT (DSCT) system. Eur Radiol 16(2):256–268
25. Gaa J, Rummeny EJ, Seemann MD (2004) Whole-body imaging with PET/MRI. Eur J Med Res 9(6):309–312
26. Gambhir SS, Berman DS, Ziffer J et al (2009) A novel high-sensitivity rapid-acquisition single-photon cardiac imaging camera. J Nucl Med 50(4):635–643
27. Green DM, Swets JA (1974) Signal detection theory and psychophysics. Krieger Publishing, New York
28. Gutierrez D, Monnin P, Valley JF et al (2005) A strategy to qualify the performance of radiographic monitors. Radiat Prot Dosimetry 114(1–3):192–197
29. Gutierrez D, Schmidt S, Denys A et al (2007) CT-automatic exposure control devices: what are their performances? Nucl Instrum Methods Phys Res A 580:990–995
30. Hart D, Wall BF (2002) Radiation exposure of the UK population from medical and dental X-ray examinations. Publication NRPB-W4. National Radiological Protection Board, Chilton
31. Hasegawa B, Zaidi H (2006) Dual-modality imaging: more than the sum of its components. In: Zaidi H (ed) Quantitative analysis in nuclear medicine imaging. Springer, New York
32. Hofmann M, Pichler B, Schölkopf B et al (2009) Towards quantitative PET/MRI: a review of MR-based attenuation correction techniques. Eur J Nucl Med Mol Imaging 36(Suppl 1):93–104

33. Holte S, Ostertag H, Kesselberg M (1987) A preliminary evaluation of a dual crystal positron camera. J Comput Assist Tomogr 11(4):691–697
34. Hounsfield GN (1973) Computerized transverse axial scanning (tomography) 1. Description of system. Br J Radiol 46(552):1016–1022
35. Huang B, Law MW, Khong PL (2009) Whole-body PET/CT scanning: estimation of radiation dose and cancer risk. Radiology 251(1):166–174
36. Hurter F, Driffield VC (1890) Photochemical investigations and a new method of determination of the sensitiveness of photographic plates. J Soc Chem Indian 9:455
37. International Commission on Non-Ionizing Radiation Protection (1998) Guidelines for limiting exposure to time-varying electric, magnetic, and electromagnetic fields (up to 300 GHz). Health Phys 74(4):494–522
38. International Commission on Non-Ionizing Radiation Protection (2004) Medical magnetic resonance (MR) procedures: protection of patients. Health Phys 87(2):197–216
39. International Electrotechnical Commission (2003) Medical electrical equipment – characteristics of digital X-ray imaging devices. Part 1: determination of the detective quantum efficiency. International Electrotechnical Commission, Geneva, 62220–1
40. Kachelriess M, Watzke O, Kalender WA (2001) Generalized multi-dimensional adaptive filtering for conventional and spiral single-slice, multi-slice, and cone-beam CT. Med Phys 28(4):475–490
41. Kalender WA, Seissler W, Klotz E et al (1990) Spiral volumetric CT with single-breath-hold technique, continuous transport, and continuous scanner rotation. Radiology 176(1):181–183
42. Kalra MK, Maher MM, Toth TL et al (2004) Techniques and applications of automatic tube current modulation for CT. Radiology 233(3):649–657
43. Kalra MK, Wittram C, Maher MM et al (2003) Can noise reduction filters improve low-radiation-dose chest CT images? Pilot study. Radiology 228(1):257–264
44. Katsura M, Matsuda I, Akahane M et al (2012) Model-based iterative reconstruction technique for radiation dose reduction in chest CT: comparison with the adaptive statistical iterative reconstruction technique. Eur Radiol 22(8):1613–1623
45. Kimme-Smith C, Wang J, DeBruhl N et al (1994) Mammograms obtained with rhodium vs molybdenum anodes: contrast and dose differences. AJR Am J Roentgenol 162(6): 1313–1317
46. Kinahan PE, Hasegawa BH, Beyer T (2003) X-ray-based attenuation correction for positron emission tomography/computed tomography scanners. Semin Nucl Med 33(3):166–179
47. Kuhls-Gilcrist A, Jain A, Bednarek DR et al (2010) Accurate MTF measurement in digital radiography using noise response. Med Phys 37(2):724–735
48. Lcdley RS, Chiro GD, Luessenhop AJ et al (1974) Computerized transaxial X-ray tomography of the human body. Science 186:207–212
49. Lewellen TK (1998) Time-of-flight PET. Semin Nucl Med 28(3):268–275
50. Mariani G, Bruselli L, Duatti A et al (2008) Is PET always an advantage versus planar and SPECT imaging? Eur J Nucl Med Mol Imaging 35(8):1560–1565
51. Marin D, Nelson RC, Schindera ST et al (2010) Low-tube-voltage, high-tube-current multidetector abdominal CT: improved image quality and decreased radiation dose with adaptive statistical iterative reconstruction algorithm-initial clinical experience. Radiology 254(1):145–153
52. Matsumoto Y, Masuda T, Imada N et al (2012) Examination of the chest computed tomography scan condition optimization in consideration of the influence of the position of the arms. Nihon Hoshasen Gijutsu Gakkai Zasshi 68(7):851–856
53. Melcher CL, Schweitzer JS (1992) Cerium-doped lutetium oxyorthosilicate: a fast, efficient new scintillator. IEEE Trans Nucl Sci 39:502–505
54. Metz C (2008) ROC analysis in medical imaging: a tutorial review of the literature. Radiol Phys Technol 1(1):2–12
55. Metz CE (1986) ROC methodology in radiologic imaging. Invest Radiol 21(9):720–733
56. Monnin P, Gutierrez D, Bulling S et al (2007) A comparison of the performance of digital mammography systems. Med Phys 34(3):906–914

57. Monnin P, Gutierrez D, Bulling S et al (2005) A comparison of the imaging characteristics of the new Kodak Hyper Speed G film with the current T-MAT G/RA film and the CR 9000 system. Phys Med Biol 50(19):4541–4552
58. Morgan RH, Hodges PC (1945) An evaluation of automatic exposure control equipment in photofluorography. Radiology 45:588–593
59. Moses WW (2003) Time of flight in PET revisited. IEEE Trans Nucl Sci 50(5):1325–1330
60. Pan X, Sidky EY, Vannier M (2009) Why do commercial CT scanners still employ traditional, filtered back-projection for image reconstruction? Inverse Probl 25(12):123009
61. Pichler BJ, Wehrl HF, Kolb A et al (2008) Positron emission tomography/magnetic resonance imaging: the next generation of multimodality imaging? Semin Nucl Med 38(3):199–208
62. Radon J (1917) On the determination of functions from their integrals along certain manifolds. Ber Saechs Akad Wiss 69:262–277
63. Reader AJ, Zaidi H (2007) Advances in PET image reconstruction. PET Clin 2:173–190
64. Rhodes DJ, Hruska CB, Phillips SW et al (2011) Dedicated dual-head gamma imaging for breast cancer screening in women with mammographically dense breasts. Radiology 258(1):106–118
65. Rose A (1948) The sensitivity performance of the human eye on an absolute scale. J Opt Soc Am 38:196–208
66. Rossmann K, Wiley BE (1970) The central problem in the study of radiographic image quality. Radiology 96(1):113–118
67. Samei E, Badano A, Chakraborty D et al (2005) Assessment of display performance for medical imaging systems: executive summary of AAPM TG18 report. Med Phys 32(4):1205–1225
68. Samei E, Flynn MJ, Reimann DA (1998) A method for measuring the presampled MTF of digital radiographic systems using an edge test device. Med Phys 25(1):102–113
69. Schlyer D, Rooney W, Woody C et al (2004) Development of a simultaneous PET/MRI scanner. In: IEEE Nucl Sci Symp Conf Rec 3419–3421
70. Seguchi S, Aoyama T, Koyama S et al (2010) Patient radiation dose in prospectively gated axial CT coronary angiography and retrospectively gated helical technique with a 320-detector row CT scanner. Med Phys 37(11):5579–5585
71. Shaw R (1963) The equivalent quantum efficiency of the photographic process. J Photogr Sci 11:199–204
72. Smith-Bindman R, Lipson J, Marcus R et al (2009) Radiation dose associated with common computed tomography examinations and the associated lifetime attributable risk of cancer. Arch Intern Med 169(22):2078–2086
73. Townsend DW, Beyer T (2002) A combined PET/CT scanner: the path to true image fusion. Br J Radiol 75:S24–S30
74. Verdun FR, Theumann N, Poletti PA (2006) Impact of the introduction of 16-row MDCT on image quality and patient dose: phantom study and multi-centre survey. Eur Radiol 16(12):2866–2874
75. Wagner RF (1978) Decision theory and the signal-to-noise ratio of Otto Schade. Photogr Sci Eng 22:41–46
76. Wagner RF, Weaver KE (1972) An assortment of image quality indexes for radiographic film-screen combinations – can they be resolved? Appl Opt Instr Med Proc SPIE 35:83–94
77. Watson CC, Casey ME, Bendriem B et al (2005) Optimizing injected dose in clinical PET by accurately modeling the counting-rate response functions specific to individual patient scans. J Nucl Med 46(11):1825–1834
78. Webb S (1992) The physics of medical imaging. Institute of Physics, London
79. Weber MJ, Monchamp RR (1973) Luminescence of Bi4Ge3O12: spectral and decay properties. J Appl Phys 44:5495–5499
80. Xia T, Alessio AM, De Man B et al (2012) Ultra-low dose CT attenuation correction for PET/CT. Phys Med Biol 57(2):309–328
81. Xu J, Mahesh M, Tsui BM (2009) Is iterative reconstruction ready for MDCT? J Am Coll Radiol 6(4):274–276

82. Zaidi H (ed) (2006) Quantitative analysis in nuclear medicine imaging. Springer, New York
83. Zaidi H (2007) Is MRI-guided attenuation correction a viable option for dual-modality PET/MR imaging? Radiology 244(3):639–642
84. Zaidi H (2007) Is radionuclide transmission scanning obsolete for dual-modality PET/CT systems? Eur J Nucl Med Mol Imaging 34(6):815–818
85. Zaidi H, Del Guerra A (2011) An outlook on future design of hybrid PET/MRI systems. Med Phys 38(10):5667–5689
86. Zaidi H, Hasegawa B (2003) Determination of the attenuation map in emission tomography. J Nucl Med 44(2):291–315

Chapter 11
System for Reporting and Analysing Incidents

Catherine Mandel and William Runciman

Abstract Incident reporting is a key safety tool in high-risk sectors, including healthcare. To be effective, an incident reporting system must be well constructed and reporting encouraged and supported. The presence of a fair and just culture in the workplace, where reporting is seen as a means of improving patient care, and not a tool to punish others, encourages open and honest reporting. This chapter outlines the rationales, benefits, issues, and features essential for an incident reporting system for radiology and medical imaging by using the Radiology Events Register (RaER) as an example. The challenges limiting incident reporting and the possible solutions are also presented.

Keywords Incident report • Near miss • Patient safety • Radiology error • Risk management

1 Introduction

Primum non nocere, 'first, do no harm', is the underlying tenet of healthcare. The vast majority of healthcare practitioners do not go to work intending to do harm; however, no individual, machine or system is infallible and every individual does things that could, or do, result in harm. Errors are inevitable. By knowing what happens and why, it is possible to put systems into place to minimize frequency,

C. Mandel (✉)
Cancer Imaging, Peter MacCallum Cancer Centre, St Andrew's Place,
3002 East Melbourne, VIC, Australia
e-mail: catherine.mandel@petermac.org

W. Runciman
School of Psychology, Social Work and Social Policy, University of South Australia,
City East Campus, North Terrace, 5000 Adelaide, South Australia, Australia

L. Lau and K.-H. Ng (eds.), *Radiological Safety and Quality: Paradigms in Leadership and Innovation*, DOI 10.1007/978-94-007-7256-4_11,
© Springer Science+Business Media Dordrecht 2014

maximize detection and minimize impact. One way to detect incidents, including near misses, is by using a dedicated incident reporting system. An incident reporting system collects data, analyses causes, classifies events, develops solutions and advocates their use.

Incident reporting is a way to detect real or potential harm. A well-constructed and utilized incident reporting system is able to detect near misses, which in other high-risk industries account for most of the reported incidents. Incident reporting is well established in many high-risk industries, such as aviation, oil, gas, nuclear power, and rail; and has been shown to improve safety and reduce adverse events [31]. Although incident reporting in healthcare has been in existence for decades, its routine use as a quality improvement and safety tool is less well accepted, despite this being a core component of patient safety improvement systems [7].

Examples of safety improvements in healthcare that have been brought about by the use of incident reporting are the routine use of pulse oximetry and capnography in general anaesthesia [27]. Perhaps the most significant benefit is an increased awareness of risk and safety, which is not easily measured.

2   What Are Incidents and Why Do They Happen

A patient safety incident, or incident, is defined as 'an event or circumstance which could have resulted, or did result, in unnecessary harm to a patient' [30]. Incidents in healthcare are like icebergs: the visible tips are the incidents that result in harm; the majority, equivalent to the submerged parts of the icebergs, are near misses i.e. incidents that did not reach the patient.

Most, if not all, incidents are the result of a sequence of events; but each event in isolation would not have resulted in harm. A near miss is when the combination of circumstances results in the potential harm being detected and mitigated before an adverse outcome occurs. This scenario is well described by Reason's Swiss cheese model (Fig. 11.1) [24, 25]. Whilst individual actions may have a significant impact on safety, it is the culture and working environment that affect the thinking and behavior of the practitioners, and encourage or constrain safe practice.

3   Why Report Incidents

Errors are prevalent in healthcare. For example, up to 18,000 Australian patients died from healthcare mistakes and over 16 % of admissions were associated with an adverse event [27]. The Institute of Medicine in the USA published 'To err is human' [11], in which it was reported that between 44,000 and 98,000 Americans died each year as a result of preventable medical errors, more than the number of deaths from common malignancies such as breast cancer. A key finding of this report is that most of the harm was due to system issues rather than bad or incompetent practitioners or

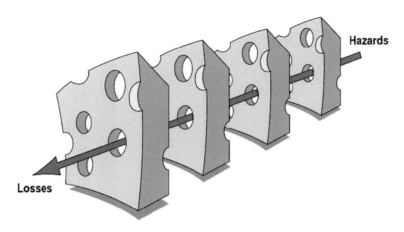

Fig. 11.1 Swiss cheese model. The slices of cheese represent the barriers in place to prevent harm but no preventative strategy is totally reliable. The holes in the cheese represent deficiencies in the barriers. By having several preventative strategies in place it is more likely that a potential harm will be detected and mitigated before it reaches the patient. Occasionally, however, the holes line up and the patient is harmed. Incident reports can be used to analyze these incidents and the knowledge gained can be used to strengthen the defences (Reproduced with adaptation from 'Reason J (2000) Human error: models and management. BMJ 320(7237):768–70', with permission from the BMJ Publishing Group Ltd)

other staff. The report recommended that the American government should '... develop, disseminate, and evaluate tools for identifying and analyzing errors ...'; and '... identify and learn from errors by (1) developing a nationwide public mandatory reporting system and (2) encouraging healthcare organizations and practitioners to develop and participate in voluntary reporting systems ...'. This report further commented that '... (voluntary) systems can focus on a much broader set of errors, mainly on those that do no harm or result in minimal harm, thus helping to detect the system weaknesses that should be fixed before the occurrence of any serious harm, thereby providing rich information to the healthcare organizations in support of their quality improvement efforts...'.

In the United Kingdom approximately 900,000 incidents are reported within the National Health Service (NHS) every year and approximately 2,000 of these result in death [15]. These figures demonstrate that error is common in healthcare. In order to understand errors, reduce incidents and improve patient safety, it is important to capture the data on all incidents including near misses, rather than just those resulting in harm. Incident reporting is more likely to capture this type of data than other methods of studying patient safety, errors and incidents.

The practice of radiology and medical imaging can be reviewed and audited more easily than other specialties as images, reports, and other data produced are recorded and retained. However, the scope of this data is very limited because the vast majority of incidents including all near misses and much of the information about the underlying causes are not captured. In order to obtain this information, including the data that enables the identification of the underlying causes, a more comprehensive approach such as an incident reporting system is needed.

4 Components of an Incident Reporting System

The National Patient Safety Agency (NPSA) in the United Kingdom, published 'Seven Steps to Patient Safety' [16, 17] and describes the following elements as the 'circle of safety':

1. Reporting;
2. Analysis;
3. Solution development;
4. Implementation;
5. Audit and monitoring;
6. Feedback; and
7. Reporting.

4.1 Reporting System

A good incident reporting system should: be independent, trustworthy, relevant, non-punitive, accessible to all, and user-friendly; encourage entry of honest, thoughtful narrative and data; and have peer involvement. The design must balance the need for adequate information against the time required. Unfortunately, under reporting is widespread [18, 27].

Scope

An incident reporting system for radiology and medical imaging includes all the events along the patient journey. A reportable event is not limited to the acquisition and interpretation of images or performance of procedures, i.e. events relating to patient preparation, referral communication, result delivery etc. should also be recorded. These events can occur outside a radiology department. In a British study of anesthetic equipment failure, nine incidents occurred in remote sites, mostly in radiology [6]. These anesthetic incidents should also be part of a radiology incident database.

Common Taxonomy

A common language for patient safety needs to be used in the classification, analysis and dissemination of the findings. This facilitates the sharing of data and the lessons learned [28]. The World Health Organization (WHO) has developed a conceptual framework [31] that is currently being developed into a classification. Common terminology enables computer-aided data analysis. This is important when analyzing large volumes of data and detecting trends [14].

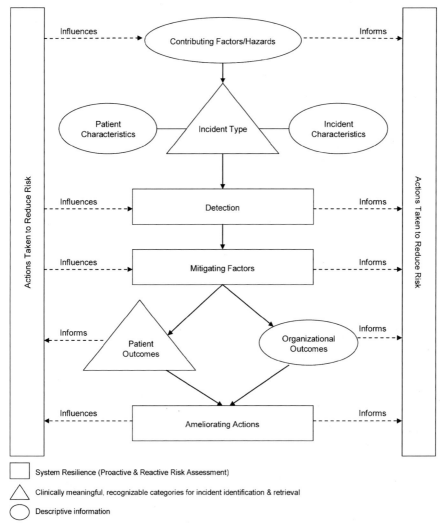

System Resilience (Proactive & Reactive Risk Assessment)

Clinically meaningful, recognizable categories for incident identification & retrieval

Descriptive information

The solid lines represent the semantic relationships between the classes. The dotted lines represent the flow of information.

Fig. 11.2 The WHO Conceptual Framework for the International Classification for Patient Safety. This diagram shows the stages in the development and detection of incidents. This information, when captured in an incident report, can be used to analyze, classify and group incidents and devise strategies to prevent, detect and mitigate further occurrences of incidents (Reproduced from 'Chapter 2, page 8 of Conceptual Framework for the International Classification for Patient Safety (2009) WHO, Geneva' with permission from the World Health Organization)

WHO Classification Framework

A diagram illustrating the WHO conceptual framework for the classification of patient safety is shown in Fig. 11.2 [38]. For any incident there are usually several contributing factors or hazards, some of which may be mitigated by actions taken to reduce risk. The outcome, whether a near miss or harm occurring to the patient or organization will depend on when the incident is detected and the effectiveness of the mitigating and ameliorating factors. Information from all stages is useful in developing strategies and tools to prevent further incidents.

Some examples of the contributing factors to a misidentification incident (e.g. procedure performed on a wrong patient) could include: poor communication between staff, wrong name on request form, two patients with the same or similar name, and inability to communicate due to coma, a minor, dementia or language barrier etc.

Patient characteristics are demographic data, reason(s) for seeking care and the primary diagnosis.

Actions taken to reduce the risk could include a robust patient identification policy, e.g. asking the patient to actively identify himself or herself rather than asking 'Are you John Smith?'

An example of detection and mitigating factors is the cross-checking of patient identification before a procedure. These are secondary and tertiary prevention strategies.

Patient outcomes could be a near miss, a delay in receiving the correct procedure, or receiving an unnecessary procedure and/or radiation exposure.

Organizational outcomes could include root cause analysis leading to sanction or review for incorrectly irradiating a patient.

Ameliorating actions could include an apology to the patient and treatment of any harm sustained.

Actions taken to reduce risk after the event could be the development of a robust patient identification policy and the education of staff and patients to ensure proper implementation.

Data Fields

The following fields are suggested as the essential data required for a reporting system [21]:

1. The discovery:

   (a) Who discovered the incident
   (b) How was it discovered

2. The event:

   (a) What happened: type of event and narrative
   (b) Where in the care process did the event occur and at what stage was it discovered
   (c) When did it occur
   (d) Who was involved (type of staff, not necessarily named)
   (e) Why did it happen: primary and secondary causes
   (f) Risk assessment: severity, probability and preventability
   (g) Narrative of the event including the contributing factors and ancillary information

(h) Product information
(i) Patient demographics, diagnosis, procedure and co-morbidities

The other important elements are the contributing and mitigating factors, consequences, modifiers and prevention strategies.

Paper Or Electronic Forms

Traditional incident reports have been paper documents that needed to be read, collated, classified and analyzed. This was time-consuming and prone to problems, especially when interpreting illegible handwriting. Whilst there is a role for this format in those settings without internet access, its use is becoming less common.

Electronic or on-line report filing is common in healthcare. Electronic data entry has many advantages, not least computer-aided analysis, classification and data search. Another advantage is the ability to pose questions that are specific to the type of incident. For example, different information would be required for a patient identification incident compared with an equipment failure incident. On-line reporting allows incidents to be reported from anywhere and at any time. Making incident reporting easy and contemporaneous is important.

As with all computer-based processes, stringent attention to data security including encryption before transmission and storage, and data back up/duplication to safeguard against system failure are required. The ability to use the reporting system on a variety of platforms and software is critical.

The use of 'pick-lists' and forcing functions ensure that essential data are entered. An ability to enter free narrative text is important [18], because this is an important source of information about what happened, the events before the incident, what actions reduced the harm and what actions could potentially have prevented the incident.

A system may be custom-designed or an existing proven program may be modified to meet the requirements of radiology and medical imaging. One such programme is the Advanced Incident Management System AIMS™ software [2], which is used for the Radiology Event Register (RaER) database [24].

In some industries such as aviation, reports can be filed by telephone initially. A final written document is still needed and this can be done by fax. Staffing a call centre increases cost but this option increases accessibility especially if other means are not readily available.

A Collaborative Or Stand-Alone Database

An incident reporting system for radiology and medical imaging can be part of a large collaborative international, national or hospital-wide reporting system or a stand-alone, dedicated database.

A national database such as the National Reporting and Learning System (NRLS) in the United Kingdom obtains data directly from the local risk management systems (LRMS) used in hospitals and other healthcare organizations. Data is entered once and is available in two systems. It is possible to report incidents directly to the NRLS, including by patients and carers. The data is shared with other organizations when the incidents are relevant, e.g. incidents involving medical devices or drug reactions. Such system has the benefit of single data entry from any field of healthcare but the data unique to radiology may not be specifically recorded. Vendors of risk management systems are able to include relevant datasets in their products so they can interact with larger systems [16, 17].

An international incident reporting model for radiology and medical imaging is RaER, a peer-led and radiology-specific database that collects data from Australia and New Zealand [24]. Such a system is not a substitute for reporting incidents into a hospital database. Ideally data should be available in both systems after a single entry.

Whilst a stand-alone, facility-based reporting system can collect data about the common problems, it may not detect trends or support analysis. Since many different incidents occur infrequently, it is important to share and aggregate this data for an analysis of the risks and trends, as is the case with adverse drug reactions.

An incident collection agency must operate independently from an enforcement authority, maintain patient and practitioner confidentiality, comply with legal requirements and preferably be endorsed by the stakeholders. Local access to a reporting system is essential as this facilitates the timely collection of data. The dispersion of reports amongst the different systems, e.g. hospital, national, radiation regulatory body etc. should be avoided [30].

Voluntary Or Mandatory Reporting

Ideally all incidents should be recorded in an incident reporting database; in reality many are not [18, 27]. As voluntary reporting has varying degrees of success, the reporting of serious incidents is often mandated or a legal requirement. In aviation the reporting of serious incidents is required by law. Whilst this is also meant to be the case in healthcare, e.g. in the notification of wrong side, wrong site and wrong patient operations etc., it does not always occur. There are many reasons why this may not happen. Improving the safety culture and acceptance of incident reporting increase reporting rates. The ability of a system to accept voluntary reports about any aspect of patient care or any other perceived risk within an organization will give the greatest opportunity to detect potential threats and vulnerabilities within systems and will improve patient safety [21].

Known, Confidential Or Anonymous Reporting

Reporting an incident can cause anxiety about possible backlash from colleagues, employers or other organizations. In some cases reports have been used punitively [27]. The person reporting an incident may fear being blamed for doing something wrong, even if the actions were not deliberate; of being made a scapegoat i.e. blame the person rather than fix the system; of backlash from other individuals or organizations who may be scrutinized or found to be imperfect as a result of the report. Protecting the identity of the person reporting an incident is therefore critical. This could be achieved by confidential reporting either by legislation or permitting anonymous reporting of incidents [29].

Confidentiality has a clear advantage: the person who reported an incident can be asked to provide more information. This method is used in aviation and in the NRLS. In aviation the accident investigation and regulatory bodies are separate entities. This is important, as the investigating body cannot have a punitive role. Legislation describes what can and cannot be disclosed to others outside the investigating authority.

If an organization does not have a just culture i.e. one that is based on fairness and justice, and one that seeks to balance the need to learn from mistakes and the need to take disciplinary action [38]; there may be hesitation to provide full and frank details or to report at all. With confidential reporting, individuals are accountable for their actions. Whilst confidential reporting provides the best opportunity to gather extra information this is only possible where a fair and just culture exists and there is clear protection from punitive action be it from litigation, employers or colleagues.

The extent to which an individual's identity is kept confidential varies. In some systems the reporter's managers or those able to influence the reporter's career or employment will not see the report. In others the report is sent to the reporter's line manager. In those practices without a robust just culture this will inhibit reporting, as staff will fear punitive actions.

An anonymous reporting system gives the reporter an added layer of protection: without detailed forensic analysis of IP address, fax number or handwriting it is not possible to know the reporter's identity. There are, however, major disadvantages, in particular more detailed information cannot be obtained. Fears of malicious reporting have been raised if the reporter is neither identifiable nor accountable. This has not been the experience in reviewing and analyzing the reports in the RaER database or in other large professionally based reporting systems [36].

Regardless of the chosen method, any identifying material about staff and patient must be removed before a report is analyzed, sent to another party, published as a safety alert or used in any other way.

Who Should and Who Does Report

All healthcare stakeholders need to be able to report incidents [16]: not just clinical staff such as doctors, nurses and radiographers but also clerical and other non-clinical staff. Patients and carers also need to be able to report incidents: these may be gathered by feedback surveys or correspondence to the healthcare facility or patient advocates.

The reporting rates by nurses are typically higher than doctors [13, 27]. This is probably a reflection of the differences in training and culture. This does, however, lead to bias in the type of incidents reported and data recorded [27]. This is particularly true in anonymous reporting, as there is no way of getting more information to fill in any gaps and explore the causative factors.

Accountability

Whichever system is used to collect data individuals must be accountable for their actions [17]. This needs to be done in a way that does not deter reporting or learning from incidents. This can be a difficult outcome to achieve, hence the need for a just culture.

Ethics

Incident reporting is a quality improvement activity and in many jurisdictions is subject to statutory immunity. Open disclosure is good practice and is best performed independently of the report: anonymous reports cannot always be linked to individual patients.

Resources

Funding is necessary to establish and maintain the infrastructure for the database, including the collection, analysis and classification of data; development of solutions; and dissemination of guidance, tools, and educational material. The cost of this is far less than that needed for litigation and further patient care due to harm caused by incidents.

4.2   Barriers and Enablers

Fear, poor awareness and understanding, administrative issues, and the efforts required affect incident reporting [16, 37]. Simpler forms, easier access to and confidence in the reporting system, stakeholder awareness and education in incident reporting, and a non-punitive environment increase incident reporting [4].

A study of over 200 Japanese teaching hospitals noted that incident reporting by medical practitioners was more common when online reporting was possible, the time required to complete the report was short, and education about safety and incident reporting and dedicated patient-safety staff were employed [8].

Seven factors were described as either barriers or enablers to incident reporting: system access; system user-friendliness; perceived system confidentiality and security; staff training in and knowledge of incident reporting; feedback following incident reporting; workplace safety and incident reporting culture; and value or scepticism towards incident reporting [4]. Training that was perceived as good or helpful increased the rate of reporting. Security, accessibility and a local safety culture were important enablers.

Not recognizing errors, wrongly assuming near misses and minor events are too trivial to report, and mistakenly believing that certain events should not be reported, result in under-reporting. Poor understanding of the terms used in incident reporting forms or software can decrease reporting [18].

Fear

Fear of being blamed for an error or its adverse consequences, or perceived as incompetent by peers are reasons for not reporting incidents. Fear of being perceived as telling tales, penalized, or sued; and anonymity not being respected hinder reporting [21]. The language used in incident reporting forms and in follow-up questions must be carefully chosen to avoid suggestion of blame and to reinforce the value of reporting. Interestingly, immunity from punishment did not affect but the time required to complete the incident form did affect reports filed by Japanese nurses [8].

Organizational Issues

The organizational barriers include a lack of feedback and acknowledgement following incident reporting [16] and a perceived mismatch between an organization's response and the severity of the incident outcome. Feeling that suggestions to improve safety are being ignored hinders reporting. Hierarchical rather than team-oriented cultures are associated with lower reporting rates [37]. The NPSA noted that reports by doctors are over-represented in the incidents entered directly into NRLS [18]. This suggests there are barriers to reporting locally and the extra protection offered by an independent reporting agency is important, at least until the doctors are more confident that the workplace culture is just and non-punitive.

Leadership

For incident reporting to become and stay as a part of the culture, i.e. 'the way we do things around here', it is essential to have the support of, and active promotion by, an organization's leadership [16]. Whilst managers and directors need to promote reporting, the active participation of opinion leaders or champions is vital.

Likelihood versus *consequence*	Insignificant	Minor	Moderate	Major	Extreme
Almost certain	○	○	○	○	○
Likely	○	○	○	○	○
Occasionally	○	○	○	○	○
Unlikely	○	○	○	○	○
Rare	○	○	○	○	○

Fig. 11.3 Risk evaluation. The risk for an incident is evaluated by taking into account of the likelihood i.e. probability shown in rows, and the consequence i.e. severity of damage shown under columns. The risks are grouped as extreme (black), high (red), moderate (orange) and low (green). Risk evaluation enables the identification of incidents with high risk and the prioritization of corrective actions

Incentives

Incentives for reporting incidents can be used to encourage participation. Continuing professional development (CPD) credits are offered to radiologists in Australia and New Zealand whenever a report is filed in the RaER database. Radiology trainees are required to file incident reports as part of their training. From a broader perspective, another key motivation is a contribution to better patient safety by the sharing of information anonymously with other practitioners and thus minimizing future errors [5].

4.3   *Closing the Loop*

Collecting data is ineffective if it is not used to improve patient safety. Incidents need to be analyzed and classified; and solutions developed, communicated and evaluated [30].

Data Analysis

A classification system is needed to analyze incident data. The use of a common system helps data sharing and comparison. This should be constructed to meet the specific needs of the clinical field. For example, RaER uses a dedicated classification based on AIMS, which in turn uses the WHO's international classification for patient safety [31].

The risk posed by an incident should be evaluated according to its severity and likelihood (Fig. 11.3). A risk matrix is a commonly used tool [35]. This maps the severity of impact against the likelihood of occurrence and can help to prioritize analysis and development of solutions.

When analyzing incidents, three key steps need to be reviewed [34]:

1. The events preceding and leading up to the incident;
2. The problems and errors that occurred during the incident; and
3. The factors contributing to the incident.

By identifying the contributory factors and systems issues it is possible to develop solutions to improve quality and safety. Whilst some incidents are common, others are very rare in any setting. By pooling the incident data, rare events are readily recognized rather than being regarded as 'one off' events. This is one way in which rare adverse drug reactions are detected.

Solution Development

Some solutions may be apparent from the narrative text; others may need input from experts from other fields or an application of solutions used elsewhere. The individuals who reported the incidents are often aware of the ways in which these events could have been avoided. Common themes may be detected by examining the narrative. The contribution of radiology, patient safety and human factors experts in the development of solutions for incidents in radiology and medical imaging will most likely result in solutions that will work in practice.

Implementation Strategies

The effective implementation of preventive and corrective actions needs clear communication of workable and practical safety solutions. Clear communication about the rationales for the patient safety strategy, education on how to use the solutions, and implementation assistance all contribute to the uptake of a new policy or procedure. Sometimes it is necessary to mandate changes by legislation. Medicolegal risk is a powerful motivator.

Following an analysis of the radiology incidents in the NRLS, the NPSA in conjunction with the Royal College of Radiologists have modified the WHO's surgical safety checklist for use in radiology [19, 20]. Key items included in this checklist include the confirmation of:

1. Patient identification, procedure side and site checks;
2. Compliance with radiation exposure regulations;
3. Renal function, allergies, bleeding risk factors, antibiotic and DVT prophylaxis;
4. Imaging reviewed, equipment available;
5. Recording of procedure and implantable devices, all equipment accounted for;
6. Specimens labelled.

Additional requirements include team introductions, briefings and debriefings and the marking of the procedure site.

Using a variety of education strategies simultaneously recognizes the fact that individuals learn in different ways. These strategies can include email, letters, visual or audio broadcasts or recordings, workshops and seminars. The contents might contain case studies, data analysis and safety improvement solutions. The NPSA issues 'safety alert broadcasts'. Medical defence organizations send newsletters and some hold workshops and seminars. Informal communication of information conveyed through professional networks is not sufficiently robust and cannot guarantee that everyone will be informed.

Audit and Monitoring

The effectiveness of any new process or policy must be assessed. A suitable period of time for education and implementation is necessary before audit is performed. The results of audit and ongoing monitoring can determine how effective a strategy is in the workplace and what, if anything, needs to be changed to achieve the intended safety improvement.

Feedback

Timely feedback showing how incident reporting is improving healthcare is vital [16]. An acknowledgement of reports informs the reporters that their efforts and concerns are valued. Feedback about the effectiveness and user-friendliness of safety improvement solutions is invaluable and can often explain why the uptake has been satisfactory or less than expected.

Closing the Loop

Continued incident reporting closes the loop. An increase in incident reporting is associated with a timely acknowledgement of reports and the development and implementation of solutions. Ongoing incident reporting helps to identify new threats, detect recurrence of old threats and encourage the reflection on patient safety and, therefore, safer practice.

5   Benefits

Better use of resources, reduced costs, increased responsiveness and pre-empting complaints are cited as benefits [16]. These benefits are in addition to safer care, which can improve staff morale.

6 Limitations

The limitations of incident reporting systems include under-reporting and reporting bias. Under-reporting is inherent in any voluntary system. By its very nature, incident reporting will not capture all events. In one review, under-reporting has been estimated to range from 50 % to 96 % [8]. Another review noted near misses are widely under-reported both in its databases and the literature and concluded that it is impossible to determine the magnitude or frequency of any problem [27].

When compared with a review of medical records, incident reporting is more likely to find near misses or incidents resulting in no or little harm than catastrophic incidents, and is unlikely to include known side effects and complications [22].

Another bias is that some incident types, such as medical device failures, are more likely to be reported to other agencies.

The reporting of incidents varies between staff groups. The data collected depends on the design of the reporting system and this can reflect the background of those who set up the database. The details may not be as complete as, for example, root cause analyses [16]. Unless there are robust links between the different databases it is necessary to report related incidents more than once: this may result in under-reporting in each individual database. Incident reporting is subject to hindsight bias and lost information [32]: a reporting system needs to be designed to minimize these risks.

7 Reporting Systems

7.1 Radiology Event Register (RaER)

The Radiology Events Register [9, 24] is a radiology incident reporting system established in 2006 as part of the Royal Australian and New Zealand College of Radiologists' (RANZCR) Quality Use of Diagnostic Imaging Program [12]. The Australian Patient Safety Foundation (APSF), an organization with a long history of incident report collection and analysis, manages this database.

The aims of RaER are to:

1. Collect data on errors that occur in radiology and medical imaging;
2. Classify and analyze this data; thus gaining an understanding of the types of incidents, the contributing factors, the mitigating factors and outcomes; and
3. Disseminate this information to improve patient safety.

The model chosen was on-line, anonymous and open to anyone, professional or lay, who had experienced an incident in radiology and medical imaging.

As incident reporting is not a part of the culture in radiology, an extensive education campaign was conducted to demonstrate the system, its ease of use and value. Modifications were made, based on feedback from the users. RaER is peer-led and driven and is not connected to any regulatory body or employer. The on-line reporting feature means that it is easy to report an incident from anywhere with internet access.

The RaER database does not replace other hospital-based or agency-based incident reporting systems. Ideally, there would be links between these databases and the relevant information would need to be entered only once.

7.2   Other Reporting Systems

The first specialty-specific incident reporting databases were in anesthesia. Given the potential for catastrophic harm to occur very quickly to a patient, this is, perhaps, not surprising.

The Radiation Oncology Safety Information System (ROSIS) is a voluntary international incident reporting system for radiation oncology developed under the auspices of the European Society of Therapeutic Radiology and Oncology in 2001 [23]. French practitioners are legally required to report over-exposures in radiotherapy [3]. Based on more than 7,500 incidents the WHO classified the radiotherapy processes and outlined the procedural risks and controls.

In addition to RaER, examples of other incident reporting systems for radiology are: Radiology Events and Discrepancies (READ), Safety in Radiological Procedures Program (SAFRAD) and General Radiology Improvement Database (GRID). The Royal College of Radiologists (RCR) established READ: access is restricted to RCR members and fellows via password-protected login [33]. The International Atomic Energy Agency launched SAFRAD as a reporting system for over-exposures in fluoroscopy [10]. It informs the practitioners of the causes and the controls. The General Radiology Improvement Database is an initiative from the American College of Radiology as part of the National Radiology Data Registry (NRDR). It collects and aggregates data such as turnaround times and incident rates and uses these data to establish benchmarks and to improve practice [1].

There are several databases devoted to incidents in general practice. Most hospitals and healthcare systems have an incident reporting system. This ranges from small systems based in individual hospitals to nation-wide systems such as NRLS set up in England and Wales by the NHS in 2003. During 2009, the NRLS had recorded a total of 16,112 incidents involving radiology; 15,239 of these resulted in no or low patient harm [20].

8   Other Methods of Error Detection

There are many other ways to detect error, each with it own strengths and weaknesses. As these methods often gather different sorts of information they are complementary to incident reporting. Examples include: morbidity and mortality meetings, audit, medical record review, complaint analysis, prospective risk assessment, confidential review, observational study, organizational review and blind double reporting in radiology [30].

9 Challenges and Innovations

An incident reporting culture is not widespread in medicine and radiology in particular, resulting in low reporting rates [9]. Despite strong promotion, the logging of incidents has not been an overwhelming success. The use of innovative approaches by including the requirement to log incidents during training, i.e. 'educate the young and regulate the old' and by awarding CPD credits for incident reporting hopefully will change the reporting culture and rate over time.

On-going improvements to the incident reporting system are essential, e.g. by making the data fields in the incident form less ambiguous, more intuitive and user-friendly, thus facilitating reporting and minimizing the time required. Any hurdles encountered during report filing will only discourage reporting in the future. Systems where relevant incidents entered into a database and are automatically shared with other craft-specific databases, after removal of identifying information, would be invaluable. The technology to do this is currently available [16]. Finally, the development of a just culture based on fairness and social justice in the workplace is necessary to encourage the reporting of incidents.

10 Conclusion

There are many actions used to improve quality and to minimize error in radiology and medical imaging, e.g. by strengthening awareness, research, education, and infrastructure. Past experience has laid a strong foundation and provides guidance to future actions meeting the needs of the different stakeholders. Radiologists have a responsibility to provide quality radiology and medical imaging procedures and a duty to ensure safe and effective patient care. By leading the charge through advocacy for, and participation in, incident reporting and collaborating with other experts in patient safety, radiologists will ensure good medical practice and position the specialty at the forefront of quality and safety. As radiology and medical imaging play a pivotal role in the management of almost all in-patients and many outpatients, effective implementation of incident reporting will bring real benefits to the community.

References

1. American College of Radiology (2012) General Radiology Improvement Database (GRID). Available at: http://www.acr.org/SecondaryMainMenuCategories/quality_safety. Accessed 31 Dec 2012
2. Australian Patient Safety Foundation (2012) Advanced Incident Management System (AIMS™). Available at: http://www.apsf.net.au/about.php. Accessed 31 Dec 2012
3. Autorité de sûreté nucléaire (2012) Available at: http://www.french-nuclear-safety.fr. Accessed 31 Dec 2012
4. Braithewaite J, Westbrook M, Tavaglia J et al (2010) Cultural and associated enablers of, and barriers to, adverse incident reporting. Qual Saf Health Care 19(3):229–233
5. Brun A (2005) Preliminary results of an anonymous internet-based reporting system for critical incidents in ambulatory primary care. Ther Umsch 62(3):175–178
6. Cassidy CJ, Smith A, Arnot-Smith J (2011) Critical incident reports concerning anaesthetic equipment: analysis of the UK National Reporting and Learning System (NRLS) data from 2006–2008. Anaesthesia 66(10):879–888
7. Department of Health (2000) An organisation with a memory. The Stationery Office, London
8. Fukuda H, Imanaka Y, Hirose M et al (2010) Impact of system-level activities and reporting design on the number of incident reports for patient safety. Qual Saf Health Care 19(2):122–127
9. Galloway H, Hibbert P, Agar A (2009) Radiology events register progress report – second phase. Available at: http://www.ranzcr.edu.au/quality-a-safety/qudi/past-projects/quality-services-accredited-providers. Accessed 8 May 2012
10. International Atomic Energy Agency (2012) Safety in radiological procedures (SAFRAD). Available at: https://rpop.iaea.org/SAFRAD/About.aspx. Accessed 31 Dec 2013
11. Kohn LT, Corrigan JM, Donaldson MS (eds) (2000) To err is human: building a safer health system. National Academies Press, Washington, DC
12. Lau LSW (2007) The Australian national quality program in diagnostic imaging and interventional radiology. J Am Coll Radiol 4(11):849–855
13. Lawton R, Parker D (2002) Barriers to incident reporting in a healthcare system. Qual Saf Health Care 11(1):15–18
14. Magrabi F, Ong MS, Runciman W et al (2012) Using FDA reports to inform a classification for health information technology safety problems. J Am Med Inform Assoc 19(1):45–53
15. National Audit Office (2009) Patient safety – sixth report of session 2008–2009. The Stationery Office. London
16. National Patient Safety Agency (2004a) Seven steps to patient safety: an overview guide for NHS staff Second print. National Patient Safety Agency, London
17. National Patient Safety Agency (2004b) Seven steps to patient safety. The full reference guide. National Patient Safety Agency, London
18. National Patient Safety Agency (2005) Building a memory: preventing harm, reducing risks and improving patient safety. The first report of the National Reporting and Learning System and of the Patient Safety Observatory. National Patient Safety Agency, London
19. National Patient Safety Agency (2010a) WHO surgical safety checklist for radiological intervention only. In: National Patient Safety Agency "How to guide" five steps to safer surgery. Available at: http://www.nrls.npsa.nhs.uk/EasySiteWeb/getresource.axd?AssetID=74068&type=full&servicetype=Attachment. Accessed 5 May 2012
20. National Patient Safety Agency (2010b) NPSA issue Europe's first interventional radiology surgical checklist. Available at: http://www.npsa.nhs.uk/corporate/news/npsa-issue-europes-first-interventional-radiology-surgical-checklist/?vAction=fntDown. Accessed 5 May 2012
21. O'Beirne M, Sterling P, Reid R et al (2010) Safety learning system development – incident reporting component for family practice. Qual Saf Health Care 19(3):252–257
22. O'Neil AC, Petersen LA, Cook EF et al (1993) Physician reporting compared with medical-record review to identify adverse medical events. Ann Intern Med 119(5):370–376

23. Radiation Oncology Safety Information System (2012) Available at: http://www.rosis.info/. Accessed 30 Dec 2012
24. Radiology Events Register (2012) Available at: http://www.raer.org/index.htm. Accessed 8 May 2012
25. Reason J (1990) Human error. Cambridge University Press, New York
26. Reason J (2000) Human error: models and management. BMJ 320(7237):768–770
27. Ricci M, Goldman AP, de Leval MR et al (2004) Pitfalls of adverse event reporting in paediatric cardiac intensive care. Arch Dis Child 89(9):856–859
28. Runciman WB, Webb RK, Helps SC et al (2000) A comparison of iatrogenic injury studies in Australia and the USA. II: Reviewer behavior and quality of care. Int J Qual Health Care 12(5):379–388
29. Runciman B, Merry A, Smith AM (2001) Improving patients' safety by gathering information. Anonymous reporting has an important role. BMJ 323(7308):298
30. Runciman WB, Williamson JA, Deakin A et al (2006) An integrated framework for safety, quality and risk management: an information and incident management system based on a universal patient safety classification. Qual Saf Health Care 15(Suppl 1):i82–i90
31. Runciman W, Hibbert P, Thomson R et al (2009) Towards an international classification for patient safety: key concepts and terms. Int J Qual Saf Health Care 21(1):18–26
32. Simon A, Lee RC, Cooke DL et al (2005) Institutional medical incident reporting systems: a review. Health Technology Assessment Unit, Alberta Heritage Foundation for Medical (AHFMR) Research, Alberta
33. Spencer PA (2012) Shared learning can help minimize errors. Available at: http://www.auntminnieeurope.com/index.aspx?sec=nws&sub=rad&pag=dis&ItemID=607379. Accessed 19 Jan 2013
34. The Royal College of Radiologists, Society and College of Radiographers, Institute of Physics and Engineering in Medicine, National Patient Safety Agency, British Institute of Radiology (2008) Towards safer radiotherapy. The Royal College of Radiologists, London
35. VA National Center for Patient Safety (2011) VHA national patient safety improvement handbook. Department of Veterans Affairs, Washington, DC. Available at: http://www1.va.gov/vhapublications/ViewPublication.asp?pub_ID=2389. Accessed 30 Dec 2012
36. Webb RK, Currie M, Morgan CA et al (1993) The Australian incident monitoring study – an analysis of 2000 incident reports. Anaesth Intensive Care 21(5):520–528
37. Wolf ZR, Hughes RG (2008) Error reporting and disclosure. In: Hughes GH (ed) Patient safety and quality: an evidence-based handbook for nurses. Agency for Healthcare Research and Quality, Rockville
38. World Health Organization (2009) More than words. Conceptual framework for the international classification for patient safety version 1.1. WHO, Geneva

Part III
Infrastructure Improvements

Chapter 12
Promoting Public Awareness and Communicating Radiation Safety

Theocharis Berris and Madan M. Rehani

Abstract Information and communication technologies have rapidly advanced in the last decade. Moreover, communication through the internet has become two-ways in recent years. People who were only able to receive information through web pages are now able to publish their own information or even criticisms in real time. The advent of social media platforms is facilitating a more direct and real-time communication between the source of information and the users. This new feature poses some challenges to the organizations involved in educating professionals and disseminating information to patients and public about radiation safety. There are indications that harnessing the power of these new communication tools could be effective in reaching those who are interested and in raising awareness about radiation safety in medicine. However, this is a relatively new approach and data about its effectiveness are scarce. Nevertheless, many organizations are already exploring the advantages and the power of modern real time, communication. In this chapter a brief review of the international literature on the advantages, disadvantages and pitfalls of the new media is provided. Information about the use of the internet and social media by the Radiation Protection of Patients Unit of the International Atomic Energy Agency is also presented. These data might be used as a guide for other organizations, interested in using such media for raising public awareness.

Keywords Advertising • Blogs • Facebook • Internet • LinkedIn • Marketing • Non-profit organizations • Patients • Professionals • Professional societies • Public • Radiation safety • Raising awareness • Social media • Twitter • Web 2.0

T. Berris (✉) • M.M. Rehani
Radiation Protection of Patients Unit, International Atomic Energy Agency,
PO Box 100, A 1400, Vienna, Austria
e-mail: t.berris@iaea.org; theocharisberris@gmail.com; madan.rehani@gmail.com

L. Lau and K.-H. Ng (eds.), *Radiological Safety and Quality: Paradigms in Leadership and Innovation*, DOI 10.1007/978-94-007-7256-4_12,
© Springer Science+Business Media Dordrecht 2014

Abbreviations

AAPM	American Association of Physicists in Medicine
ACC	American College of Cardiology
ACR	American College of Radiology
ARRS	American Roentgen Ray Society
ASRT	American Society of Radiologic Technologists
ASTRO	American Society for Radiation Oncology
DEXA	Dual energy X ray absorptiometry
ESC	European Society of Cardiology
ESR	European Society of Radiology
HIV	Human immunodeficiency virus
http	Hypertext Transfer Protocol
https	Hypertext Transfer Protocol Secure
IAEA	International Atomic Energy Agency
ICRP	International Commission on Radiological Protection
ROI	Return on investment
RPOP	Radiation Protection of Patients
RSNA	Radiological Society of North America
SIR	Society of Interventional Radiology
SM	Social media
SNM	Society of Nuclear Medicine
UNSCEAR	United Nations Scientific Committee on the Effects of Atomic Radiation
URL	Uniform resource locator
WoM	Word of mouth

1 Introduction

Radiation safety has become a public issue in the last decade with increasing attention being drawn to the carcinogenic potential of repeated CT scans and more recently to the accidental over-exposures in CT procedures [16–18].

Unfortunate as it may be, safety receives attention through accidents. If there is a single accident in nuclear reactor in 20 years, the whole world's attention gets focused on that event. However, in day-to-day work, the nuclear reactors are safe for the workers and the public. On the contrary, in medicine, accidents such as skin injuries to patients from over-exposures following CT procedures occur more often than they are widely perceived because they are not so vividly publicized. Nevertheless, they form a driving force for the escalation of interest from patients and public at large.

The media has always played a significant role in modulating perception. In the past, there was a distinct divide between information obtained through the news media and peer-reviewed scientific publications for medical and scientific professionals. The credibility associated with non peer-reviewed scientific material published in the public news media was not high. With time there has been major change primarily because of the internet. Websites are accessible to anyone, be it scientific, professional or public. There is an increasing tendency by scientific and professional organizations to offer more and more information to the public. Even though in general, full text of scientific publications is available against payment, there are growing numbers of organizations that make full text publications freely available. Communication through the public media is becoming an important activity of most organizations. Scientific and professional organizations want to reach the public rather than to restrict their audience to professionals. Besides an organization's own website, there are a number of other channels to go public. Wikipedia and a large number of social media (SM) platforms are examples in this context. Not only professional organizations have to change strategies to handle public information, traditional public media (newspaper, TV channels etc.) have to deal with a growing dependence of the public on the internet and SM. Recently, a very important transition has been observed: the traditional professional reservations about the quality of information presented in the public media are easing. This shift in thinking has compelled most professional organizations to get involved with the public media. How to disseminate scientific and technical information to the public is a challenge faced by various organizations.

According to the United Nations Scientific Committee on Effects of Atomic Radiation (UNSCEAR) 3.6 billion diagnostic x-ray examinations are performed worldwide annually [29]. Almost everyone at some time in their life undergoes an x-ray examination. The rate of increase in CT use has been almost 10 % per year in many countries. Probably there was never a time in history since the discovery of x-rays, when a number of individual patients underwent more than 10 CT scans within a few years with resultant effective radiation dose to the individual of about 100 mSv or higher [25, 28]. Not all of these examinations are medically justified. Over-utilization of imaging is increasingly being recognized in US [10, 24]. A number of reasons have been cited for this, such as defensive medicine, patient's wishes, financial interest, self-referral, health system factors, industry, use of media and lack of awareness.

These situations call for the use of appropriate means to achieve a more effective dissemination of information and to improve public understanding. The key issue is not to lose sight of the proven benefits of medical imaging, nor to create undue concern of radiation but to propagate the message of rational use and radiation protection.

The International Atomic Energy Agency (IAEA) has been involved in promoting public awareness and communicating the message of radiation safety in medicine. The dominant source has been its public website on radiation protection of patients [14]. In addition, recently, it has embarked on use of SM via Facebook and Twitter. This chapter covers the experience gained and provides some basic information on these communication tools.

2   Radiation Protection of Patients (RPOP) Website

The RPOP website [14] was released in September 2006 and has become an indispensible source of information on medical radiation protection worldwide. It is accessed by more than 190 states and currently receives more than 10 million hits through 0.24 million visits annually. The characteristics of the website that have made it unique and popular are by:

1. Providing regular updates: couple of times a month;
2. Providing information in simple language and in non-technical way;
3. Using: catchy questions that invite attention; daily practice scenarios; closed questions as far as possible so that a concise sentence in the first line gives a feel of the rest of the answer text; and short and meaningful answers;
4. Grouping of information: under professional categories and covering adequately not only traditional fields like radiology, nuclear medicine, radiotherapy but also those areas that are normally not covered in other websites such as: use of X rays in cardiology, gastroenterology, orthopedic surgery, urology, dual energy X ray absorptiometry (DEXA) and dental radiology;
5. Providing specific information for children and pregnant women;
6. Offering separate sections for patients and public;
7. Providing free download of huge amount of training material as power point slides duly supported by major international organizations; and
8. Making available free download of most IAEA publications in this field.

These features have made the RPOP website so popular that it appears on first page of Google, Yahoo! and other search engines when searching by using: radiation protection. RPOP appears on the first page of search results also when the string radiation protection is followed by any of the following terms: radiology, cardiology, orthopedics, gastroenterology, fluoroscopy, computed tomography, mammography and so on. Such unique features mean its web address does not need to be remembered. Just by adding the above search terms the web address can be found among the search results of the first page.

The credibility of information is an important requirement and there are mechanisms in place to ensure this. A large number of experts from many countries are involved in reviewing information before it is published on the website. The training material is reviewed and endorsed by international organizations and professional bodies. There is an international action plan on radiation protection of patients under which organizations cooperate [26].

The statistics on visits has shown that more than 50 % of visitors to RPOP website are through search engines and that was the driving force to improve search optimization. Furthermore, search terms like fluoroscopy, dental, PET/CT, and training material were found to be used often. Accordingly, the contents on these topics were strengthened. Also actions were taken to facilitate search by adding metadata tags to terms in the pages.

Some lessons that may be of interest to others are: providing audience specific information rather than organization specific, keeping information non-technical, short and simple; making valuable material such as presentation slides of training material available for free with freedom to modify rather than files which cannot be edited.

3 The Role of Social Media in Raising Awareness About Radiation Protection in Medicine

3.1 Web 2.0 and Social Media Platforms

In the early days of the internet, communication was one-way. Information was directed from organizations, authorities and firms to the users. The users were effectively consumers of the published information. Nowadays the users may circulate in real time, pieces of information they have created themselves. In fact there are whole websites, which base their content generation on interactions among users. This interaction leaves room for advertisers to participate and create revenues for platforms hosting user-generated material. For instance, Facebook, the largest social networking platform, provides networking services for free to its users. Like other similar platforms, Facebook makes its profits by selling targeted advertisements to online marketers who want to take advantage of Facebook's vast databases of personal preferences in order to reach potential customers. Blogs, SM platforms, gaming online communities and other platforms where content is constantly modified by user interaction signifies the Web 2.0 era [19].

As of March 27, 2013 the dominant international SM platforms based on a combination of average daily visitors and pageviews were Facebook Inc. USA, YouTube LLC USA, Wikipedia Foundation Inc. USA, Twitter Inc. USA, and LinkedIn Corporation USA, ranking within the top 15 places (2nd, 3rd, 6th, 11th and 13th place respectively), within the top 500 global sites list maintained by the web information company Alexa Internet Inc. USA [1]. Facebook as of December 2012 had over 1 billion monthly active users (users who have returned to the site in the last 30 days) [8]. Twitter has currently over 500 million registered users with less than half of them really "tweeting" [11]. It has been mentioned in the literature that twitter registered users were 175–200 million in 2011 [7, 27]. Another important social networking platform is Google+ attached to Google Inc. USA services. It had 500 million registered users as of December 2012 [32]. It should be noted that according to Alexa, Google has been the most visited website in the world. Separate Google+ statistics are not provided [1]. Another social platform more targeted to business social networking interactions is LinkedIn. With 200 million members in 200 countries as of December 31, 2012 [22], LinkedIn is one of the big players in the field of SM and occupied the 13th position in the top 500 global sites list maintained by Alexa on March 27, 2013. There are other smaller SM platforms, which have been successful in some parts of the world but their success has not spread everywhere. For instance, Orkut (owned by Google Inc.) is a platform that has been successful in Brazil and India [33]. MySpace USA is another platform that has been very popular in the past but failed to spread as much as other platforms in recent years. From the above platforms, Facebook, Twitter, Google+ and LinkedIn are the most widespread international platforms that allow direct exchange of various kinds of information among users. A more detailed description of the operation of SM platforms has been described elsewhere [7]. In the following paragraphs, an analysis is provided on the way organizations and businesses utilize SM platforms to raise awareness or market their products. A review of the available information is provided on how effective the leverage of SM is in achieving their goals. Furthermore, the experience gained by the authors in the use of SM, namely Facebook and Twitter accounts of the RPOP Unit of the IAEA is shared.

3.2 The Potential of SM Platforms to Improve Communication Between Organizations and Individuals

Possible effectiveness of SM in improving communication between organizations and targeted public is based on the creation of a buzz around brands products or services, as is the case in marketing in general. The difference is that the new media offer real-time interactivity and users may swiftly set-up worldwide conversations about the brand, product, or service in question. They can share their experiences and opinions with the same audience that marketers are trying to reach. On the downside for marketers is that users may also instantly criticize or even unveil exaggerated marketing claims in no time. Kaplan and Haenlein [20] argue power of SM can be described in terms of the "viral" spread of digital word of mouth (WoM). The previous arguments hint that if the pros and cons of SM are used correctly there is potential gain for organizations. However, as is the case with any marketing campaign, SM campaigns have a cost either monetary or in human resources, so proof of its marketing effectiveness is needed. Marketers need to know the return on investment (ROI) in order to regulate their SM budget [31]. Numerous examples of successful and unsuccessful SM campaigns organized by various brands are discussed in the literature [9, 20, 31]. Success is probably defined differently for different entities. Specifically for the RPOP unit of the IAEA success of the campaign would be to reach and inform more people about radiation safety in medicine. The campaign is ongoing and initial fears that uncontrolled delivery of scientific information may create credibility issues have not been substantiated as discipline has been exhibited over time.

Other organizations are also using SM for increasing awareness. Tables 12.1 and 12.2 provide a listing of the web addresses of Facebook pages and Twitter accounts of major organizations involved in radiation safety and healthcare. When Facebook allowed organizations to setup pages in April 2006, 4,000 organizations registered

Table 12.1 URLs for Facebook pages of major organizations in radiation safety and healthcare sectors

Organizations	URLs for Facebook pages
American College of Cardiology (ACC)	http://www.facebook.com/AmericanCollegeofCardiology
American College of Radiology (ACR)	http://www.facebook.com/AmericanCollegeOfRadiology
American Roentgen Ray Society (ARRS)	http://www.facebook.com/AmericanRoentgenRaySociety
American Society for Radiation Oncology (ASTRO)	http://www.facebook.com/pages/American-Society-for-Radiation-Oncology/35768312349
American Society of Radiologic Technologists (ASRT)	http://www.facebook.com/MyASRT
European Society of Cardiology (ESC)	http://www.facebook.com/europeansocietyofcardiology
Image Gently	http://www.facebook.com/pages/Image-Gently/146714032030578
Image Wisely	http://www.facebook.com/ImageWisely
International Atomic Energy Agency (IAEA)	http://www.facebook.com/iaeaorg
myESR (European Society of Radiology)	http://www.facebook.com/MyESR
Radiation Protection of Patients (RPOP)	http://www.facebook.com/rpop.iaea.org
Radiological Society of North America (RSNA)	http://www.facebook.com/RSNAfans
RadiologyInfo.org	http://www.facebook.com/radiologyinfo
Society of Interventional Radiology (SIR)	http://www.facebook.com/SocietyOfInterventionalRadiology
Society of Nuclear Medicine (SNM)	http://www.facebook.com/pages/SNM-Society-of-Nuclear-Medicine/55278318943

Table 12.2 URLs for Twitter accounts of major organizations in radiation safety and healthcare sectors

Organizations	URLs for Twitter accounts
American Association of Physicists in Medicine (AAPM)	http://twitter.com/aapmHQ
American College of Cardiology (ACC)	http://twitter.com/ACCinTouch
American College of Radiology (ACR)	http://twitter.com/RadiologyACR
American Roentgen Ray Society (ARRS)	http://twitter.com/ARRS_Radiology
American Society for Radiation Oncology (ASTRO)	http://twitter.com/ASTRO_org
American Society of Radiologic Technologists (ASRT)	http://twitter.com/ASRT
European Society of Cardiology (ESC)	http://twitter.com/escardio
myESR (European Society of Radiology)	http://www.twitter.com/myESR
Image Wisely	http://twitter.com/ImageWisely
International Atomic Energy Agency (IAEA)	http://twitter.com/iaeaorg
International Commission on Radiological Protection (ICRP)	http://twitter.com/ICRP
Radiation Protection of Patients (RPOP)	http://twitter.com/rpop_iaea
Radiological Society of North America (RSNA)	http://twitter.com/RSNA
Radiologyinfo.org	http://twitter.com/radiologyinfo_
Society of Nuclear Medicine (SNM)	http://twitter.com/SNM_MI

within the first 2 weeks. However, the data regarding the use of SM by non-profit organizations are scarce [30]. An example of successful use of SM by an international non-profit social marketing company has been published in the literature [23]. A company dedicated to HIV prevention and family planning, launched an integrated strategy, which heavily utilized SM to promote the use of condoms in Turkey. Traditional media were also used but in a lesser extent, due to budgetary constraints. The study showed that SM leveraging definitely contributed to an increase in condom sales. The cost of investment per sold condom was $0.14 for the internet investment, while the overall ratio including traditional and SM was $0.36 per sold condom.

3.3 Guidance for Proper Utilization of SM

The right way to manage SM campaigns is not straightforward. This is probably due to the complexity of social interactions among users. Human behavior is unpredictable in many cases and thus, the outcome of a campaign is uncertain. The literature is full of examples of campaigns that backfired. Case studies of successful and unsuccessful SM campaigns have been extensively discussed in the literature cited in this chapter [9, 20, 31]. Despite the difficulty of assessing the outcome of SM campaigns beforehand, it is possible to give some general guidance on SM use.

- Be social. Try to be spontaneous and communicative but without sacrificing the quality and seriousness of the discussed subjects.
- Be true and original. The users know that people can make mistakes. Informed users (and users are becoming more informed by the minute) would more easily tolerate an organization genuinely admitting a mistake than a clumsy effort to pretend that everything is fine. The latter could trigger a wave of negative "viral" feedback.
- Post useful material. Users value bits of information that they relate to as well as tools they may use and benefit from. In case they find something useful they share it with their online contacts creating waves of positive online conversations.
- Post interactive engaging material. Probably due to their nature based on fast communication, SM platforms are more suitable for publishing interactive material. Such material has more potential to engage users who are quickly consuming large amounts of information.
- Post material that will spark conversations and take part in the conversation. The material needs to be stimulating in order to have potential to create "viral" conversations. Caution should be used with controversial issues as they may backfire [20].
- Leverage your website's power. A good SM effort needs to be supported by an active, regularly updated website. Blog-like pages having the social element already built-in could be particularly good in maintaining the interest of users.

Weinberg and Pehlivan [31] argue that the SM platform itself plays a role in the type and depth of information that may be provided. In turn the depth of information is related with the best use of the platform in question. The RPOP experience has shown that the available information has been communicated more effectively through the social channels of Facebook as shown by the data presented further below in this chapter.

3.4 Possible Problems of SM Use for Advertising Purposes

Privacy Issues

Sensitive personal information published on SM such as phone numbers, home addresses and human relationship information may be used for malicious purposes. There have been reports of peoples' houses being robbed because they gave information that they were away through their online channels [2]. Hackers are always plotting new ways of stealing data. The SM platforms are ideal places for setting up social engineering plots in order to steal the data that they are interested in [5]. Sometimes a person may be targeted from hackers who have taken control of an online friend's account. However, despite the numerous online threats, the users of SM services are still increasing. Researchers have been wondering how much people value their privacy. Boyd and Hargittai [4] pointed out that even young people 18–19 years old who were considered not to be very interested in maintaining their privacy, were in fact concerned and tried to manage their privacy settings, especially during 2009–2010 when Facebook privacy was challenged strongly by privacy advocate groups. The struggle to increase safety online against threats posed by hacking is ongoing and this is probably why the SM platforms very often make changes to their privacy settings. Except for the technological security requirements they also need to balance ease of use and facilitation of social interaction. After all, social interaction is what creates their content and generates their revenue. Constant change in privacy settings although justified, is probably making the life of users more difficult. For example, Facebook has been criticized for the complexity of its privacy settings [7].

The above-mentioned threats are also concerning organizations. Hackers would be interested in compromising the SM channels of some organizations. If they succeed they may damage the reputation of the organization. Organizations intending to use SM should also keep in mind that users value their privacy. Organizations should always be careful in the handling of sensitive data of users who are affiliated with them in any way including the use of third party custom-made applications that access personal data according to user preferences. After major concerns about data security [21] Facebook announced the intention to use secure protocols (Hypertext Transfer Protocol Secure -https) for communication between its servers and users browsers as a default. For organizations developing third party applications for Facebook this means that third party applications will also have to run on secure servers.

Legal Issues

Social media are a reflection of human social interactions in the digital world. As such they are not immune from other issues in real social life. Different perceptions may lead to conflict. In turn conflict may lead to disputes that may even need to be solved in the court of law. A series of important legal concerns may arise by the use of SM from organizations. This is one of the reasons why disclaimers are used on many websites or SM pages of organizations and businesses. Especially in healthcare, liability is a very important issue that could create problems if not addressed properly. Coffield and Joiner [6] addressed a scenario where a healthcare professional provided medical advice through traceable SM channels while off-duty. Is professional liability present in such cases? Another important issue is the use of SM within organizations. Businesses find it difficult to regulate the use of SM by their employees. There are already reports of employees losing their jobs because they were using SM during working hours or made comments, which were deemed as inappropriate by their employers [3].

3.5 The RPOP Experience in Utilizing SM Platforms for Awareness Raising About Radiation Safety

Facebook

The Facebook page of RPOP was launched in August 2010. It had 213 "fans" on February 21, 2011 and reached 245 "fans" on April 13, 2011. No advertising or other means of promotion were used except the occasional posting of information about radiation safety in medicine on the page's "wall" or on the "walls" of other similar content pages. On April 13, 2011 Facebook "like" buttons were added to the RPOP website, which received about 1 million hits per month at the time. The page's "fans" increased to 1,923 as of March 27, 2013 (Fig. 12.1), averaging a mean growth rate of 2.3 new "fans" per day after April 13, 2011. Facebook provides its users with "Facebook Insights", a tool that registers information about the people who like the page, how many people commented on the page's posts and how many times a post appears in fans' newsfeeds. However, the only really substantial data are: (a) the total number of "fans" which represents the number of people that can be reached by posted updates and (b) feedback (likes, comments) that posts receive,

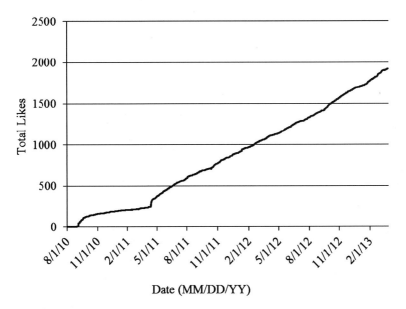

Date (MM/DD/YY)

Fig. 12.1 "Fans" of the RPOP Facebook page. The gradient change corresponds to the addition of "like button" to the RPOP website. Total percentage increase in "fans" from February 21 to March 27, 2013 was 803 %

which shows how successful the posted material was. The more feedback a post receives, the better. However, there is no reference on how much feedback is enough for a post to be considered "good". It also depends on the total number of fans of the page. Posts on RPOP Facebook page receive comparable number of "likes" as posts released by pages with similar sized "fan" base at the time such as Image Gently [12], 1,531 "fans" as of March 27, 2013 and Image Wisely [13], 1,252 "fans" as of March 27, 2013.

Cooperation between the main Facebook page of the IAEA, which has a wider outreach among the public at large and the IAEA RPOP Facebook page, which has mostly audience interested in medical radiation protection, has been established. The larger fan base of IAEA's main Facebook page, currently exceeding 40,000 fans, is an effective way to reach people who are interested in radiation use and safety.

On late September 2011, the "Ask an expert in Radiation Protection" campaign, which was organized in cooperation with the IAEA Facebook page, was concluded. In this campaign, an "event" was created on the IAEA Facebook page on September 5, 2011 and Fans were encouraged to ask (on the "event" thread) questions pertaining to radiation protection in medicine that they might have (https://www.facebook.com/iaeaorg?ref=ts#!/event.php?eid=100561103384094). Daily updates during the following week were posted on the IAEA Facebook page prompting users to ask their questions. Users asked additional questions in the form of comments under each post. All questions were pooled and assessed. The most relevant questions were answered in a video interview by radiation safety experts in a public event organized by the IAEA and RPOP unit on September 26, 2011. The videos with the experts answering a total of 12 selected questions were finally published on RPOP website and Facebook page as well as the IAEA Facebook page. The results indicated that these efforts were very successful. One hundred and forty users were linked to the event page and many more asked their questions under each post while keeping the conversation to a high scientific level. Each post on the IAEA Facebook page received an average of 35,000 "impressions" (number of people who saw these posts) and 50 "likes". The videos received 15,000 "impressions" in the following 3 weeks after posting on 27 September, 2011. According to the Division of Public Information of the IAEA, which handles the IAEA Facebook page, the success of the campaign was unprecedented [15].

In 27 September 2012 an interactive quiz regarding basic radiation protection was posted on the RPOP Facebook page. The weekly total reach (the number of unique users who have seen any content associated with the RPOP page) increased by 600 %. The post was also shared by the IAEA Facebook page. More than 5,000 users interacted with this post through the IAEA Facebook page.

Twitter

RPOP had 87 Twitter "followers" as of February 21, 2011. They reached about 130 just before the addition of the "follow button" on the RPOP website. A steady increase was observed after an initial steep increase in followers in the first few days after the "follow button" was added. "Followers" reached 619 on March 27, 2013. Twitter does not provide any integrated tools to monitor progress. The quality of posts could possibly be assessed by the number of re-tweets they get and the inclusion of the Twitter profile in lists that other users compile. Twitter does not keep records for re-tweets. This profile is comparable to other organizations with similar populations of "followers", e.g. Image Wisely, which had 495 "followers", as of March 27, 2013.

Contribution of Social Media to the RPOP Website Traffic

According to Google Analytics statistics for the RPOP website, Facebook was the first referring site for the RPOP website traffic during the period April 13, 2011 to March 27, 2013. In the same time frame Twitter was 21st in the same list. For comparison, during the period August 2010 to February 2011, Facebook was 6th in the same list of referring sites while Twitter was in the in the 45th position. According to the above data, it is clear that SM have been gaining a more prominent position among the referring sites for the RPOP website.

Future Directions

The RPOP unit of the IAEA is constantly striving to produce high quality content on radiation safety in medicine. Public engagement with SM may be improved by using social applications already available for Facebook and Twitter platforms. The use of Facebook "iframe" feature will allow the development of more custom-made interactive, educating applications such as quizzes and enhanced multimedia experiences such as videos. Work is in progress and time will be needed since developing application requires network resources and software support.

Future plans also include cooperation with other organizations, which are interested in radiation protection of patients. Since SM leveraging is a new kind of endeavor for everyone in the field, maturity will take time. However, based on a consideration of the number of possible professional bodies and non-profit organizations in radiation related healthcare which could use SM effectively to communicate information to interested people, there is a strong potential for SM to be an even more dominant channel for communication in the future.

4 Epilogue

The digital media using the internet as a channel of communication seems to be ever increasing. Meaningful guidelines for information dissemination will likely emerge for organizations, which intend to maintain or expand their influence. Users are becoming more proficient in the use of the online media. With technological developments and increasing globalization, it is very probable that the pace of change will accelerate. This poses a standing challenge for organizations to keep up with these changes.

The credibility of open online sources of information seems to be increasing. Considering that many of these sources are maintained and administered by large professional organizations, usually including many experts on various subjects, even traditional peer-reviewing processes may need to be updated. The future possibly holds for open, online real time peer reviewing by large numbers of experts on the subject. Theoretically this could lead to even better quality of scientific information.

Researchers have argued that viral marketing may eventually lose its momentum and may become an extinct way of communication in the future as has happened before with other trends [20]. However, SM advertising is still at its early stages and we would assume that there is still enough time for organizations and firms to take advantage of its strengths. The above analysis shows that organizations may leverage SM in order to achieve their promotional goals. Experience has shown that by following simple rules such as by producing decent content, being conversational and avoid being too professional would contribute to a successful SM campaign and strengthen an organization's communication strategy. However, this new paradigm in online information exchange allows everyone to create his or her own content without complete moderation. Therefore, caution must be used in order to avoid unfavorable situations and to utilize SM in the organization's best interest.

References

1. Alexa Internet, Inc. (2013) Alexa top 500 global sites. http://www.alexa.com/topsites. Accessed 27 Mar 2013
2. Axon S (2010) Don't get robbed: burglars use Facebook to pick targets. Mashable. http://mashable.com/2010/09/11/facebook-places-burglars/. Accessed 27 Mar 2013
3. British Broadcasting Corporation (2009) "Ill" worker fired over Facebook. BBC News. http://news.bbc.co.uk/2/hi/technology/8018329.stm. Accessed 27 Mar 2013
4. Boyd D, Hargittai E (2010) Facebook privacy settings: who cares? First Monday (Online) http://www.uic.edu/htbin/cgiwrap/bin/ojs/index.php/fm/article/view/3086/2589. Accessed 27 Mar 2013
5. Charles A (2009) Facebook hit by phishing attack. The Guardian. http://www.guardian.co.uk/technology/2009/apr/30/facebook-phishing-scam. Accessed 27 Mar 2013
6. Coffield RL, Joiner JE (2010) Risky business: treating tweeting the symptoms of social media. AHLA connections. http://www.pagegangster.com//p/DqwTo/1/. Accessed 27 Mar 2013
7. Duffy M (2011) Facebook, Twitter and LinkedIn, Oh My! Making sense and good use of social media. Am J Nurs 111:56–59
8. Facebook, Inc. (2013) Key facts. http://newsroom.fb.com/Key-Facts. Accessed 27 Mar 2013
9. Hanna R, Rohm A, Crittenden VL (2011) We're all connected: the power of the social media ecosystem. Bus Horiz 54:265–273
10. Hendee WR, Becker GJ, Borgstede JP et al (2010) Addressing overutilization in medical imaging. Radiology 257:240–245
11. Holt R (2013) Twitter in numbers. The telegraph. http://www.telegraph.co.uk/technology/twitter/9945505/Twitter-in-numbers.html. Accessed 27 Mar 2013
12. Image Gently (2013) The alliance for radiation safety in paediatric imaging. http://www.pedrad.org/associations/5364/ig/. Accessed 27 Mar 2013
13. Image Wisely (2013) Radiation safety in adult medical imaging. http://www.imagewisely.org. Accessed 27 Mar 2013
14. International Atomic Energy Agency (2006) Radiation Protection of Patients https://rpop.iaea.org/RPoP/RPoP/Content/index.htm Accessed 27 March 2013
15. International Atomic Energy Agency (2011) Radiation protection of patients: use of social media to achieve interaction with the public on medical radiation protection. https://rpop.iaea.org/RPOP/RPoP/Content/News/social_media-interaction-radiation-proection.htm. Accessed 27 Mar 2013
16. International Atomic Energy Agency (2013a) Radiation protection of patients: radiation risks in CT examinations. https://rpop.iaea.org/RPOP/RPoP/Content/ArchivedNews/radiation-risks-CT-examinations.htm. Accessed 27 Mar 2013
17. International Atomic Energy Agency (2013b) Radiation protection of patients: new era in CT scanning. https://rpop.iaea.org/RPOP/RPoP/Content/News/new-era-ct-scanning.htm. Accessed 27 Mar 2013
18. International Atomic Energy Agency (2013c) Radiation protection of patients: continued occurrence of over-radiation in brain CT. https://rpop.iaea.org/RPOP/RPoP/Content/News/overradiation-brain-CT.htm. Accessed 27 Mar 2013
19. Kaplan AM, Haenlein M (2010) Users of the world unite! The challenges and opportunities of social media. Bus Horiz 53:59–68
20. Kaplan AM, Haenlein M (2011) Two hearts in three-quarter time: how to waltz the social media/viral marketing dance. Bus Horiz 54:253–263

21. King C (2011) Facebook goes HTTPS. Palo Alto networks. http://www.paloaltonetworks.com/researchcenter/2011/01/facebook-goes-https/. Accessed 27 Mar 2013
22. LinkedIn Corporation (2013) About us. http://press.linkedin.com/about. Accessed 27 Mar 2013
23. Purdy CH (2011) Using the internet and social media to promote condom use in Turkey. Reprod Health Matter 19(37):157–165
24. Rehani B (2011) Imaging overutilization: is enough being done already? Biomed Imaging Interv J. doi:10.2349/biij.7.1.e6
25. Rehani MM, Frush DP (2011) Patient exposure tracking: the IAEA smart card project. Radiat prot dosim. doi:10.1093/rpd/ncr300
26. Rehani MM, Holmberg O, Ortiz López P et al (2011) International action plan on the radiation protection of patients. Radiat Prot Dosim. doi:10.1093/rpd/ncr258
27. Shiels M (2011) Twitter co-founder Jack Dorsey Rejoins Company. British Broadcasting Corporation (BBC). http://www.bbc.co.uk/news/business-12889048. Accessed 27 Mar 2013
28. Sodickson A, Baeyens PF, Andriole KP et al (2009) Recurrent CT, cumulative radiation exposure, and associated radiation-induced cancer risks from CT of adults. Radiology 251(1):175–184
29. United Nations Scientific Committee on the Effects of Atomic Radiation, Sources and Effects of Ionizing Radiation (2008) UNSCEAR report 2008. United Nations, New York
30. Waters RD, Burnett E, Lamm A et al (2009) Engaging stakeholders through social networking: how non-profit organizations are using Facebook. Public Relat Rev 35:102–106
31. Weinberg BD, Pehlivan E (2011) Social spending: managing the social media mix. Bus Horiz 54:275–282
32. Wikipedia (2013a) Google+. http://en.wikipedia.org/wiki/Google%2B. Accessed 27 Mar 2013
33. Wikipedia (2013b) Orkut. http://en.wikipedia.org/wiki/Orkut. Accessed 27 Mar 2013

Chapter 13
Sustaining Praxis: Quality Assurance of Education and Training Programmes of the IAEA's Division of Human Health

Soveacha Ros, Thomas N. B. Pascual, and Rethy K. Chhem

Abstract In this article, we examine quality assurance as a collective action, utilizing the concepts of "praxis" and total quality management as the theoretical framework, which is consistent with our determined efforts to implement the "learning organization" vision. Drawing on the experience of the Division of Human Health of the International Atomic Energy Agency (IAEA) in assuring quality of education and training programmes, we explore practical opportunities for sustaining praxis within the Division of Human Health to ensure the on going provision of quality education and training resources in radiation medicine to the IAEA's Member States. The consistency between praxis and total quality management informs curriculum development by creating and sustaining organizational quality culture within the Division. This study is based on consultations and reports from December 2009 to August 2012, extended observations, focused group capacity building sessions, and day-to-day informal communications from key informants within the Division of Human Health. This paper aims to contribute to the theoretical and practical dialogue of quality assurance as implemented in a learning organization where praxis is continuously facilitated and implemented.

Keywords Praxis • Quality assurance • Radiation medicine • Total quality management

1 Introduction

The International Atomic Energy Agency (IAEA) was established in 1957 as the "Atoms for Peace" organization. In addition to its globally known role as the United Nation's "nuclear watchdog", the IAEA serves its 154 Member States in numerous

S. Ros (✉) • T.N.B. Pascual • R.K. Chhem
Division of Human Health, International Atomic Energy Agency, A-1400 Vienna, Austria
e-mail: S.Ros@iaea.org; T.Pascual@iaea.org; r.chhem@iaea.org

L. Lau and K.-H. Ng (eds.), *Radiological Safety and Quality: Paradigms in Leadership and Innovation*, DOI 10.1007/978-94-007-7256-4_13,
© Springer Science+Business Media Dordrecht 2014

aspects such as the use of nuclear techniques and applications in human health [17]. The IAEA is not a university. The IAEA verifies, informs, and educates the world on nuclear-related matters. The Division of Human Health (NAHU) within the IAEA's Department of Nuclear Sciences and Applications delivers global education and training resources in radiation medicine, facilitating the peaceful uses of nuclear techniques to prevent, detect and treat non-communicable and communicable diseases. The NAHU implements its programmes through four administrative sections: (1) nuclear medicine and diagnostic imaging that employs unsealed radionuclides in the diagnosis and treatment of diseases, including radiologic studies (2) applied radiation biology and radiotherapy that uses ionizing radiation in treatment of cancer, (3) dosimetry and medical radiation physics that assures the quality use of radiation in medicine for safety and effectiveness reasons; and (4) nutritional and health-related environmental studies that strengthens the Member States' know-how in applying stable isotope techniques to combat malnutrition.

NAHU establishes and sustains itself as a learning organization. A learning organization [26, 28] is one that is committed to inspire the learning of its staff and to improve this by continuously reflecting on the learning processes, including the social, political, and technological impacts. The Division promotes reflective practice through an institutional quality assurance framework (Fig. 13.1) known as the Educational Quality Assurance for Adult Learners (EQUAAL) to assure the quality of its education and training activities [22]. This framework focuses on the institution's quality culture while embracing local, regional, and global quality assurance practices [23]. The NAHU team is committed to meeting the needs of the learners and the organization when developing education and training resources. This approach is consistent with an organization putting praxis into practice [19]. As a learning organization, NAHU introduces and sustains the notion of praxis into the process of assuring quality of its education and training programmes.

The EQUAAL framework serves NAHU in the strengthening of its quality culture by bringing all professional staff together on a regular basis to share best practice in education and training, to jointly promote capacity building and curriculum development, and to execute its contextual education programmes. To implement this framework, the Division utilizes the expertise from a "resource triangle", consisting of: content experts, i.e. NAHU's professional staff members; process experts, i.e. educationalists; and information and technology experts (Fig. 13.2). By focusing on the needs of the learners, the staff members and experts work together as a team to strengthen NAHU's education and training programmes: e-learning and m-learning educational resources, coordinated research projects, conferences, symposiums, workshops, technical meetings, regional training courses, train the trainers courses, scientific visits, and fellowships.

Providing education and training programmes is the most commonly used and perhaps most tangible means to improve quality in radiology and radiation safety [20]. However, the content and delivery methodology of the education resources inevitably are outdated over time. NAHU management demonstrated leadership by

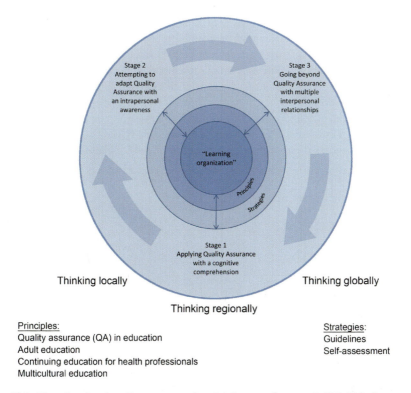

Principles:
Quality assurance (QA) in education
Adult education
Continuing education for health professionals
Multicultural education

Strategies:
Guidelines
Self-assessment

Fig. 13.1 The educational quality assurance for adult learners framework [22]. This framework focuses on the institution's quality culture while embracing local, regional, and global quality assurance practices [23]

Fig. 13.2 The NAHU's *resource triangle* team includes content experts, i.e. NAHU's professional staff members; process experts, i.e. educationalists; and information and technology experts

undertaking a comprehensive review of its education and training resources, advocating infrastructure strengthening, applying innovative process improvements by using praxis, total quality management (TQM) and continuous quality improvement (CQI), evaluating outcome and implementing refinements. These actions support NAHU's commitment to improve radiation safety and quality radiology practice in the IAEA Member States.

To provide an account of NAHU's experience in assuring quality of its education and training programmes, we first define praxis in NAHU's context. Next, we explore the process of sustaining praxis within NAHU. Then, we investigate how praxis and TQM inform NAHU's curriculum development process. Finally, we examine how a praxis-oriented quality improvement process strengthens NAHU's institutional quality culture.

2   Defining Praxis

Praxis can have different meanings in academic and social contexts. In some European cultures, praxis can be equated to "practice" as used in the English language. In Marxian beliefs, praxis can refer to an action that transforms a society. In this discussion, praxis refers to the on-going quality assurance processes and actions applied by NAHU's *resource triangle*. The *resource triangle*'s team members are mindful of their actions, which are professionally and morally committed by a tradition of quality [19] and are informed by the EQUAAL framework [22]. The team does not only care about the quality of the education and training resources in radiation medicine but also about the outcome of its actions and the resources of the Member States. Based on the *resource triangle's* perspective, the quality assurance of education and training programmes is a praxis-driven and collective action. This means that the *resource triangle* undertakes what it considers as the most meaningful actions to benefit the learners who are the radiation medicine specialists from the Member States. In this regards, praxis refers to the professional and moral commitments of the *resource triangle* team when undertaking actions to assure the quality of its education and training programmes. This contextualized praxis that is "morally-committed, and oriented and informed by tradition" [19] is applicable to NAHU's four professional fields: nuclear medicine and diagnostic imaging, applied radiation biology and radiotherapy, dosimetry and medical radiation physics, and nutritional and health-related environmental studies.

These four professional fields have tradition in enhancing the capabilities of the Member States to respond to the needs associated with the prevention, analysis and management of diseases through the application of nuclear techniques under the EQUAAL framework, which explicitly applies educational principles and strategies to the Division's education and training activities [3]. These principles comprise of quality assurance in education, notion of andragogy, transformative learning, continuing education for health specialists, and multicultural education. The strategies include internal guidelines and self-assessment tools [22]. The principles guide the *resource triangle* team when developing curriculum for the education and training programmes. The *resource triangle* plays the role as praxis-oriented decision agents [19] who prioritize the needs of the learners into the curriculum development processes. In this case, the learners are the radiation medicine specialists in the IAEA's Member States.

3 Sustaining Praxis

In education and training, as partially influenced by the manufacturing industries, praxis is also aligned with the concept of TQM, i.e. learner-driven teaching and curriculum development [7]. This alignment deliberately raises an awareness of moral duties among NAHU's *resource triangle* when developing education and training resources. It is not only about the quality of the resources, but it is also about NAHU's accountability: it should be a lifelong quality assurance process. Due to IAEA's mandatory staff rotation policy, an institutional memory has been adopted within NAHU to assist with the sustainability of the praxis culture.

These continuing quality assurance actions in education and training programmes are complementary to the learning organization vision and guide the *resource triangle* team when developing and implementing morally meaningful actions for the Member States. The Division of Human Health champions praxis by taking actions to sustain it. By sustaining praxis, NAHU can maintain its status as a sustainable learning organization despite rapid technological advances.

To sustain praxis, NAHU applies the principles of TQM into the curriculum development processes. Implementing TQM needs dedicated staff [5] with an awareness of the "praxis-oriented self" [10], which encourages and uses on going learning to capacity building thus translating the learning organization vision into a sustainable reality. To implement this initiative, NAHU strengthens its infrastructure by allocating resources and appointing experts to the *resource triangle*. On a bi-monthly basis, NAHU's staff members meet to share best practices, build professional capacity and sustain individual and institutional praxis. Freed and colleagues [13] perceive TQM as the principles used to transform a shared institutional tradition and to achieve an institutional vision by employing outcome-based measurement and team collaboration.

4 Praxis-Oriented TQM

From university to healthcare settings, TQM has been a topic of debate. Universities first experienced the implementation of TQM in administration [1, 16]. Some professors viewed TQM as a fad [14]. Some faculty members were reluctant to incorporate TQM since they thought it could jeopardize their authority over the students [12]. TQM promotes a learner-centered teaching and learning approach. Some faculty members enjoyed having full control over how they could design their teaching methodologies, for example. Some faculty members questioned whether the TQM philosophy, sourced from factories and businesses, would actually fit in higher education contexts [2]. This controversy resulted in the re-naming of TQM as Continuous Quality Improvement (CQI) by some academic groups [11]. Some scholars find TQM as useful tools to improve academic services and programmes, particularly in higher education institutions that have already implemented quality principles and strategies within their institutional framework [8, 24]. TQM has been reflected in NAHU's EQUAAL framework. This reflection resonates with the assertion that the principles of TQM do contribute to quality improvement in an education setting [6].

The *resource triangle* incorporates praxis-oriented TQM to assist with curriculum development. Its members agree on the quality assurance principles and strategies that have been implemented to ensure the quality of the education and training programmes. Quality assurance is a process of developing principles, strategies, and tools to deliver the desired quality outcome [15]. TQM incorporates the quality assurance processes and further cultivates it by applying on-going improvements [25]. Based on this concept, the IAEA's Division of Human Health incorporates TQM into its tradition of ensuring quality of the education and training programmes and strengthens the existing efforts by CQI. Thus, NAHU is committed to CQI and sustaining an institutional quality culture under a praxis-oriented TQM.

5 NAHU's Approach to Curriculum Development

Based on the praxis-oriented approach to TQM, NAHU incorporates CQI into the curriculum development process using an integrated, systematic, and structured approach to quality assurance [9]. Many of NAHU's consultations and reports in curriculum development are in keeping with this CQI approach. Next, we describe how NAHU's *resource triangle* relates each philosophical component of this CQI approach to the curriculum development process.

First, the quality assurance of the education and training programmes is integrated because NAHU has clearly defined the roles and responsibilities of the *resource triangle* and integrated quality assurance into the Division's regular and budgetary work plan. To inform the continuous process of quality improvement, it is important to define who is responsible for the development and revision of the evaluation tools, the analysis and report of surveys, and the use of the accumulated data for liability commitments. It is essential to share the evaluation results in the Division's regular forum to learn about best practices. This bi-monthly forum is internally known as the NAHU Campus meeting. This meeting serves as a brainstorming session that allows free flow of ideas and proposed actions. Some of these proposals can ultimately lead to policy actions within the Division. These actions are facilitated by praxis in a sustainable learning organization such as NAHU.

Second, the quality assurance approach is systematic because NAHU implements quality assurance to all of its education and training programmes, especially to the curriculum development process. The Division systematically engages its *resource triangle* and other in-house key stakeholders. For the education and training programmes, the different stakeholder groups inform curriculum development. The NAHU's *resource triangle*, for example, collates the needs of the learners by an in-house mechanism. Combining this data from the Member States with the team's teaching and learning experience, the *resource triangle* develops a curriculum to meet the needs of the learners, e.g. by conducting regional training courses as required by the learners from a specific geographical location. As another example, the NAHU is conducting a tracer study to evaluate the impact of its programmes in applied radiation biology and radiotherapy. In this study, the targeted groups provide feedback on the relevance and usefulness of the NAHU programmes they have attended. The data from this tracer study is used to strengthen future curriculum design as driven by the praxis principle [4, 21].

Third, NAHU's quality assurance of education and training programmes is structural in nature because the four administrative sections are unique in their own professional field. This means that each section shall sustain praxis differently depending on the needs of their learners. Within NAHU, there are generic principles and strategies for curriculum development embedded in the EQUAAL framework [22], but each administrative section can adapt these to suit its own requirement.

A well-developed curriculum provides an integrated framework to disseminate knowledge. The task for the experts is to define the learning objectives and the specific learning outcomes for each education activity. When designing an activity, the *resource triangle* ensures that it meets the needs and requirements of the learners. This praxis-oriented CQI principle is applicable to a stand-alone short course or to a module in a more extensive programme. The *resource triangle* applies the same principle to the planning of a conference or an individual fellowship. For the program's sustainability, the *resource triangle* places a strong emphasis on the evaluation and feedback from the learners of these education and training activities.

It is important to ensure the sustainability of the pool of local educators residing in the Member States who are more aware of their own cultural, social, and professional needs. These culturally and socially informed educators serve as "praxis ambassadors" by transferring their knowledge and know-how to the learners thus sustaining praxis beyond the direct reach of NAHU's education and training programmes. In this regard, the *resource triangle* team members may be described as "praxis architects" [18, 27] who reflectively co-learn from each other and from the participants of these education and training programmes. The combination of the praxis and TQM concepts informs curriculum development and sustains organizational quality culture, thus enabling NAHU to better serve the IAEA's Member States.

6 Conclusion

We explore the process of applying the praxis concept into the quality assurance of education and training programmes of the IAEA's Division of Human Health. The contextualized praxis refers to the continuous processes and actions of quality assurance as endorsed by NAHU's resource triangle, which are morally and professionally driven by NAHU's tradition of quality. This quality culture is further enhanced by the EQUAAL framework. Praxis plays an important role in ensuring the quality of NAHU's education and training programmes by adding value to the EQUAAL framework. Since 2008, NAHU has been implementing and sustaining praxis in its work plans by introducing the learning organization vision into all education and training programmes. The learning organization vision is consistent with the praxis and TQM concepts, which jointly enhance the curriculum development process. The praxis process is most invaluable to NAHU and has strengthened its quality culture. This successful model is adaptable and applicable to other organizations by strengthening institutional quality culture and improving the development and implementation of radiation safety and quality radiology actions.

References

1. Beaver W (1994) Is TQM appropriate for the classroom? College Teaching 42(3):111–114
2. Birnbaum R (2001) Management fads in higher education: where they come from, what they do, why they fail. Jossey-Bass, San Francisco
3. Chhem RK, Hibbert KM, Van Deven T (eds) (2009) Radiology education: the scholarship of teaching and learning. Springer, Berlin
4. Cousins JB, Leithwood KA (1986) Current empirical research on evaluation utilization. Review of Educational Research 56(3):331–364
5. Daily BF, Bishop JW (2003) TQM workforce factors and employee involvement: the pivotal role of teamwork. Journal of Managerial Issues 15:393–412
6. DeJager HJ, Nieuwenhuis FJ (2005) Linkages between total quality management and the outcomes-based approach in an education environment. Quality in Higher Education 11:251–260
7. Deming WE (1986) Out of the crisis. M.I.T, Centre for Advanced Engineering Study, Cambridge, MA
8. Dew J (2007) Quality goes to college. Quality Progress 40:45–48
9. Dolmans D, Wolfhagen H, Scherbier A (2003) From quality assurance to total quality management: how can quality assurance result in continuous improvement in health professions education? Education for Health 16(2):210–217
10. Edwards-Groves C (2008) The praxis-oriented self: continuing (self-) education. In: Kemmis S, Smith TJ (eds) Enabling praxis: challenges for education. Sense Publishing, Rotterdam
11. El-Khawas E (1993) Campus trends. American Council on Education, Washington, DC
12. Ensby M, Mahmoodi F (1997) Using the Baldrige award criteria in college classrooms. Quality Progress 30:85–91
13. Freed JA, Klugman MR, Fife JD (1997) A culture for academic excellence: implementing the quality principles in higher education. ASHE-ERIC higher education report 25(1) George Washington University, Washington, DC
14. Hansen WL (1993) Bringing total quality improvement into the college classroom. Higher Education 25:259–279
15. Harvey L, Green D (1993) Defining quality. Assessment and Evaluation in Higher Education 18(1):9–34
16. Horine JE (1993) Improving the educational system through Deming's system theory. Educational forum 58:30–35

17. International Atomic Energy Agency (2010) More than a "watchdog": Director General Yukiya Amano addresses 54th general conference. http://www.iaea.org/newscenter/news/2010/gc54opens.html. Accessed 19 Oct 2011
18. Kemmis S, Grootenboer P (2008) Situating praxis in practice: practice architectures and the cultural, social and material conditions for practice. In: Kemmis S, Smith TJ (eds) Enabling praxis: challenges for education. Sense Publishing, Rotterdam
19. Kemmis S, Smith TJ (eds) (2008) Enabling praxis: challenges for education. Sense Publishing, Rotterdam
20. Lau LSW, Perez MR, Applegate KE et al (2011) Global quality imaging: improvement actions. J Am Coll Radiol 8(5):330–334
21. Leviton LC, Hughes EFX (1981) Research on the utilization of evaluations: a review and synthesis. Eval Rev 5(4):525–548
22. Ros S, Pascual TN, Chhem RK et al (2012, in press). Quality e-learning resources for global radiation medicine specialists: the IAEA experience. Paper presented at the 19th international conference on learning, The Institute of Education, University of London, London
23. Ros S (2010) Implementing quality assurance at Royal University of Phnom Penh, Cambodia: perceptions, practices and challenges. Ed.D. Dissertation, Northern Illinois University, ProQuest (AAT 3439629)
24. Ruben BD (2007) Higher education assessment: linking accreditation standards and the Malcolm Baldrige criteria. New Dir Higher Educ 137:59–69
25. Sallis E (1996) Total quality management in education, 3rd edn. Kogan Page, London
26. Senge PM (1990) The fifth discipline. Century Business, London
27. Smith TJ (2008) Fostering a praxis stance in pre-service teacher education. In: Kemmis S, Smith TJ (eds) Enabling praxis: challenges for education. Sense Publishing, Rotterdam
28. Thomas K, Allen S (2006) The learning organization: a meta-analysis of themes in literature. The Learning Organization 13(2):123–39

Chapter 14
Professional and Regulatory Collaborations to Ensure Competency

Lizbeth M. Kenny

Abstract To maximize the benefit that diagnostic and interventional radiology (radiology) can bring to both patients and health system, a highly competent service is required. Radiology contributes substantially to the effective diagnosis and treatment of patients. The radiology community has a clear professional responsibility to develop and maintain standards of training, care, and practice to ensure the capability and competence of the radiology team. A competent system would position radiology to make the maximum impact on care. Radiology leaders need to advocate for practitioner competency and quality use of radiology. As radiology is complex, its interaction with the referrers must be effective and efficient and should make use of modern information technology. In particular, a system of guiding referrers ought to be in place to minimize inappropriate investigations and facilitate its cost-effective use. Effective collaboration between the leaders of the profession, facility managers, regulators, and policy makers and the application of innovative actions are necessary if patients and referring physicians are to receive maximum benefit. Such a system is likely to be highly cost effective. This requires acceptance that radiology is a critical enabler of good care and that competence is required in every facet and level of the complex system in which it operates. Strong professional leadership is required to set the principles and standard of care and to promote quality use. Ultimately, these actions require the support from and participation of the practitioners to ensure their successful implementation, thus contributing to better patient outcome.

Keywords Appropriateness of requesting • Collaboration • Competence • Diagnostic imaging • Facilities • Interventional radiology • Policy makers • Professional leadership • Regulators

L.M. Kenny (✉)
Royal Brisbane and Women's Hospital, Herston, QLD, Australia

Central Integrated Regional Cancer Service, Queensland Health, 362 Gold Creek Road, Brookfield, QLD 4069, Australia
e-mail: lizkenny@bigpond.net.au

L. Lau and K.-H. Ng (eds.), *Radiological Safety and Quality: Paradigms in Leadership and Innovation*, DOI 10.1007/978-94-007-7256-4_14,
© Springer Science+Business Media Dordrecht 2014

1 Introduction

To maximize the benefit to individual patients and the overall health system that diagnostic and interventional radiology can bring, a highly competent system is required. The radiology community has a clear professional responsibility to develop standards of training, care, and practice to ensure the capability and the competence of radiologists underpinning the delivery of high quality imaging. A competent system would position radiology to make the maximum impact on care in every regard and leaders of radiology need to advocate for the competence of and the quality use of radiology within the entire medical system. Such a system would ensure that the right procedure was requested and performed correctly and safely the first time, in the right facility with a competent radiology team, with maximum benefit to the patient and system and with minimum risk to the patient and staff. Such a system is likely to be highly cost effective in every respect. This requires acceptance that diagnostic imaging is a critical enabler of good care for many people and that competence is required in every facet and level of the complex system in which it operates. Strong professional leadership is required to set the principles and standard of care and to promote the quality use of diagnostic imaging. Collaboration with regulators and policy makers must be robust to bring about the changes required to enable much of what is required.

2 Background

Competency could be considered the ability to do something successfully or efficiently [13] or the ability of an individual or system to perform a job properly.

Diagnostic imaging and interventional radiology has become an integral part of medical care, so much so that it is often taken for granted. The use of diagnostic imaging has gone hand in hand with escalating medical costs in the western world and the increased burden of medical radiation exposure to patients. The increasing financial burden of health care is a major issue in every country that has a significant health care system. Consideration may not be given to the importance of governance or oversight of the totality of diagnostic imaging with regards to its utilization, the impact of imaging on patient care, its impact on the system and its potential to manage and contain escalating health care costs and iatrogenic harm to patients. A highly competent system would address all of these elements relevant to diagnostic imaging not just the individual competence of radiologists and facilities.

From a global perspective, a variety of professional leadership and governance exists influencing the requesting and practice of diagnostic imaging and it's positioning within the health care system. In general, the profession has taken the lead on the promotion of quality in the practice of diagnostic imaging. Although generalizations do not hold true in all circumstances, in some countries such as Australia, the ordering or requesting of imaging is mostly at the discretion of the requesting clinician. In Europe given the EU Council Directive [3], the duty of care as to whether to comply with such a request lies significantly with the radiologist. In the USA and Israel a third party payer may determine which procedures will be reimbursed, hence ultimately influencing the choice.

On this background, strong and determined professional leadership is vital in setting the standards of care, promoting the benefit of quality diagnostic imaging and influencing both regulators and politicians alike, in order to achieve best outcomes for patients. Planned collaboration between the profession setting the standards and the regulators helping to implement quality practice seems an ideal partnership in achieving this outcome.

3 A Competent System

Competence in radiology is important, not just of the radiologists themselves but also of the system within which they operate. Radiologists operate within a complex radiology system, which in turn sits within a highly complex system of medical care where there are various regulations and politics imparting significant influence. To consider the competence of one aspect of the system be it staff or procedures has great merit but is insufficient without considering the system as a whole.

The minimum components of such a system would include competency-based training of radiologists and other members of the radiology team, compliance of conduct and practice within a quality framework, proper education, support and guidance for referring practitioners, capable facilities and regulatory, IT and financial systems that support quality practice.

In an ideal system, which is designed to maximize the benefit that diagnostic imaging can bring, the following steps would be considered, led by the profession and managed collaboratively with funders and regulators.

- The most appropriate procedure(s) is requested and performed the first time in order to provide an accurate diagnosis and maximize the benefit of care provided; minimize radiation dose to patients; reduce the waste of unnecessary or unhelpful procedures often associated with multiple physician appointments; minimize the waste of precious time and resources; minimize the delay in providing appropriate care to those in need; and reduce the likelihood of spurious diagnoses leading to potentially harmful or meddlesome treatment.
- The requesting physician either has an educated understanding of what procedure will answer the clinical question being asked or that they are guided into requesting the best procedure(s).
- The imaging facility is properly equipped.
- Good IT system management is in place.
- The imaging procedure is acquired correctly with the lowest reasonable radiation exposure to the patient and able to be sent anywhere within the system to the most appropriate radiology team.

- The system can retrieve prior imaging records to avoid unnecessary duplication and to guide procedure selection and interpretation.
- The radiology team has access to all relevant clinical information and prior imaging.
- The radiology team is highly competent to acquire and give an expert specialist opinion on the imaging findings and its meaning in the context of the patient's illness or clinical question being asked.
- Opinion and results were delivered to the requestor and available in a clinically relevant timeframe.
- Imaging and radiologist opinions are available to wherever the patient and their physician are located.
- Feedback and audit loops are built-in to monitor utilization, compliance with appropriate requesting, impact on care and continuous improvement.

When considering competency the individual radiologist, the team, the referrer, the facility and the system are intimately intertwined and all of significance.

4 The Radiologist

4.1 Competency-Based Training

The design and governance of training programs varies across the world. Many training programs, whether they are professional college-based or university-based, have adopted the CanMEDS 2005 framework [14] in designing competency-based training and assessment programs. In developing this framework, the Royal College of Physicians and Surgeons of Canada has shown extraordinary professional leadership, with its influence extending well beyond Canada, as shown by its adoption in whole or in part across much of the globe. The framework is designed to respect the medical expert role as critical and central to requirements, but expands the skills required to make the medical expert more effective. The core competencies include:

- Medical Expert;
- Communicator;
- Collaborator;
- Manager;
- Health Advocate;
- Scholar; and
- Professional

Not only have they been adopted in some training programs but also in programs of continuing medical education or maintenance of competence and certification.

In a number of countries, national training programs are regulated. An example of professional and regulatory collaboration to ensure competency is the system used in Australia and New Zealand. The CanMEDS principles are used to guide the development of the core competency content of the training program by the fellows of the Royal Australian and New Zealand College of Radiologists in an exhaustive and constantly evolving process [22]. The RANZCR conducts the examinations and awards fellowship. The Australian Medical Council (AMC [2]) has responsibility for developing guidelines, policy and procedures for the accreditation of specialist medical education and training programs; overseeing the AMC's program of accreditation; and encouraging improvements in postgraduate medical education that respond to evolving health needs and practices, and educational and scientific developments.

The RANZCR was the first Learned College to partner with the AMC to accredit its education and examination process, which ultimately led the change from a curriculum-based to a competency-based training and assessment program. Although initially voluntary, from 1 July 2010, the process of accreditation of training programs was made mandatory. In Australia the *Health Practitioner Regulation National Law Act 2009* makes the accreditation of specialist training programs an element of the process for approval of programs for the purposes of specialist registration. The RANZCR in turn accredits hospitals and radiology facilities as having the necessary infrastructure and competence to train their trainees.

While this is an Australian process, the Medical Council of New Zealand uses AMC accreditation reports to assist it to make decisions about recognizing medical training programs in New Zealand. The AMC works with the Medical Council of New Zealand in reviewing bi-national training programs. Similarly, the Medical Board of Australia's registration standards indicate that continuing professional development programs that meet AMC accreditation requirements also meet the Board's continuing professional development requirements.

Hence like many countries, there are multiple layers of regulation. However, the standards and contents are owned and driven by the profession and their professional body. Due to the leadership shown and the collaboration with the regulators, it is reasonable to assume that with the awarding of fellowship, radiologists are well trained and have competency that extends beyond the medical expert role. The support of the regulators makes it more likely that there is confidence in the training program and that training institutions and employers comply with requirements. *This collaboration is wholesome and of benefit to all.*

4.2 Maintenance of Competence

Most developed countries or states, require either in a voluntary or mandatory manner, some evidence that medical specialists are maintaining and growing knowledge and skills. Different countries and systems coin different terms, all of which have similar goals. These include Continuing Medical Education (CME), Continuing Professional Development (CPD), Maintenance of Professional Standards (MOPS), and Maintenance of Certification (MOC). Although many countries and professionals use these terms interchangeably, the European Society of Radiology describes CME in Diagnostic Radiology (ESR [4]) as a programme of educational activities to guarantee the maintenance and upgrading of knowledge, skills and competence following completion of postgraduate training. They describe CME as an ethical and moral obligation for each radiologist throughout his / her professional career, in order to maintain the highest possible professional standards; and CPD as dealing with the acquisition and maintenance of new knowledge, skills, attitudes and values, in order to ensure that radiologists are able to maintain and improve their competence to practice in their specialist area, as well as in any area of additional responsibility in education or management to improve the overall quality of service to patients.

The support for and mandating of such programs also varies around the world. Many professional bodies and societies have rightly taken the professional lead in designing and governing the elements and composition of CPD programs for their members or fellows. Some Colleges such as the RANZCR use the CanMEDS roles to determine the reasonable elements within the CPD program requirements. Compliance with CPD programs are often then either mandated by medical boards or similar organizations as a pre-requisite to ongoing specialist registration or recognition. Many institutions also directly use CPD compliance as part of their credentialing and employment arrangements.

Ultimately this is perhaps the ideal arrangement with the profession taking the lead in standards and practice and the regulators ensuring some form of compliance. CPD in and of itself does not solely imply or measure competence in practice but is one of the important tools of measurement of quality that contributes to assessment as a whole. This is a part of professional practice taken seriously by the professional bodies and learned Colleges and consequently it carries credibility within the regulatory and general community.

An important development in the USA and in the UK has been the introduction of limited licensing arrangements. However, each of these two countries has chosen a different approach. The American Board of Medical Specialties (ABMS) and the American Board of Radiology (ABR) have introduced a process of Maintenance of Certification (MOC). The ABR now awards time-limited certification and requires its new diplomates to enroll into a process that has a 10-year cycle and is based on examinations taken during that period. The process evaluates six essential competencies on a continuous basis, similar in nature to the CanMEDS principles and is designed to document the professional development of each diplomate. The ABR considers this process to be a comprehensive vehicle through which all diplomates can provide assurance to the public and the radiological community that they are incorporating new information into their practice and are delivering excellent medical care. The process has been designed and led by the radiologists.

In contradistinction to the American system, the process of revalidation and recertification in the UK has been imposed on the medical profession by the government in the aftermath of the Shipman affair. (Harold Shipman was a General Practitioner who murdered hundreds of his own patients). However, after taking the political decision to follow this route, the government delegated the task to the regulator – The General Medical Council (GMC), which is independent of government and has substantial input from the medical profession. The GMC consulted with all the medical Royal Colleges, each of which set the standards for the specialties it controls. The Royal College of Radiologists was particularly proactive in this process and many of the other Colleges used method adopted by the RCR as a template for their own specialties. This is based on four components: Continuous Professional Development, Multisource Feedback (a process in which anonymous feedback is obtained on a doctor's practice from colleagues and patients), a robust annual appraisal process and evidence of good practice e.g. by providing patient outcomes. The main differences between the USA and UK systems are that the latter does not utilize examinations and has a 5-year cycle. The RCR has developed online tools to guide and support its fellows through the revalidation process.

5 The Radiology Team

Other key members of the radiology team include radiographers and medical physicists. In many parts of the world, specific training programs, accreditation of training programs within institutions or universities and maintenance of certification or similar programs are available for these practitioners [16, 18–20, 24].

The principles are the same as for any other professionals and are modeled along similar lines, detailing the competencies required for registration and ongoing continuing professional development. Just as important as the individual qualifications and skills but very hard to measure is the ability to interact, communicate and function as a team across all of the radiology disciplines. In real life this is a complex challenge. Few programs if any can really address this issue in a meaningful way but awareness of the importance of a well functioning team is an important leadership skill.

6 The Referrers

When one considers how imaging is requested, it is a complex and varied situation. Referrers to radiology can either be at arms-length or not. Non arms-length referral, i.e. self-referral, has inbuilt biases that despite the highest moral behavior can lead to over ordering, misuse or overuse of imaging. In countries like Australia, only arms-length referral to radiologists is acceptable as one step to limit unnecessary procedures and expenditure.

For the referring clinicians the greatest challenge is in knowing what to order, when and how to maximize the benefit to the patient and their overall management. The impact of diagnostic imaging on the care of patients is massive and this can be good and bad. Imaging can be requested out of prior expert knowledge, habit, guideline pathways, decision support systems, ignorance, arrogance, what is allowed by health care management organizations or driven by medico-legal requirements.

When an imaging procedure is unable to contribute to either a diagnosis or a management decision then not only is it unhelpful but positively harmful. It adds unwanted radiation exposure, may suggest a spurious path in care and may delay the correct diagnosis and management. It may also result in multiple visits to the referring physician with additional expenses to the patient or to the system as yet more and more tests are requested when the answer is not forthcoming. Knowing what tests to order is a complex business and no one is better placed than the radiologist themselves who has a deep understanding of the pathology of diseases and imaging pathways to diagnosis.

In order for this to work well, there needs to be an understanding between radiology and other clinical staff. Education in radiology needs to be considered at all levels from medical school through to that of internship, residency and consultant practice. Multidisciplinary team meetings are an outstanding way to demonstrate the benefit of good imaging in diagnosis and decision making show-case the benefit of good radiology and are an excellent teaching environment.

For leaders in radiology engagement with the referrers is a complex but critically important issue. When one's livelihood depends on the ordering practices of referring clinicians or in a corporatized practice where more activity will lead to a greater financial return to shareholders, internal conflict naturally exists. This is the reality of life in many states and countries. However the compounding growth of diagnostic imaging in developed countries is unable to be sustained and is a likely major contributor to the escalating cost of health care not just the actual cost of imaging per se. It requires a holistic review of where diagnostic imaging sits in health care as a whole and a conscious decision to bring it well forward in the decision making process to maximally influence good care. We simply must be in a position to do the right imaging procedure the first time with maximum potential benefit to patients. The importance of this principle cannot be underestimated and requires committed, clear and unwavering professional leadership. To effect this will require close collaboration with regulators, policy makers and funders as well as participation of the practitioners.

Unless the referring clinicians are extremely well educated or regulated, guidelines and decision support systems with feedback loops can make a tremendous difference to the utilization of diagnostic imaging. Guidelines have been developed by a number of groups and these include but are not limited to the guidelines such as the ACR Appropriateness Criteria® [17] and Diagnostic Imaging Pathways [12].

The use and benefit of an integrated decision support has been well demonstrated by The Massachusetts General Hospital [15]. Their system, which is part of a holistic radiology management approach, has made a major impact on the previously compounding growth rate within their group of facilities. Of critical importance, it has broad support and buy in from the clinical community who have contributed to and continue to modify the system. There has been buy in also from funders who do not require pre-authorization by clinicians within this system. Feedback loops, which allow an assessment of utilization and compliance, are a key modifier of both the decision support algorithms and of clinician ordering behavior.

Another method is simply to limit who can access requests for high-end radiology such as MRI. This is likely to be of little benefit to patients and may result in unnecessary radiation exposure and delay in diagnosis.

7 The Facility

Professional standards are precisely that and the responsibility of setting professional standards lies with the profession. They must be developed and owned by the profession and not by the regulators. Standards however can then be used within a quality and safety framework either in their entirety or in part. Accreditation programs and systems that are designed to look at a part or the whole of the radiology system are usually facility-based and may be assessed in isolation from the rest of the overall facility. The central focus of all accreditation programs should be on safety and quality first and foremost.

In most developed countries facility accreditation systems exist either for training, practice or reimbursement purposes. These can be whole of country or state-based programs. Both the ACR and the RANZCR have to their credit developed gold standard facility accreditation programs focusing on quality and safety assessing a full suite of practice standards. The RCR in conjunction with the Society and College of Radiographers has developed the Imaging Services Accreditation Scheme. Likewise similar programs have been developed in Canada [1]. All of these schemes have been led by the profession. The Colleges or professional bodies then partner with accrediting bodies to actually perform the accreditation process.

By way of example the ACR accreditation program is one of the programs used by The Joint Commission [21] in their Imaging Services Accreditation scheme, which facilitates reimbursement through the Medicare and Medicaid accredit practices. The RANZCR/NATA Medical Imaging Accreditation Program [8] is pitched at a higher level of accreditation than that required by the Australian government (The Diagnostic Imaging Accreditation Scheme) in order for patients to be eligible for Medicare reimbursement. In both countries compliance with the basic accreditation schemes are linked to national reimbursement – either to facilities or to patients.

Facility accreditation programs are not uniquely owned by the professional and learned Colleges and the need for the profession and the regulators to come together is of paramount importance. The accreditation process is normally carried out by accreditation bodies rather than the professional Colleges. It remains of great importance however that professional standards be owned by the profession and this is not always apparent. The use of a subset of standards in accreditation programs would seem entirely appropriate and such programs may also incorporate other practice standards.

8 The System

This is the most complex issue and few countries or states have truly considered the pivotal role of radiology and why getting it right is so important in the management of patients in the acute and chronic health care settings. States such as Minnesota,

USA and complex facilities such as the Massachusetts General Hospital have attempted this on a relatively encompassing scale.

The relationships are complex. When radiology is either considered as a source of major income for a facility or an expensive burden it is difficult then to regard it as an enabler of the system. If however we focus on the extraordinary diagnostic and interventional capability of radiologists to perform procedures guided by relevant prior imaging and proper clinical information; if there was a direct relationship between radiologists and other clinicians; if the system was equipped to manage the IT complexities to allow good management and audit of process; and if funding rewarded the providers of high quality radiology, then it is possible that the current runaway expenditure of health care in most countries could be brought under some semblance of control. The focus on quality imaging is an important key [11]. Just restricting access to imaging across the board or simply reducing the reimbursement of procedures does not in any way reward quality or improve patient outcome. The impact of radiological diagnosis on getting answers to complex problems quickly is likely to be underestimated, e.g. by reducing the number of visits and unnecessary procedures and reducing the number of iatrogenic complications as a consequence of choosing incorrect treatment. The impact of radiology to the prevention of these undesirable scenarios is hard to estimate.

Some countries such as Australia have attempted to take a holistic view at the whole of system issues. Unique programs such as the Quality Use of Diagnostic Imaging Program (QUDI) [9, 23] are a case in point. The initial phase of the QUDI program was established by collaboration between the Australian Government and the RANZCR and aimed to develop a framework for a systems- and evidence-based approach to the appropriate and quality use of diagnostic imaging services in Australia. Funding has been made available between 2004 and 2011.

This profession led and managed, government funded program considered projects across safety, effectiveness, appropriateness, acceptability, access and efficiency. The projects were organized into four streams of work organized around consumers, referrers, radiologists and economic sustainability. A number of the projects have been completed and some are continuing in an ongoing fashion. This has produced information and data that has changed aspects of practice. Some examples include:

- A standard set of diagnostic imaging referral forms;
- CT dose optimization;
- Development of a Radiology Events Register (RaER);
- Diagnostic imaging reporting guidelines;
- CT criteria for low back pain;
- Promoting public awareness of the appropriate use of diagnostic imaging;
- Role evolution;
- Teleradiology;
- Professional supervision and reporting standards for radiologists;
- Closing the loop; and
- Radiologists in multidisciplinary breast cancer teams

Incorporation of the findings of these various projects into daily practice is ongoing. QUDI has helped to improve the quality and safety of service delivery of diagnostic imaging in Australia across a number of key areas. Collaboration between the profession and government was critical to the QUDI program being enabled.

At the international level, UN agencies and professional organizations such as the International Atomic Energy Agency, World Health Organization, International Society of Radiology, International Society of Radiographers and Radiological Technologists and International Quality Radiology Network [5, 6] have major roles to play in focusing on the quality use of diagnostic imaging in a global sense. As an example, the WHO launched a Global Initiative on Radiation Safety in Health Care Settings to mobilize the health sector in the safer use of radiation in medicine [25].

9 Summary

The relationship between the profession – those who set the standard, and the regulator – those that regulate the system is indeed complex. To have a common and agreed vision is of great value and needs to be centered on the sustainable and quality use of diagnostic imaging, i.e. imaging that will make a real difference to patients and their care and imaging that will provide them with the lowest reasonable radiation exposure. Individual and system competency underpin quality and safety in radiology. Quality and safety in radiology is a continuum requiring on-going efforts, innovations and collaboration from all stakeholders [7]. Appreciating that the repositioning of radiology within the system and rewarding quality can help to manage the whole of the health care system, and having a clear understanding of the difference between owning and being responsible for a professional standard and using them in a regulatory framework, will advance this common goal for all the stakeholders. These are the challenges of professional leadership [10]. The need for innovation in thinking and funding mechanisms that reward quality is clear. Above all else, these changes should be professionally led, actively and willingly participated in by the practitioners, sustainable and of benefit to the community at large.

References

1. Accreditation Canada (2012) Accreditation Canada's Diagnostic Imaging standard. http://www.accreditation.ca/accreditation-programs/qmentum/standards/diagnostic-maging/. Accessed 31 Jan 2012
2. Australian Medical Council (AMC) (2012) Assessing specialist medical education and training. http://www.amc.org.au/index.php/ar/sme. Accessed 31 Jan 2012
3. Council of the European Union (1997) Council Directive 97/43/Euratom of 30 June 1997 on health protection of individuals against the dangers of ionizing radiation in relation to medical exposure, and repealing Directive 84/466/EURATOM. OJEC 1997/L 180/22

4. European Society of Radiology (ESR) (2012) Definition of CME/CPD. http://www.myesr.org/cms/website.php?id=/en/education_training/cme.htm. Accessed 31 Jan 2012
5. International Radiology Quality Network (IRQN) (2012) http://www.irqn.org/. Accessed 31 Jan 2012
6. Lau LSW (2004) International radiology quality network. J Am Coll Radiol 1(11):867–870
7. Lau LSW (2006) A continuum of quality improvement in radiology. J Am Coll Radiol 3:233–239
8. Lau LSW (2007) The design and implementation of the RANZCR/NATA Accreditation Program for Australian Radiology Practices. J Am Coll Radiol 4(10):730–738
9. Lau LSW (2007) The Australian national quality program in diagnostic imaging and interventional radiology. J Am Coll Radiol 4(11):849–855
10. Lau LSW (2007) Leadership and management in quality radiology. Biomed Imaging Interv J 3(3):e21
11. Lau LSW, Perez MR, Applegate KE et al (2011) Global quality imaging: improvement actions. J Am Coll Radiol 8(5):330–334
12. Mendelson RM, Bairstow PJ (2009) Imaging pathways: will they be well trodden or less traveled? J Am Coll Radiol 6(3):160–166
13. Oxford Dictionaries (2010) Oxford dictionary of English. Oxford University Press, Oxford
14. Royal College of Physicians and Surgeons of Canada (RCPSC) CanMEDS Framework (2005) http://www.royalcollege.ca/portal/page/portal/rc/canmeds/framework. Accessed 1 Oct 2013
15. Sistrom CL, Dang PA, Weilburg JB et al (2009) Effect of computerized order entry with integrated decision support on the growth of outpatient procedure volumes: seven-year time series analysis. Radiology 251:147–155
16. The American Association of Physicists in Medicine (AAPM) Education (2012) http://www.aapm.org/education/default.asp. Accessed 31 Jan 2012
17. The American College of Radiology (ACR) Appropriateness Criteria® (2012) http://www.acr.org/Quality-Safety/Appropriateness-Criteria. Accessed 1 Oct 2013
18. The American Registry of Radiologic Technologists (ARRT) Radiography Certification (2012) https://www.arrt.org/Certification/Radiography/. Accessed 31 Jan 2012
19. The Australasian College of Physicists Scientists and Engineers in Medicine (ACPSEM) (2012) ACPSEM CPD program. http://acpsem.org.au/index.php/education/teap-cpd. Accessed 31 Jan 2012
20. The Australian Institute of Radiography (AIR) (2012) Continuing professional development. http://www.air.asn.au/cpd.php. Accessed 31 Jan 2012
21. The Joint Commission (2012) http://www.jointcommission.org/. Accessed 31 Jan 2012
22. The Royal Australian and New Zealand College of Radiologists (RANZCR-1) (2012) CPD overview. http://www.ranzcr.edu.au/cpd/overview. Accessed 31 Jan 2012
23. The Royal Australian and New Zealand College of Radiologists (RANZCR-2) (2012) The quality use of diagnostic imaging program. http://www.ranzcr.edu.au/quality-a-safety/qudi. Accessed 31 Jan 2012
24. The Society and College of Radiographers (SOR) (2012) Education accreditation. http://www.sor.org/learning/education-accreditation/. Accessed 31 Jan 2012
25. World Health Organization (WHO) Global Initiative on Radiation Safety in Health Care Settings (2012) http://www.who.int/ionizing_radiation/about/med_exposure/en/index1.html. Accessed 6 Feb 2012
26. Ferber AL, O'Reilly Herrera A (2013) Teaching privilege through an intersectional lens. In: Kim C (ed) Deconstructing privilege: teaching and learning as allies in the classroom. Routledge, New York

Chapter 15
Safety and Radiation Protection Culture

Kelly Classic, Bernard Le Guen, Kenneth Kase, and Richard Vetter

Abstract At a time of significant developments in the use of ionizing radiation in medicine and power generation, the radiation protection profession is facing the challenge of enhancing safety culture throughout the world.

As the voice of radiation protection professionals, the International Radiation Protection Association (IRPA) has initiated a process for promoting radiation protection and safety culture. This chapter presents general theories of organizational culture and discussions from three IRPA workshops organized in Europe, Asia and the United States.

We present an outline of how these theories correspond to specific values and beliefs underlying a safety culture. We posit that there is no difference in the basis of a radiation protection culture among the various radiation-using sectors, whether it is medicine, nuclear power generation or industry.

Radiation protection focuses on people and behavior (culture) to prevent harm to individuals when radiation or radioactive materials are being used; radiation safety focuses on system design to allow the use of hazardous equipment or materials without harming individuals and the environment. A radiation protection culture is an underlying requirement to successful implementation of radiation safety.

K. Classic (✉) • R. Vetter
Radiation Safety, Mayo Clinic, MSB B-28, Rochester, MN 55905, USA
e-mail: classic.kelly@mayo.edu; rvetter@mayo.edu

B. Le Guen
Generation Nuclear Power Plant Operations, EDF, Site Cap Ampere 1, place Pleyel, Saint-Denis Cedex 93282, France
e-mail: bernard.le-guen@edf.fr

K. Kase
International Radiation Protection Association, 10 Place Adolphe Max, Paris 75009, France
e-mail: kr.kase@alumni.stanford.edu

L. Lau and K.-H. Ng (eds.), *Radiological Safety and Quality: Paradigms in Leadership and Innovation*, DOI 10.1007/978-94-007-7256-4_15,
© Springer Science+Business Media Dordrecht 2014

Keywords Behavior • Culture • Employees • Framework • IRPA • Leadership
• Management • Protection • Radiation • Regulatory • Safety • Staff • Stakeholders
• Traits • Values

1 Introduction

Often, the relationship between an organization's culture and the organization's safety culture is not fully appreciated by safety professionals[1] who, historically, rely on compliance as their primary tool to gain necessary support. A key obstacle to successful implementation of safety as a basic value has been the failure of safety professionals to consider the various cultures of the organization: *the founding culture, the workforce culture and the culture of the safety staff*. Embedding safety within the organization's existing cultures is the most effective way of delivering the performance to which we aspire.

In 2008, at the International Radiation Protection Association 12 Congress, the IRPA began looking for the mechanisms to actively involve radiation protection (RP) professionals[1] in discussions regarding the advancement of a RP culture, by defining the characteristics of a RP culture, assessing an organization's current culture against these characteristics and determining the successful features after achieving a high performing RP culture. Three open stakeholder meetings were held from 2009 to 2011 in Europe, Asia and the United States to facilitate this discussion [12]. This initiative by IRPA represents leadership and innovation in the development and integration of RP into an overall safety culture. Some of the ideas presented in the "*Draft Principles for Establishing a Radiation Protection Culture*" and the outcomes of those stakeholder meetings are presented here.

2 Organizational Culture

2.1 Culture Defined

Culture is a phenomenon that surrounds us at all times, being constantly enacted and created by our interactions with each other. Definitions of culture found in literature have in common the view that culture, as a central value in society, consists of practices that organizational members share about appropriate behavior.

---

[1] "Safety professionals" are well-educated, experienced persons who recognize, evaluate and control risks to people, property and the environment from a variety of hazards that may include radiation. "Radiation protection professionals" are well-educated, experienced persons who recognize, evaluate, and control radiation-related risks to people, property and the environment. In a healthcare facility, this would include methods to prevent, control, or reduce the exposure of staff, patients, and the public [11].

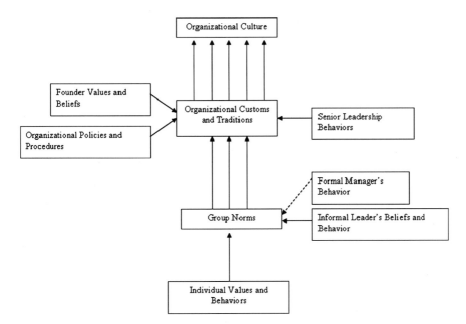

Fig. 15.1 Factors contributing to organizational culture [23]

It is that complex whole, which includes knowledge, beliefs, art, morals, law, customs, values, symbols, rituals, and any other capabilities and habits acquired by people as members of society that determines appropriate attitudes and behavior.

There is some consensus that organizational culture manifests itself in the features of organizational life. Many suggest that culture is the "way things are done around here" or that culture at a basic level is simply a common way of thinking that drives a common way of acting. Culture can be considered a system of endurance and continuation (education), transfer of knowledge and expertise to the next generation. But it's also a combination of conservation and innovation accepted by the group [4, 20–23, 29].

Schein [22] stated that an organization's culture comes from three sources: (1) beliefs, values, and assumptions of the founders of an organization, (2) learning experiences of group members as the organization evolves, and (3) beliefs, values, and assumptions brought in by new members and leaders. The most important source of culture is the impact of the founders. Figure 15.1 depicts the interaction of these factors and their place in the formation of an organizational culture [23].

Organizational culture, therefore, is the pattern of basic assumptions invented, discovered or developed by a group who have shared significant problems, solved them, and observed the effects of their solutions. If the solutions have worked well enough, they then are considered valid. Once considered valid, the assumptions are taught to new members as the correct way to perceive, think, and feel in relation to those problems. The longer we live in a given culture and the older the culture, the more it will influence our perceptions, thoughts, values, feelings and behavior.

2.2   Cultural Influence

Institutions are being challenged to look at their operations and to find more efficient ways of doing business. Forces include competition, customer satisfaction, perceived value, market share and the need to remain profitable. Organizational structure institutionalizes how people interact with each other, how communication flows and how power relationships are defined. It also reflects the value-based choices made by the institution [25].

Corporate culture is believed to be key to long-lasting, positive change and high performance. The real culture in some organizations is far removed from the desired culture and can be damaging; employees who do not share the organizational values may jeopardize the performance of the institution. A poor corporate culture fosters poor employee morale, which leads to low profits and inadequate safety performance. If an organization is not performing to expectations, chances are that real, as opposed to desired, corporate culture is subversively working in the background, blocking even the best strategies and efforts [17].

Most large organizations have a dominant culture and numerous sets of subcultures [9]. A dominant culture expresses the core values that are shared by a majority of the organization's members. It is this shared meaning aspect of culture that makes it such a potent device for guiding and shaping behavior [20]. A culture can convey a sense of identity for the organization's members and facilitate commitment to something larger than one's individual self-interest. Culture increases the consistency of employee behavior; a positive safety culture influences how workers behave in the workplace [6, 29].

Any effort to change behaviors and systems in an organization demands a look into the corporate culture to find hidden rules that govern employee behavior. If the organization learns that the hidden culture sends the wrong message, it must work to change that culture. One of the culture sources mentioned by Schein includes what he believed to be the primary basis for organizational culture: *underlying assumptions* [22].

Basic underlying assumptions of group members are the invisible but identifiable reasons why group members perceive, think, and feel the way they do. These assumptions impact workers' methods of problem solving and how relationships are created. Problems in organizations come from cumulative results of basic underlying assumptions and the fact that behaviors are often widely separated in time and space from results. The problem, simply put, is that people behave according to the immediate consequences they experience and, if negative results are far into the future, the [negative] behavior becomes part of the underlying assumption, "It must be okay because nothing bad has happened." However, when people operate from true commitment, they feel a profound personal ownership and responsibility for the success of the organization and for accomplishing its strategic direction. The cultures of an organization, as well as the larger societal context within which the organization exists, create the influential environment for individual values.

3   Safety and Organizational Success

3.1   Regulatory Environment

The field of safety is highly regulatory. In addition to its regulatory responsibility, an organization may also find their safety programs under scrutiny by accrediting agencies. The dominant paradigm in safety has been and continues to be enforcing compliance [5, 21].

In the United States, safety has passed through several eras since the early 1900s [21]:

- The inspection era. This is likely due to the obvious poor working conditions present at the time (1912–1933).
- The unsafe act and condition era. This led to two approaches: addressing work conditions and training employees in safe working habits.
- The industrial hygiene era. This led to the control of occupational illnesses: inspecting and improving physical conditions, focusing on worker behavior and improving environmental conditions that lead to disease.
- The noise era. Primarily due to the first hearing loss claim being accepted [in court].
- In the 1950s and 1960s safety was 'managed' with quality control, statistical sampling, etc.
- In the 1970s, improvements in safety were at a standstill with technical, scientific people analyzing outcomes. The idea that workers play an important role in safety had still not been recognized. Also in the 1970s, the Occupational Safety and Health Agency (OSHA) passed safety legislation focusing even more attention on documentation, inspection and control of the physical environment. This led to a further downgrade of the human role and pushed safety management back into the 'inspection era.'

Organizations typically first come to terms with the need for a safety program because of regulatory requirements. Safety professionals are then hired for the ultimate goal of compliance or, possibly for some institutions, to show regulators that steps are being taken to follow regulations. This often is the basis from which both organizational leadership and safety professionals work: to keep the institution free from fines, reduce legal liability and maintain some form of compliance.

3.2   Safety Culture and the Leadership Role

The concept of a RP safety culture developed primarily from the Chernobyl nuclear reactor accident. The results of an investigation into what occurred at Chernobyl led the International Safety Advisory Group to coin the phrase 'safety culture' because of the lack of worker concern about safety matters and shortcomings in the organizational culture [1, 3, 14, 15]. This, then, led to the International Atomic Energy Agency lending an initial broad definition of safety culture as "*that assembly of characteristics and attitudes in organizations and individuals which establish that, as an overriding priority, nuclear plant safety issues receive the attention warranted by their significance*" [10].

Others have defined safety culture as "*individuals guided in their behavior by their joint belief in the importance of safety, and their shared understanding that every member willingly upholds the group safety norms and will support other members to do the same*" [4, 13, 16]. In a self-sustaining positive safety culture, employees not only feel responsible for their own safety, they feel responsible for their peers' safety and the organizational culture supports them acting on that responsibility. Safety culture is a subculture within the overall organizational culture and is strongly influenced by the overarching organizational culture [6, 9, 27].

Like corporate culture, there are visible and invisible components in a safety culture. The visible components are the leadership, symbols, rituals, values, stories and myths. The invisible components are the norms and assumptions and these, more than policy and procedures, determine safety performance [19]. It ultimately is the workers' value of safety for themselves and the organization's prioritization of safety that dominates safety performance [13]. This led to the U.S. Nuclear Regulatory Commission's (NRC) new definition for nuclear safety culture: "*Nuclear Safety Culture is the core values and behaviors resulting from a collective commitment by leaders and individuals to emphasize safety over competing goals (e.g., production vs. safety, schedule vs. safety, and cost of the effort vs. safety) to ensure protection of people and the environment*" [28].

Findings from surveys suggest that management actions and supervisory support may have the largest impact on positive safety performance [6, 21]. Any safety effort is doomed to failure unless management leads and supports the effort [7]. If actual experiences shatter an employee's confidence in the organization's concern for safety, many employees shift from a 'want to' attitude to a 'have to' attitude regarding safety policies. This attitude usually prompts people to violate or abide by the policies depending on who's watching. The more widespread the 'have to' attitude, the more likely it will become a cultural trait. In a 'have to' safety culture, employees believe safety is the responsibility of the organization [26].

However, all staff can be directed towards an operational focus, and more specifically, ongoing reliability, human performance, and organizational effectiveness. This will lead to the development of a "field culture" in addition to the "science, engineering or medical culture" to anticipate problems and to obtain the commitment of all employees. Safety culture is a learned way of life. It is achieved only when it is instilled and practiced by top management and top management is trusted by the workforce [26]. Safety culture must be an ongoing dialogue among safety professionals, organizational management and the workforce, and between the organization and all relevant stakeholders (regulators, vendors, patients, public, etc.). Leaders play a key role through their presence in the field to coach workers and focus all staff on the operational Safety (and Radiation Protection) Culture. The values of an organization are based on what is important to its leaders. If the leaders

Table 15.1 Traits of various types of safety cultures (Adapted from [18])

Pathological	Reactive	Calculative	Proactive	Generative
Compliance but little else	Worry about costs	Focus on current problems	Benchmark and adapt	Benchmark and involve all organizational levels
Audit after accidents	Put a quick solution in place with no follow-up	Regular audits of known hazard areas	Audits are positive and provide help	Continuous informal search for non-obvious issues
No safety planning	Safety planning based on past issues	Emphasis on hazard analysis	Planning is standard practice	Planning based on anticipation of problems and review of process
Training is necessary evil	Training as consequence of accident	Testing of knowledge	Ongoing on-the-job assessments	Development is a process not an event
Punishment for failure	Disincentives for poor performance	Lip service for positive safety performance	Some rewards for safe behavior	Strong safety performance is in itself rewarding
Employee fired after accident	Accident reports not forwarded	Management does not want to hear about accidents	Management disappointed in accident	Top management seen on the floor after an accident to make sure workers are okay
Safety costs money	Can afford preventive maintenance	Safety and profitability juggled not balanced	Money counts but safety is right up there	A safe environment makes money

emphasize safety, those who report to them will build structures that are likely to perpetuate behaviors that satisfy the leader's emphasis. As these behaviors spread, the culture evolves to take on this flavor.

3.3   Safety and Organizational Success

Research has indicated that within organizations, a better safety culture directly correlates to positive safety performance. Because safety is important to the welfare of staff and makes good business sense, a positive safety culture is essential to a successful organization [4]. A belief that there is a positive link between safety and the bottom line can initiate a process causing safety to become a basic value [5]. Parker and colleagues offered their view on a topology suggesting five levels of a safety culture [18]. Table 15.1 shares some of the traits within these five levels.

A generative safety program status requires a profound understanding by management and employees of the relevant safety culture principles, methods and tools needed. All companies have safety values, whether or not they have been identified.

Table 15.2 Key personal and organizational traits of a radiation protection safety culture (Adapted from [28])

Leadership safety values and actions	Problem identification and resolution	Personal accountability
Leaders demonstrate commitment to safety in their decisions and behaviors	Issues potentially impacting safety are promptly identified, evaluated, addressed and corrected commensurate with their significance	All individuals take personal responsibility for safety
Work processes	**Continuous learning**	**Environment for raising concerns**
Planning and controlling work activities is implemented so safety is maintained or enhanced	Opportunities to learn safety methodologies are sought out and implemented	Personnel feel free to raise safety concerns without fear of retaliation, intimidation, harassment or discrimination
Effective safety communication	**Respectful work environment**	**Questioning attitude**
Communications focus on safety, because the local and regional contexts are part of the narrative; the narrative must be in a common language, understood by all	Trust and respect permeate the organization	Individuals continually challenge the existing conditions and activities so discrepancies that might result in error or inappropriate action are identified

Although, basic safety values don't need to be formally identified and documented, no organization can achieve positive and significant safety results unless such values are in place [5]. Safety excellence is not the result of one singular strategy. Generic solutions or universal answers cannot be given because no one best way exists. Safety excellence is the outcome of a strategy continuum – one that addresses an organization's regulatory, technical, engineering, organizational, behavioral, managerial and cultural loss sources [8].

The United States Nuclear Regulatory Commission [28] has proposed nine traits to develop safety culture (Table 15.2). It should be noted that these traits are not necessarily subject to inspection and were not developed for that purpose. Experience has shown that certain traits are present in a positive safety culture. A trait, in this case, is a pattern of thinking, feeling, and behaving that emphasizes safety, particularly in goal conflict situations, e.g., production vs. safety, schedule vs. safety, and cost of the effort vs. safety.

3.4 Safety and Employee Behavior

Employees may think of safety in terms of limits and not getting caught. This thinking tends to generate personal injunctions of restraint – the shall-nots. Such reasoning is fine insofar as it provides the brakes for certain obvious kinds of misbehavior such as falsifying records or willfully disobeying laws. But employees also need intellectual tools for more positive thinking to open up recognition of the safest choices. A framework and psychological attitude have to be adopted that puts safety interests first in order to transform the attributes of excellence into productive actions. Basic values guide the personal behavior of individual group members and influence their decisions regarding safety. If a group believes job hazards are not significant or injuries are an expected part of the job, group members are less likely to develop safety as a basic value. If a group believes hazards are significant then they will be more rigorous about their safety practices. Where safety is a basic value, workers understand what is expected of them and spend little time deciding whether they should perform safely [5].

Shaping a safety culture and shaping employee perceptions go hand in hand. Because short cuts may be attempted and can often be taken without any immediate risk, workers need to have the value of safe behavior demonstrated to them; a demonstration of the value of the extra time spent doing the task correctly and safely. The manner in which the message of safety is demonstrated to employees is significant; it can cause a positive or defensive reaction. If the message is sent that safety is important, but there is only a weak regulatory compliance system in place and leadership isn't modeling the behavior, the message received – whether intended or not – is that the organization doesn't really care and doesn't feel responsible for worker injuries [6, 18, 26]. Research on modeling shows that leaders must exhibit the desired safety behavior to maximize worker acceptance. Modeling correct safety behavior is an ideal way to lead the safety effort [7].

In his book *"The Fifth Discipline"*, Senge [24] describes five disciplines for a learning organization: systems thinking, shared vision, mental models, personal mastery and team learning. Two of these disciplines, in particular, have direct application to implementation of a positive safety culture. The first discipline, *"systems thinking"*, includes policy resistance, leverage and the idea that structure influences behavior. Systems thinking means looking at all influences on the task at hand not just a focus on the task itself. For the safety profession, this translates to a global view of the organization. With what is the organization dealing? What influences the organization's decisions? Within this global view, some details can be added. Which of the influences can safety impact? What are those impacts – good and bad? Where are the safety 'levers'? Systems thinking for safety professionals will not only influence the ability to find the best and longest lasting solution but can also put this group alongside counterparts who often have the ear of the CEO. If safety professionals use global business impacts to market safety and a safety culture, they are much more likely to be looked upon as a strategic business partner rather than, simply, the cost of compliance.

The third discipline, *"mental models"*, describes our deeply held internal images of how the world works. These images limit us to familiar ways of thinking and acting. They are our assumptions, theories and simple generalizations. Our mental models influence what we see and how we interpret events. Failure to see our mental models and to examine why they exist, constrains our ability to find the real problem and define appropriate solutions. If we try to examine our own mental models, we often rationalize their correctness; if someone else begins to examine or question our mental models, we often become defensive to insulate them from being scrutinized. To suggest to safety professionals that compliance with regulations need not be the ultimate goal is the equivalent of telling them there is no longer a reason to go to work. For many, compliance is their reason for working.

4 Implementation Framework for an RP Culture

4.1 Strategies

Organizations with a record of effective safety practices are better able to invest in people than those who must divert funds into such nonproductive costs as medical claims, environmental cleanups and safety related litigation [2]. Learning organization disciplines can be applied to create a RP safety culture. Application of these principles is the foundation to making safety a basic value of the organization. A framework to begin such a process is presented in the following bullets:

- The vision. Concern for the welfare of staff and consumers are the purpose of a radiation safety program. Compliance may be a portion of the overall strategy, but compliance cannot be the vision; compliance will not create nor maintain a safety culture. Compliance is not inspirational and alone is not enough to motivate employees.
- Ownership. All staff must believe that their behavior can impact their own protection and the protection of their co-workers from injury and illness. RP professionals must take responsibility for RP programs delegated to them by senior leadership. RP professionals are stewards of the RP culture and every worker is individually responsible for maximizing protection and minimizing risk in the work environment.
- Inspirational leadership. To implement and maintain an RP culture, organizational leaders assisted by all staff, must articulate what is missing and the changes that are required to embed RP into their organization. There must be a shared vision of RP and RP must be a top priority and a basic value. The behaviors of leaders acting by example of this new RP safety culture are necessary.
- Continuous thinking and learning. Thinking about the organization's objectives, protection of workers and safe work behaviors, and their relationship must be stimulated at all levels. Stories about successful solving of RP issues must be acknowledged and passed on.
- Mentoring and coaching. Key leaders who firmly believe in the nature of RP as a strategic business partner need to be available to provide mentoring and coaching to future leaders.
- Individually and collectively. To successfully achieve a culture of RP and safety, there must be a blend of individualism to allow for creativity and innovation and collectivism for the implementation of interdependent solutions to complex problems.

- Reward. Changes must be introduced into daily practices following adequate education to foster safety within the learning organization. Individuals and departments must be praised for their RP efforts rather than rewarded for short-term, unsafe gains in productivity.
- Management by exception. Wherever possible, leaders at all levels must be encouraged to incorporate RP and safety into their decisions. Leaders use active management by exception – observing behavior and actively performing one-on-one intervention for unsafe behaviors.
- Bureaucracies rooted out. The structure of the organization must be flexible, encourage openness and provide opportunity. Safe work practices must be built into all operations and activities. There can be no respect for divisional boundaries. Procedures and policies that reinforce unsafe behaviors must be rescinded or rewritten. Leaders who continue to support unsafe practices through actions such as rewards for short-term productivity gains must be worked out of the organization.
- Linked to success. A safety culture needs to be intertwined and aligned with organizational and personal success. Safety outcomes must be clearly and explicitly linked with organizational needs and strategic goals, which in turn are related to job behaviors, job skills and job success.

The leadership exhibited by managers or team leaders through their presence in the workplace, staff coaching, and the reinforcement of operational focus will contribute to ongoing reliability, human performance, and organizational effectiveness. It will help develop a "field culture" in addition to the existing "engineering culture" to anticipate problems and to obtain the commitment of all staff. Figure 15.2 shows how a RP safety culture might fit within a healthcare organization.

4.2   Practices: Actions for Radiation Protection Professionals

There are potential obstacles at the onset of fully implementing a RP safety culture. Some of the more common include agreeing on a common definition, concepts and parameters for RP safety culture; overcoming the viewpoint that RP and safety are the responsibilities of the RP professionals; and disagreement over levels of risk posed by certain activities. However, when performed regularly, we believe some practices will mesh the RP safety culture into the institutional culture. As stewards of the RP safety culture, RP professionals must foster the collective performance of these practices because they are highly dependent upon each other. The following are examples of useful actions for RP professionals:

- Maintain a high level of technical competency within the RP program. Individuals who run the organization are expected to be highly proficient in their area of expertise. Safety professionals are no different. Being certified in the applicable safety field and keeping abreast of trends and best practices are expected.

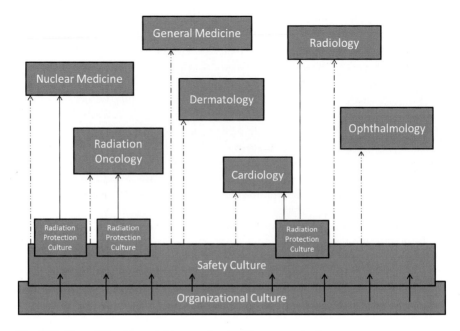

Fig. 15.2 How a RP culture might appear in a healthcare organization

- Know the pressures the organization is facing. Talk to organizational leaders. With what are they dealing on a day-to-day basis? What do they think the future will bring to the organization? Determine how the safety program can lessen those pressures.
- Talk the talk. Organizational leaders tune out compliance talk. Talk in financial terms to the Chief Financial Officer. Talk in recruitment and retention terms to the Human Resources Department Chair. Talk in terms of increased productivity and decreased costs to the Chief Operating Officer. Radiation protection and safety has an impact in each of these areas.
- Know the global RP safety culture picture. Few safety professionals realize how safety and RP can be the foundation for nearly every activity at the institution. Assurance of protection from hazards and a safe work environment can increase job satisfaction, increase productivity, increase retention and reduce recruitment efforts. Safety is important not only in the work environment, but also in the design of products, equipment and facilities, and in the provision of services. Use these examples when talking the talk.
- Find RP advocates. It may not be obvious at times, but there are formal and informal leaders within the organization who are RP and safety advocates. Building a network of these advocates, working with them on RP issues and showing support for them in their efforts will ultimately turn into support for a RP safety culture.
- Maintain compliance. RP programs should achieve more than the minimum level of compliance; however, achieving compliance is not the only goal.

Go beyond minimal compliance where there is obvious value. Benchmark best practices but don't just follow them – make them yours and make them better where they add value.

- Identify problems within the RP safety culture, don't gloss over them and don't hide the truth. Carefully decide what might be done and with whom to discuss those plans. Use the global view.

- Be proactive. As hard as it is to get the day-to-day tasks done, the future is always challenging for an organization to do better. By having conversations with leaders about the future of the organization, RP professionals can decide how best to position themselves for future impacts. Not only must RP professionals look at organizational pressures but they must be well prepared to respond to proposed regulations and to determine how to implement them with maximum value for and minimal impact on the institution.

- Be a mentor. RP professionals must be mentored into a culture of protection and mentored out of compliance thinking. RP leadership must share the global picture, explain why organizational pressures must be considered even in the RP department and encourage opportunities to participate in the creation of a [new] RP safety culture. Organizational leadership must also be mentored into a culture of protection and out of believing safety is necessary only because of regulations.

- Build a compelling vision. What will the RP safety culture look like in 10 years? Will it be a strategic business partner in an organization where safety is an integral part of work? Will it be incorporated into all aspects of the decision making process? The organization must be able to "see" and "live" the RP vision. Workers must believe they play a key role in advancing RP and safety at their organization. Compliance is not a compelling vision.

- RP professionals and professional societies must participate in, contribute to and foster an environment that promotes dialogue and disseminates information among the profession, other stakeholders, institutions, organizations, companies (private and public) and society as a whole.

5 Summary

The development of a RP safety culture gives visibility to the fundamentals of safety (science and values); promotes radiation risk awareness; shares responsibility among practitioners, operators, management and regulators; maintains the safety heritage for the next generation of workers; and improves the quality and effectiveness of a safety program.

These practices can be combined with daily tasks eventually evolving into the strategies outlined earlier. All of the outlined practices and strategies have as their basis Senge's five disciplines. To change established culture within an organization, the founding culture must be well understood. Thus, the first premise regarding organizational safety culture is that safety professionals must be aware of and understand their organization's founding culture and the impact it is having in today's operations. Also important is recognition of the safety culture within various workgroups to develop a program for identifying and resolving problems (corrective action program) which would provide and to permit independent external audits (transparency).

To advance a philosophy or certain beliefs, there must be a proponent group with a high stake in the outcome. We suggest that it is the RP professionals who have the highest stake in a RP safety culture.

RP professionals focus on RP issues in their daily work and understand what is at stake (workers' livelihoods, profits, etc.). They must use this information to prompt organizational leaders and co-workers toward a positive safety culture. This is only possible in a safety conscious work environment where personnel feel free to raise safety concerns without fear of retaliation, intimidation, harassment or discrimination [28].

References

1. Antonsen S (2009) Safety culture and the issue of power. Safety Sci 47:183–191
2. Blair E, Geller ES (2000) Becoming world class in HSE management. Occup Health Saf 69(9):61–79
3. Choudhry RM, Fang D, Mohamed S (2007) The nature of safety culture: a survey of the state-of-the-art. Safety Sci 45:993–1012
4. Dalling I (1997) Understanding and assessing safety culture. J Rad Prot 17:261–274
5. Earnest RE (2000) Making safety a basic value. Prof Saf 45:33–38
6. Fernandez-Muniz B, Montes-Peon JM, Vazquez-Ordas CJ (2007) Safety culture: analysis of the causal relationships between its key dimensions. J Safety Res 38(6):627–641
7. Grant JG (2000) Involving the total organization. Occup Health Saf 69(9):64–65
8. Hansen LL (2000) The architecture of safety excellence. Prof Saf 45:26–29
9. Haukelid K (2008) Theories of (safety) culture revisited – an anthropological approach. Safety Sci 46(3):413–426
10. International Atomic Energy Agency (1991) Safety culture. Safety Series No. 75-INSAG-4. IAEA, Vienna
11. International Atomic Energy Agency (2011) Radiation protection and safety of radiation sources: international basic safety standards. Interim edition. IAEA Safety Standards Series GSR Part 3 (Interim). http://www-pub.iaea.org/MTCD/Publications/PDF/p1531interim_web.pdf. Accessed 28 Mar 2012
12. International Radiation Protection Association (2012) Radiation protection culture. http://www.irpa.net/page.asp?id=179. Accessed 22 July 2013
13. Mansdorf Z (1999) Organizational culture and safety performance. Occup Hazards 61(5):109–116
14. Martinez-Corcoles M, Gracia F, Tomas I et al (2011) Leadership and employees' perceived safety behaviours in a nuclear power plant: a structural equation model. Safety Sci 49(8–9):1118–1129
15. Mengolini A, Debarberis L (2007) Safety culture enhancement through the implementation of IAEA guidelines. Reliab Eng Syst Safety 92(4):520–529
16. Merritt AC, Helmreich RL (1996) Creating and sustaining a safety culture: some practical strategies. CRM Advocate 96(1):8–12
17. O'Malley J (2000) How to create a winning corporate culture. Birm Bus J 17:22–26

18. Parker D, Lawrie M, Hudson P (2006) A framework for understanding the development of organizational safety culture. Safety Sci 44(6):551–562
19. Reiman T, Rollenhagen C (2011) Human and organization biases affecting the management of safety. Reliab Eng Syst Safety 96(10):1263–1274
20. Robbins SP (1994) Essentials of organizational behavior, 4th edn. Prentice Hall, Englewood Cliff
21. Sarkus DJ (2001) Safety and psychology: where do we go from here? Prof Saf 46:18–25
22. Schein EH (1992) Organizational culture and leadership. Jossey-Bass, San Francisco
23. Schein EH (1995) Coming to a new awareness of organizational culture. In: Kolb DA, Osland JS, Rubin IM (eds) The organizational behavior reader, 6th edn. Prentice Hall, Englewood Cliff
24. Senge PM (1990) The fifth discipline. Currency Doubleday, New York
25. Tata J, Prasad S, Thorn R (1999) The influence of organizational structure on the effectiveness of TQM programs. J Manage Issues 11:440–454
26. Torner M (2011) The "social-physiology" of safety. An integrative approach to understanding organizational psychological mechanisms behind safety performance. Safety Sci 49(8–9):1262–1269
27. Wu TC, Lin CH, Shiau SY (2010) Predicting safety culture: the roles of employer, operations manager and safety professional. J Safety Res 41(5):423–431
28. United States Nuclear Regulatory Commission (2010) Safety culture. http://www.nrc.gov/about-nrc/regulatory/enforcement/safety-culture.html. Accessed 20 Sept 2011
29. Williamson AM, Feyer A, Cairns D et al (1997) The development of a measure of safety climate: the role of safety perceptions and attitudes. Safety Sci 25(1–3):15–27

Chapter 16
Improving Quality Through Lean Process Improvement: The Example of Clinical Decision Support

Lucy Glenn and C. Craig Blackmore

Abstract The Institute of Medicine (IOM) defines quality in health care as being safe, timely, effective, efficient, patient-centered and equitable. Quality within Radiology, based on the IOM definition, can be broadly defined as safe, appropriate, effective and patient-centered. The key to actual quality improvement is to have a system in place that will both enable and sustain change. At Virginia Mason, we have adopted the Lean Manufacturing method to health care as our process improvement method for continuously improving the quality of care across the institution, including radiology. The major principles of the Lean process improvement methodology include the concepts of value, value streams, flow, pull and perfection. One of the basic tools of Lean is the concept of zero defects and mistake proofing. We have applied Lean throughout radiology to improve quality. One of the most important factors in the quality equation is appropriateness. In the rest of this chapter we will describe how we have used our Lean process improvement methods to ensure the appropriate use of imaging. There are two different imaging management methods that are currently employed, clinical decision support and preauthorization. These two methodologies are explored utilizing the Lean lens. Our implementation of clinical decision support to improve the appropriateness of imaging is an example of how the Lean process contributes to quality, safety and patient-centered radiology.

Keywords Appropriateness • Appropriate imaging • Clinical decision support • Flow • Lean in health care • Lean thinking • Mistakes vs. defects • Mistake proofing • Preauthorization • Process improvement • Pull • Quality • Six sigma • Value • Value streams

L. Glenn (✉) • C.C. Blackmore
Department of Radiology, Virginia Mason Medical Center, 1100 Ninth Avenue, Seattle, WA 98111, USA
e-mail: lucy.glenn@vmmc.org; craig.blackmore@vmmc.org

L. Lau and K.-H. Ng (eds.), *Radiological Safety and Quality: Paradigms in Leadership and Innovation*, DOI 10.1007/978-94-007-7256-4_16,
© Springer Science+Business Media Dordrecht 2014

1   Definition of Quality

The Institute of Medicine (IOM) defines quality in health care by six criteria. It should be safe, timely, efficient, effective, patient-centered, and equitable. Their seminal reports in 1999 "To Err is Human" and 2001 "Crossing the Quality Chasm" have stimulated the quality movement in health care in the United States [8, 9]. These same principles can be applied to defining quality within Radiology.

Quality within radiology can be divided into 4 broad categories that are based loosely on the IOM definition. These are safety, appropriateness, effectiveness, and patient centricity [2, 12, 19]. Safety refers to both the environment in which the imaging studies are performed and the technique. Environment means that the radiology department should be a safe place. The patient should expect not to fall, that the providers would practice appropriate hand hygiene, that a nosocomial infection would not be acquired, and that the appropriate resources would be available should any untoward event occur. In addition, the imaging technique should be safe, radiation dose should be minimized, interventional procedures should be performed using appropriate technique, and patients should be screened and monitored regarding the use of any medications and contrast agents.

Imaging should also be appropriate. This means performing only those imaging studies that have the potential to improve patient outcome through changes in therapy. Appropriateness also encompasses ensuring that examinations that are performed are the best for the clinical situation.

Imaging should be effective, meaning both that the interpretation should be performed at the highest professional standard, and that communication should occur in a timely manner, allowing the imaging results to affect the patient's care. Finally, radiology should be patient-centered, indicating that the focus is on the patient both in terms of the experience of radiology care and the timeliness.

Having defined quality and recognized the need for quality improvement, however, is only the starting point. Often the greater challenge, and the greater need for leadership and innovation, is in affecting the changes in processes of care that will lead to improvements in quality across all domains. The key to actual quality improvement is to have a system in place that will both enable and sustain change. At Virginia Mason (VM), we have adopted the Lean Manufacturing method to health care as our process improvement method for continuously improving the quality of care across the institution, including radiology.

In this chapter, we will focus on Lean process improvement to define the pathway to achieving quality, safety and patient-centered radiology. We provide an extended example of applying this process improvement to quality through application in imaging appropriateness.

2   Process Improvement

There are many different methodologies that can be used in process improvement. Two proven strategies are Six Sigma and Lean. These have been used extensively in the manufacturing sector but only recently have they been applied in health care [4, 15].

Sigma refers to the Greek letter symbol for standard deviation. It is a measure of variation in a process. The six sigma goal is to have so little variation in a process that only at six standard deviations away from the mean will one encounter a defect. The objective is high performance, reliability and value to the end customer. There are five phases of six sigma that are referred to as "DMAIC".

1. **Define** – identify the problem and the desired process improvement goals.
2. **Measure** – map and measure the process and collect data about the current state.
3. **Analyze** – root cause analysis of the issues.
4. **Improve** – implement steps to improve the process based on the analysis.
5. **Control** – measure the new process and set up controls to ensure variances are corrected before they result in defects.

Six sigma is problem-focused with a view that variation is waste. Six sigma uses statistics to understand variation.

Lean Thinking, on the other hand, is focused on process flow and views any activity that does not add value as waste. Lean uses visual tools such as process mapping, value stream mapping, and flow charting to understand the process. There are five core concepts in Lean Thinking [20].

1. **Value** – value is defined by the customer.
2. **Value stream** – is a process flow map with areas of waste identified.
3. **Flow** – make value flow at the pull of the customer.
4. **Pull** – the downstream process must pull from the upstream process in order to create continuous flow.
5. **Perfection** – continuously improve in the pursuit of perfection.

Lean principles include: (1) a quest for zero defects with perfect first-time quality; (2) waste minimization, i.e. eliminating all activities that do not add value; and (3) one piece flow without batching.

Applying Lean principles begins with observing a process and creating a value stream map. The map diagrams and times each step in the process. At each step in the process the value added portions are identified and the waste is also explicitly called out. The process improvements are generated by the people working within the process. The front line staffs identify the issues and the solutions that are to be tried. Once the solutions are implemented, the new process is re-measured and if successful, the new standard work is created. Standardization is a key element to high reliability and quality [15].

At VM, standard work has been created for all sections of Radiology and for all staff, including radiologists. Standardization of protocols for all exams is one example of standard work. For example, all CT exams have a standard protocol unless there is a clinical exception requiring deviation from the standard. Communication of results is another example of standard work that has been created. Clear standards for communicating critical results, as well as non-urgent but clinically important results have been established, written and agreed to by all the radiologists.

3 Quality and Lean

Embedded within the Lean methodology is quality. A perfect process creates value in every step with no waste. It produces a good result every time, i.e. there is high reliability. It is available when you need it, i.e. there is open access. And the entire process is linked by continuous flow without the waste of waiting between steps.

Human factors engineering tells us that when humans are asked to repetitively perform a task there is an inevitable 2–3 % error rate. Therefore it should be expected that humans would make mistakes. Correcting these mistakes as close to the source as possible will lead to the least harm. Mistakes are easiest to fix and least harmful the closer you get to the time and place they arise. Defects are mistakes that were never fixed and the error is picked up by the customer or patient. If mistakes are fixed at their source it is possible to have a zero defect product.

Mistake-proofing is a process to correct errors before they cause harm to the patient. There are various levels of mistake-proofing. The most basic levels involve inspection. In level 1, the customer inspects and finds the error. This is obviously the least desirable form of inspection. In level 2, the successive check has taken place in the downstream department. This is slightly better than having the patient discover the mistake, but the further down the line the mistake gets propagated, the greater potential for harm and the greater the cost in terms of time and manpower to fix the mistake. In level 3, the successive check takes place within the work unit and the error is corrected more quickly. An example of this would be when a radiologist discovers that the technologist has mislabeled an image.

In level 4, there is self-check, rather than successive check, so the mistake-proofing takes place at the time of occurrence. Mistakes get corrected before they are passed down the line to the next step in the process. Checklists are a common form of self-check. Doing a procedural pause before doing an interventional procedure is an example of self-checking. Another example is using two patient identifiers to be sure the correct exam is performed on the correct patient.

In level 5, mistakes are prevented from happening in the first place. This is the ultimate solution to prevent mistakes, when restrictions are in place making it impossible to make the mistake. An example of this in health care would be the anesthesia gas hoses in the OR. Each different gas has a different color and a different connector so it is impossible to mistakenly connect the nitrous oxide to the oxygen delivery. Another example in radiology is having the PACS integrated with the voice recognition dictation system so that the report automatically opens to the patient whose images are being displayed, therefore making it impossible to dictate a report on the wrong patient because one typed in the wrong patient's ID.

4 Application of Lean to Health Care

In 2002, the executive team of Virginia Mason including the CEO, president, vice presidents and department chairs all went to Japan to learn about the Toyota Production System and Lean management. We came away from the trip convinced that the

Lean management system that had been so successful in the manufacturing industry had applicability to the health care industry. All the senior leaders went through several months of training to become Lean certified. This is the management method we have employed for the past 10 years and it has resulted in marked improvements in quality, safety, patient-centered care and our financial performance [11].

Part of the success of this Lean management process improvement approach has been the involvement of the entire organization in learning this method. Every employee goes through an introductory 1 day class about the Virginia Mason Production System (VMPS). In addition, the department chairs and vice presidents are expected to become certified Lean leaders. The next level of leadership also gets training in VMPS for leaders. There is also a central office that provides process improvement experts to operational units to help with the improvement work. There are many staff that participate in the 2 day and 5 day improvement workshops that take place throughout the organization.

We have been using this method for the past 10 years. The combination of the breadth of involvement of all people in the organization, and the continued focus on Lean, has resulted in a major cultural change within the organization. One of the major goals of the organization is quality and process improvement and our Lean management process is the method to achieve that goal.

We have applied Lean throughout radiology to improve quality. One of the most important factors in the quality equation is appropriateness. In the rest of this chapter we will describe how we have used our Lean process improvement methods to ensure the appropriate use of imaging.

5 Appropriate Imaging

A major focus on improving quality specific to radiology is to eliminate inappropriate imaging. Inappropriate utilization of imaging has been identified as a major problem both in terms of quality and cost. Estimates are that 20–40 % of imaging studies performed in the United States are unnecessary or are not the optimal examination for a given patient [6, 13]. This is particularly relevant for CT scanning where volumes are rapidly rising, [14, 18] and which is now the leading source of medical radiation in United States [7]. Unfortunately, CT, particularly in the emergency department setting, is also likely heavily over utilized. There is also considerable variability in the utilization of CT scanning even across the United States. For example there is a 50 % greater chance that an individual will get a CT scan living in Atlanta, Georgia then a similar individual living in Seattle, Washington, without concomitant improvement in health [17]. Inappropriate imaging is a form of waste, not adding value to patient care, and causes adverse direct impacts from unnecessary radiation and cost, and adverse indirect effects from induction of unnecessary therapy [10]. Accordingly, we have focused our Lean quality efforts on improving appropriate utilization of imaging.

Table 16.1 Efficacy of imaging

Level of efficacy	Definition	Measures
Technical efficacy	Ability to produce a diagnostic image	Signal to noise ratio, resolution
Diagnostic accuracy efficacy	Ability to differentiate normal from abnormal	Accuracy, sensitivity, specificity, receiver operator characteristic curves
Therapeutic efficacy	Ability to change patient management	Patient management plan before and after imaging
Patient outcome efficacy	Effect on patient-centered outcomes	Morbidity, mortality, health-related quality of life, quality adjusted life years
Societal efficacy	Benefits, costs and harms for the society	Cost-benefit analysis, cost-effectiveness analysis, cost-utility analysis, population exposure

At Virginia Mason, since 2005, we have sought to apply Lean principles to the elimination of inappropriate imaging. The first step in the process is to identify "value." However, imaging does not directly influence patient outcome, but rather has an indirect effect on outcome through modification of therapy. Thus, determination of the appropriateness, or value, of imaging is not a simple question, but is predicated upon defining effectiveness at multiple levels [5].

First, the imaging study must be capable of defining the anatomic component or pathologic process, known as technical efficacy (Table 16.1). Second, the imaging study must have diagnostic accuracy efficacy, meaning that the test must be able to differentiate between normal and disease states. This is usually measured through sensitivity, specificity, or receiver operator characteristic curves. Studies determining the technical efficacy and accuracy efficacy of imaging form the bulk of the published literature. However, accuracy itself is insufficient to define value for the patient or society. In addition, the information from the imaging test must have the potential to change patient management. This is known as therapeutic efficacy and is measured through assessment of therapeutic planning before and after an imaging test is performed. Only if there is the potential to change management can an imaging test affect outcome, and therefore have value. The fourth level of efficacy is actual change in patient outcome. This is dependent on all three of the preceding levels and is measured through the patient-centered outcomes of morbidity, mortality, and quality of life. In screening studies, patient outcome can be measured directly as the screening study is the sole determinant of which individuals undergo treatment. However, for most diagnostic imaging, because all patients undergo some form of treatment, the assessment of the direct impact of imaging on outcome is challenging. Finally, the highest level of efficacy is efficacy from the perspective of society. This means that the cost in societal resources must not exceed the benefit to the individuals. This cost effectiveness estimation must include the assessment of radiation risk to the population for imaging studies that

Current State

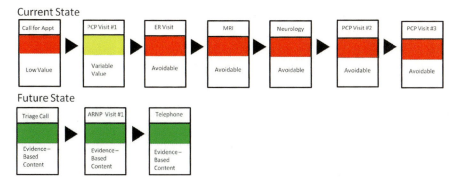

Future State

Fig. 16.1 The current state represents all the steps in the care of a patient with headache prior to development of the Lean clinical value stream. The future state is the ideal care pathway based on the elimination of all waste, only retaining those elements of the care process that add value to the patient. *Red boxes* are steps in care delivery that do not add value. *Yellow boxes* are steps of variable value. *Green boxes* represent value added steps. *ER* emergency department, *ARNP* Advanced Registered Nurse Practitioner, *PCP* primary care provider

use ionizing radiation. The benefit to an individual must be reconciled with the potential cancer induction in other members of society who may not have received the direct benefit.

To help understand if imaging adds value, we first construct Lean "value streams" for the care of individuals in specific clinical scenarios. For example, what is the entire process of care for an individual with acute uncomplicated headache? This can be defined and areas of waste identified (Fig. 16.1). For most patients with uncomplicated headache, CT or MRI does not add value, and therefore is waste and should be eliminated. Accordingly, there is need for a system to prevent ordering of unnecessary imaging, in Lean terms, to "mistake-proof" the process of image ordering.

6 Management of Appropriate Imaging

Ideally, under Lean core concepts, this imaging management system would occur without disruption to the process of patient care ("flow"), would be available when needed by the clinicians ("pull"), and would achieve "perfection" in preventing inappropriate imaging.

There are multiple approaches to imaging management, including computer decision support, preauthorization, and billing denial. In addition, imaging management systems can affect different levels of control, in a spectrum from purely advisory recommendations, to rigid rules, without exception. In understanding the system VM has employed for imaging management, it is useful to view the various approaches through the Lean lens, and evaluate whether they are based on value, value streams, flow, pull, and perfection.

6.1 Clinical Decision Support

Recently, clinical decision support systems have been hailed as an ideal mechanism for managing imaging [1]. Clinical decision support generally consists of online information integrated into the computerized physician order entry system. These systems provide information to clinicians at the time of ordering of an imaging study and so can be considered to both be in "flow," and to be "pull" systems with information available when needed. Computer decision support provides guidance regarding the appropriateness of a particular imaging study, thereby assuring "value."

Computer decision support can be used in differing ways. The information in the computer decision support system may be educational only, and non-binding to clinicians; it may be advisory, but associated with data collection for benchmarking; or it may be mandatory, as a form of "mistake-proofing" whereby inappropriate orders are blocked. Though all these approaches may lead to improvement, only the "mistake-proofing" approach, blocks all mistakes from becoming "defects" with inappropriate care being delivered. In Lean terminology, this is level 5 mistake proofing.

The final choice in computer decision support is whether to adopt a global approach or a focal approach to imaging management. Under the global approach, imaging management would be attempted for all advanced imaging studies and indications, leading to clearly tens of thousands of imaging study/indication pairs. The global system therefore has the potential of high impact in terms of reducing inappropriate imaging. However, it must be remembered that most of the indication/study pairs will be without evidence or even consensus. Unless worked through in a value stream covering the entire care process, it is difficult to establish if the imaging study is adding value. The alternative is to adopt a focal approach, whereby imaging management is limited to those areas where clinical value streams have been established based on strong evidence or at least consensus.

Starting in 2004, VM has implemented computer decision support for targeted areas where we have established value streams defining when imaging is value added. These value streams are targeted to areas of high utilization and where evidence on value of imaging exists. Currently implemented clinical conditions in our computer decision support system include headache, low back pain, sinusitis, breast concerns, and joint pain, including knee, shoulder, and hip pain. We employ *mistake-proofing* approach (level 5), whereby imaging will not occur unless an approved indication is met. The effectiveness of this system is demonstrated through a 20–25 % decrease in the rate of imaging for these selected conditions after implementation (Fig. 16.2). The clinical scenarios included in the mistake-proofing are areas with strong evidence, or at least consensus on best practice for imaging use. Hence, we are confident that the imaging we block through this system is unnecessary. Elimination of unnecessary imaging can prevent further unnecessary or even harmful care downstream; can decrease radiation exposure and decrease costs, therefore leading to an improvement in the overall quality of health care [3].

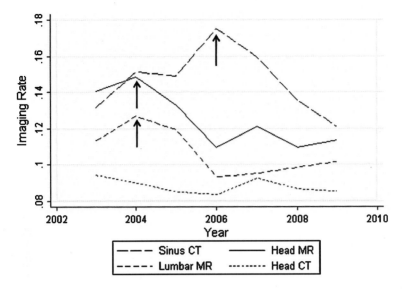

Fig. 16.2 Implementation of clinical decision support at VM occurred at differing times (*arrows*) for different Lean clinical value streams. In each case, there was a subsequent decrease in the rate of imaging for patients in these value streams. Head MR decreased by 23 %, sinus CT decreased by 27 %, and lumbar MR decreased by 23 %. Head CT serves as a control, with implementation of clinical decision support occurring subsequent to the study (Reprinted with permission from [3])

6.2 Preauthorization

There are alternatives to computer decision support for the control of imaging, specifically preauthorization schemes. Preauthorization is widespread in the United States and is currently being considered for the Medicare program that provides coverage for all individuals of age 65 and over. Preauthorization requires a phone call or interaction with a computer web portal prior to performance of imaging. This permission step generally occurs after the patient visit with a provider, hence, is not in "flow." Permission for imaging under these preauthorization schemes, are generally based on third-party proprietary criteria, which may or may not be transparent. Because the criteria for imaging are generally not available at the time clinicians attempt to order imaging studies, providers are unable to "pull" this information when needed. Finally, the preauthorization interaction is often between clerical staff of both parties, rather than providers. Hence there is no opportunity to learn, and improve toward "perfection." The systems are expensive for both payer and provider as personnel must be hired by each party to interact around the imaging coverage decision. It is also unclear if preauthorization schemes are cost-effective as the large administrative cost may outweigh the savings from decreased utilization [16].

In summary, clinical decision support implementation is an example of how Lean processes lead to improvement in quality. In the example we focus on appropriateness of imaging. However, there is overlap in aspects of quality, with the clinical decision support system also improving patient safety through decreases in unnecessary radiation. The system also addresses patient centricity, in that the outcomes of interest in developing the value streams that led to the clinical decision support system are those of importance to the patient, not simply properties of the imaging system.

7 Conclusion

Quality in radiology is the result of a process of care delivery that must be continuously improved. At Virginia Mason, we employ the Lean methodology for process improvement, with core concepts of value, streams, pull, flow, and perfection. Our implementation of clinical decision support to improve the appropriateness of imaging is an example of how the Lean process contributes to improvements in health care.

References

1. Bernardy M, Ullrich CG, Rawson JV et al (2009) Strategies for managing imaging utilization. J Am Coll Radiol 6(12):844–850
2. Blackmore CC (2007) Defining quality in radiology. J Am Coll Radiol 4(4):217–223
3. Blackmore CC, Mecklenburg RS, Kaplan GS (2011) Effectiveness of clinical decision support in controlling inappropriate imaging. J Am Coll Radiol 8(1):19–25
4. Bush RW (2007) Reducing waste in US health care systems. JAMA 297(8):871–874
5. Fryback DG, Thornbury JR (1991) The efficacy of diagnostic imaging. Med Decis Making 11(2):88–94
6. Hillman BJ (2006) Foreword. In: Medina LS, Blackmore CC (eds) Evidence based imaging: optimizing imaging for patient care. Springer, New York
7. Hricak H, Brenner DJ, Adelstein SJ et al (2011) Managing radiation use in medical imaging: a multifacted challenge. Radiology 258(3):889–905
8. Institute of Medicine (2000) To err is human: building a safer health system. National Academies Press, Washington, DC
9. Institute of Medicine (2001) Crossing the quality chasm: a new health system for the 21st century. National Academies Press, Washington, DC
10. Jarvik JG, Hollingworth W, Martin B et al (2003) Rapid magnetic resonance imaging vs radiographs for patients with low back pain: a randomized clinical trial. JAMA 289(9):2810–2818
11. Kenney C (2010) Transforming health care: Virginia Mason Medical Center's pursuit of the perfect patient experience. Productivity Press, New York
12. Lau LSW (2006) A continuum of quality in radiology. J Am Coll Radiol 3(4):233–239
13. Lehnert BE, Bree RL (2010) Analysis of appropriateness of outpatient CT and MRI referred from primary care clinics at an academic medical center: how critical is the need for improved decision support. J Am Coll Radiol 7(3):192–197

14. Levinson DR (2007) Office of the Inspector General Report: growth in advanced imaging paid under the Medicare physician fee schedule. Department of Health and Human Services, Centers for Medicare and Medicaid Services, Washington, DC
15. Miller D (ed) (2005) Innovation series 2005: going lean in health care. Institute for Healthcare Improvement, Cambridge
16. The Moran Company (2011) Assessing the budgetary implications of alternative strategies to influence utilization of diagnostic imaging services. The Moran Company, Washington, DC
17. Parker L, Levin DC, Frangos A et al (2010) Geographic variation in the utilization of noninvasive diagnostic imaging: national Medicare data, 1998–2007. AJR 194(4):1034–1039
18. Smith-Bindman R, Miglioretti DL, Larson EB (2008) Rising use of diagnostic medical imaging in a large integrated health system. Health Aff (Millwood) 27(6):1491–1502
19. Swensen SJ, Johnson CD (2010) Flying in the plane you service: patient centered radiology. J Am Coll Radiol 7(3):216–221
20. Womack JP, Jones DT (2003) Lean thinking: banish waste and create wealth in your corporation. Free Press, New York

Chapter 17
Clinical Audit and Practice Accreditation

Hannu Järvinen and Pamela Wilcox

Abstract In countries with a high level of health care, exposure to ionizing radiation from medical imaging, on average, equals to 80 % of that from natural sources. However, the dose to the patients for various procedures varies widely between facilities. The risk from ionizing radiation and the variability emphasizes the need for a systematic evaluation of radiologic facilities to improve the quality and safety of the care provided.

This chapter discusses the elements of both clinical audit and accreditation. It highlights the differences in the two types of assessment of radiologic facilities. There is an emphasis on the importance of some form of practice validation in assuring that the quality and safety of care in diagnostic imaging is optimized.

A description of the governmental regulatory environment in Europe as compared to the United States is provided. In the European Union, there is mandate for clinical audits although each country can determine the process for such audits. In the United States, there are differing mandates, e.g. federal mandates that are implemented by the Food and Drug Administration (FDA) for mammography and the Centers for Medicare and Medicaid Services (CMS) for outpatient MRI, CT, Nuclear Medicine and PET. However, much of diagnostic imaging remains unregulated.

Keywords Accreditation • Clinical audit • Continuous quality improvement • Dose • Governmental regulation in diagnostic imaging • Practice assessment • Quality control • Quality assurance • Quality management • Standards • Systematic assessment • Validation of quality

H. Järvinen (✉)
Radiation Practices Regulation, Radiation and Nuclear Safety Authority (STUK),
P.O. Box 14, FI-00881 Helsinki, Finland
e-mail: hannu.jarvinen@stuk.fi

P. Wilcox
American College of Radiology, 1891 Preston White Dr., Reston, VA 20191, USA
e-mail: pwilcox@acr.org

L. Lau and K.-H. Ng (eds.), *Radiological Safety and Quality: Paradigms in Leadership and Innovation*, DOI 10.1007/978-94-007-7256-4_17,
© Springer Science+Business Media Dordrecht 2014

1 Introduction

Medical exposure remains by far the largest artificial source of exposure to ionizing radiation and continues to grow at a remarkable rate [22]. In countries with high level of health care, exposure from medical uses now equals on average to about 80 % of that from natural sources. Irrespective of the level of health care in a country, the medical uses of radiation continue to increase as techniques develop and become more widely disseminated: about 3.6 billion radiological procedures are conducted worldwide every year. While the increased medical exposure is likely associated with increased health benefits to the population, it has been postulated that the rapid growth of new technologies, computed tomography (CT) and certain nuclear medicine studies in particular, is associated with a high rate of inappropriate or unnecessary procedures; with figures ranging from 20 % to about 80 % reported [2, 16, 17]. Furthermore, doses to patients for the same type of procedure differ widely between centres, suggesting that there is considerable scope for a better management of patient dose. Besides the increased risk for stochastic adverse effects there are nowadays also a number of cases where the non-optimized use of radiation has lead to deterministic effects such as skin damage.

The above development has stressed the importance of proper justification, optimization and quality assurance procedures in order to ensure the high quality and safety of radiological procedures. A lot of attention has been paid to the quality management in the medical use of radiation, and worldwide there has been a tendency to establish quality systems and introduce appropriate quality audits. Among the various approaches to quality auditing, two concepts, clinical audit and accreditation are gaining more and more importance. The former is a systematic examination or review of radiological procedures against agreed standards for good practices, seeking to improve the quality and the outcome of patient care, while the latter aims at assessing the competence of the facility to perform certain practices, in accordance with given standards. In this section, the basic features and differences in these concepts, the challenges and possible solutions in their implementation are reviewed.

2 Clinical Audit: A Systematic Assessment of Clinical Practice

2.1 General Features and Development

The concept of clinical audit is not a new one but has long been applied in many health care facilities. In Europe, the European Commission directive [5] introduced this concept for the assessment of medical radiological facilities. In this directive, clinical audit is defined as:

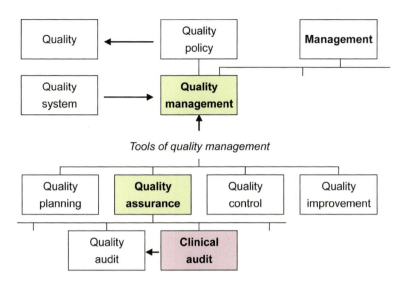

Fig. 17.1 Clinical audit in the "jungle" of quality management concepts [13]; quality management scheme based on EN ISO 8402 [10]

> a systematic examination or review of medical radiological procedures which seeks to improve the quality and the outcome of patient care, through structured review whereby radiological practices, procedures, and results are examined against agreed standards for good medical radiological procedures, with modifications of the practices where indicated and the application of new standards if necessary.

According to the directive, clinical audits shall be implemented in accordance with national procedures. Due to some confusion in the understanding of clinical audit and the high variation between the approaches of the EU Member States in its implementation, the European Commission has published specific guidelines on clinical audits [6].

The general purpose of any clinical audit is to:

- Improve the quality of patient care;
- Improve the effective use of resources;
- Enhance the provision and organization of clinical services; and
- Further professional education and training.

With these objectives, clinical audit is an integral part of the overall quality processes. In the conceptual framework of quality, clinical audit is a type of quality audit and one of the tools of quality management as shown in Fig. 17.1.

Clinical audit should cover the whole clinical pathway, and address the three main elements of the radiological practices: structure, process, and outcome. For radiological procedures, the priorities are shown in Table 17.1 [6].

Clinical audits can be of various types and levels, either by reviewing specific critical parts of the radiological process (partial audit) or assessing the whole process (comprehensive audit). Audits can also address various "depths" of the

Table 17.1 The priorities of clinical audit of radiological facilities

Structure	The mission of the unit for radiological facilities
	Lines of authorities and radiation safety responsibilities
	Staffing levels, competence and continuous professional development of staff, in particular for radiation protection
	Adequacy and quality of premises and equipment
Process	Justification and referral practices, including referral criteria
	Availability and quality of examination and treatment guidelines (protocols, procedures)
	Optimization procedures
	Patient dose and image quality in diagnostic radiology and nuclear medicine procedures, and comparison of patient dose with nationally accepted reference levels
	Procedures for dose delivery to the patient in radiotherapy (beam calibrations, accuracy of dosimetry and treatment planning)
	Quality assurance and quality control programmes
	Emergency procedures for incidents in use of radiation
	Reliability of information transfer systems
Outcome	Methods for the follow-up of outcome of examinations and treatment (short term and long term)

procedure, from generic features to details of a given procedure or treatment. Clinical audit aims at continuous improvement of the medical facilities. Therefore, it should be carried out regularly.

By its definition, clinical audit is a truly multi-disciplinary and multi-professional activity. It must be carried out by auditors with extensive knowledge and experience in the type of radiological facilities to be audited, i.e., they must generally be professionals involved in clinical work within comparable facilities. Further, the general understanding of the concept "audit" implies that the review or assessment is carried out by auditors independent of the organizational unit or facility to be audited.

Both internal audits and external audits should be implemented. These should be of equal importance and supplement each other. Internal audits are undertaken within a given health care facility by staff from the same facility, while the audit findings can also be externally reviewed. In very small health care facilities, internal audits can be replaced by self-assessments [14]. External audits involve the use of auditors who are independent of the facility to be audited. External audits are needed to remove possible "blindness" of internal experts, to recognize weaknesses of the facility and to give more universal and broader perspectives. External auditors should also possess better benchmarking skills in relation to the assessment.

Despite the fact that clinical audit has some similarities with other activities it is imperative not to confuse it with such activities as:

- Research;
- Quality control program for equipment; or
- Regulatory inspection or any other regulatory activity.

Further, clinical audit should not be confused with other external quality assessment systems [6], broadly categorized as:

- Certification by International Standards Organization (ISO);
- Accreditation;
- Professional peer review-based schemes; or
- Award seeking such as European Quality Award and their national variants, i.e. European Foundation for Quality Management (EFQM) Excellence Model.

In particular, the system of accreditation can be very close to the aims of clinical audit (see Sect. 4). To avoid the duplication of efforts, clinical audits should be developed to supplement the other activities.

The EC guidelines [6] provide a general framework to establish sustainable national systems of clinical auditing. Simultaneously with the European development, the IAEA has developed comprehensive audit programs under the term of clinical audit [8, 9]. Further, several dosimetry or quality audit programs traditionally applied in the field of radiotherapy, e.g. by the IAEA [11] or the ESTRO [7], have been recognized to form an important part of clinical audit. There are several national approaches to undertake clinical audits, e.g. the AuditLive system in the UK [19] and the Finnish nationwide organization [20].

2.2   Present and Future Challenges

Practical Organization and Financing

While the requirements and recommendations on clinical auditing of radiological facilities have been available for several years now, the practical organization and financing for the audit activities still remains one of the challenging issues.

For internal clinical audits, establishing the organization for audit within the facility is relatively straightforward. The detailed aims and objectives of the audit can be defined in the most flexible way and targeted to the expected weakness of the facilities. Little additional expenditure is usually needed but appropriate time for the preparation, execution and follow-up of the internal audits shall be reserved in the work budget of the facility.

For external clinical audits, several approaches for the practical organization are possible, with varying needs for financial input. The most straightforward but perhaps the most expensive systematic approach is to establish a special national or regional organization with audit teams to carry out clinical audits through site visits. The special organization should preferably be a non-profit organization, whereby recognized support by the radiological professional and/or scientific societies increases its general acceptance. If the special organization is a government body, the audits might be financed through the government budget. However, the general tendency in health care systems seems to assume that the health care facility requesting the clinical audit and deriving the benefits of it should also cover the costs incurred. If the special organization is a private body, special fees to cover the costs are inevitable, and provisions for clinical audits need to be made in the annual budgeting of the radiological facility. The auditors in this approach are most typically engaged for each individual audit from a pool of volunteered health care professionals based on special agreements. To ensure the full competence of the special auditing organizations, specific training to the auditors has to be arranged and the organizations should preferably be accredited by a national accreditation body.

Another approach for a systematic, regular auditing can be based on the collection and central processing of data via mail or internet. This approach relies on the local or regional support in providing the necessary information and data, and applies only to a limited part of facilities where relevant documented or measurable data is available. This approach can be cost-effective but inherently does not enable very comprehensive audits in a short time. A good example of this approach is the AuditLive program in the UK [19], e.g. national audit of the provision of MRI services [3]. Another particular approach to clinical audit is the establishment of a special Workshop to present selected clinical cases and to compare and analyze the associated clinical examination or treatment practices [18].

There are other approaches providing solutions for occasional and less systematic efforts. For clinical audits through site visits, individual "case by case" agreements are possible between a facility to be audited and another suitable facility providing the auditors (analogous to peer review activities). The drawback of this may be that the arrangement does not ensure adequate independence of the procedures, in particular if the audits are based on mutual audits between the two units. When no national systems exist, an "easy" way of starting can be by making use of international audit services; these could be very useful in providing some "model audits" in the process of developing a national organization for clinical audits. The drawback is that international services for clinical audit are not widely and extensively available, or are available only under special conditions, or for very limited applications. For example, the clinical audit service provided by the IAEA [8] is bound to the Technical Co-operation projects between the IAEA and the IAEA Member State.

For both the comprehensive site visits and more limited internet or mail based systems, a special national project can be established to undertake clinical audit in a well defined purpose but for a limited scope and timescale. An example of such a mail-based system is the assessment of the quality of CT referrals by mailed questionnaire [2]. These types of special projects can be very effective in the short term, but a clear drawback is that the project itself is only a temporary activity.

Standards of Good Practice

By definition, the criteria of good procedures (i.e. good practice) are the cornerstones for the development of clinical audits. These should be the basis of assessments regardless of the type of the audit – external, internal, comprehensive or partial.

According to the EC guidelines [6] the standards of good practice should be derived from evidence-based data, long-term experience and knowledge. In practice, these can be adopted from legal requirements, results of research, consensus statements, or recommendations by learned societies or local agreements (if other more universal reference is unavailable). International medical, scientific and professional societies could play an important role in developing such standards.

Three different levels for good practice can be distinguished:

1. The most generic criteria: applicable to all fields of radiology (diagnostic radiology, nuclear medicine and radiotherapy);
2. Criteria generic to a field of application (diagnostic radiology, nuclear medicine or radiotherapy); and
3. Criteria specific to a given procedure or treatment.

A number of publications by international or national professional societies, scientific organizations and authorities provide a basis of setting up standards of good practice, in particular for the first two level of criteria [6, 21]. For practical use in clinical audits, the criteria need to be compiled, sorted out and agreed on for the particular aims of an audit (comprehensive or partial, external or internal). The third level of criteria is the most challenging: for the specific procedure or treatment method, there is often the lack of wide consensus in the clinical community, caused by insufficient clinical evidence. This may be complicated by an inevitable varia-tion of local resources as for the availability and quality of necessary equipment and services. In such a situation, the criteria of good practice adopted should be regarded as giving only preliminary orientation, and the results of audit could then be used as a benchmarking tool to achieve improved evidence and possible adjustment of the chosen criteria (Fig. 17.2).

Motivation and Feedback

The progress of clinical auditing has sometimes been slow due to the lack of motivating actions and insufficient follow-up and reporting of the benefits. In fact, the proper follow-up and identification of the benefits is a key motivating factor to encourage the use of clinical audits.

There are a number of actions or principles, which are necessary or beneficial in order to motivate the facility and its staff to undertake internal and external clinical audits. Among other things this includes the following:

• The undertaking of internal audit, as well as the request for external clinical audit, should be endorsed by the staff at higher management level of the facility;
• Good definition of the objectives is essential: the critical areas for audit should be identified and the objectives targeted to the needs;
• Internal audits or self-assessments should be the first step when there has been no prior experience on audit and when clinical audits are introduced for the first time [14]. Another useful principle is to introduce a common program for self-assessments or internal audits, for a selected topic(s), and review and compare the results through external auditing;

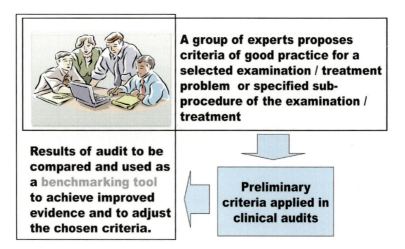

Fig. 17.2 Use of clinical audit as a benchmarking tool for the development of the criteria of good practice [21]

- Particular effort should be taken to create a motivating environment and a positive attitude before an external audit;
- The results of the audit should be well taken into consideration: the facility should respond to the recommendations with an agreed timeline for improvement, otherwise there is a risk of loosing respect and motivation of the staff for subsequent re-audits;
- Unnecessary overlap of clinical audit with other quality assessments and regulatory inspections should be avoided, otherwise it may cause frustration and feeling of wasting resources; and
- A special national or regional advisory group, or steering committee, consisting of clinical experts, who are independent of the auditing organizations, may prove useful in the overall coordination, development and implementation of clinical audit.

To maintain the stakeholders interested and motivated and to enable a proper cost/benefit analysis, the impact of clinical audits on the quality of procedures and the accountability of the auditing process should be regularly assessed. The presentation and open discussion of the findings and the benefits of clinical audits to the clinical community e.g. in national meetings, will improve the credibility of the process and promote motivation.

3   Accreditation: Assessment of Competence Against Standards

3.1   Overview of Accreditation

Accreditation can be voluntary or mandatory and promotes quality practice and programs by developing methods that assess whether facilities meet specified criteria established by the profession [12]. Accreditation in healthcare is generally based on the standards that have been developed through consensus of experts in the profession to be assessed. It is a tool for validating that the quality of care provided is adequate. The purpose of accreditation encompasses some of the same elements of a clinical audit:

• Improve the quality of patient care; and
• Further professional education and training.

It also includes demonstrating the quality of radiologic care provided to outside stakeholders including referring providers, patients, payers and governmental agencies.

In the US, accreditation is generally overseen by private non-profit organizations, although some states have taken on accreditation of mammography centers (see section "Federal Government and Accreditation"). Accreditation is granted for 3 years. At the time of renewal a complete re-assessment against all standards is required.

Certification and National Accreditation

Internationally, the International Organization for Standardization (ISO) offers a generic certification process for laboratories under ISO 17025 that can be used to assess the quality management system of a radiology facility [10]. However, it does not address the radiology-specific requirements or site inspections that are applicable in dedicated diagnostic imaging accreditation programs.

There are a number of accreditation programs that have been implemented around the world, on a voluntary or mandatory basis. The following section describes the program of the American College of Radiology. Due to space limitation, a detailed description of other programs is not possible. However, one should be aware that there are robust programs in Australia, Korea, New Zealand and the United Kingdom (UK). A program that is a partnership between the Royal Australian and New Zealand College of Radiologists and the National Association of Testing Authorities, was established in 2004 [15]. The Korean program was launched in 2005, New Zealand in 2011 and the UK in 2009. These programs are underpinned by standards agreed by the national radiological societies and other key stakeholders but are administered by independent bodies. In addition, they all have ties to their national governments of varying degrees.

ACR Accreditation

In the mid 1980s the American College of Radiology (ACR) recognized that there was wide variability in the quality and processes used in radiology facilities. The College determined that it was critical for the profession to be pro-active and establish standards as the experts in the field rather than allowing outside agencies or the government to set these standards. Furthermore, the development of accreditation programs to assess facilities against those standards was essential.

To be valid and credible within the profession and to other stakeholders the accreditation programs developed by the ACR are based on standards developed by committees of experts in the specific modality as well as those published in the literature. All programs are peer-reviewed and are educational. For providers who have deficiencies, the outcome of any accreditation evaluation should not be punitive but rather focused on providing education and guidance on how to improve. Ultimately, the impact of these programs should be appropriate decision making in diagnostic imaging and therapeutic treatment that improves the quality and efficiency of radiologic patient care. However, an inevitable and necessary aspect of any accreditation program is the denial of accreditation to those who are unable to meet the criteria after corrective action.

In compliance with recognized principles of accreditation law in the U.S. all ACR accreditation programs meet the following criteria:

- Provide a public benefit;
- Policy-making functions are independent of the parent organization;
- Voluntary (except for government mandates, see section "Elements of ACR Accreditation" on the Federal Government and accreditation);
- Available to all who meet the criteria;
- Valid, credible, reasonable, substantive, procedurally fair and able to withstand external scrutiny;
- Formal reports are provided;
- Accreditation is time-limited;
- Have an appeals process in place;
- Have a complaints process in place;
- Conflict of interest procedures are established; and
- Corrective action procedures to assist deficient providers to meet the accreditation requirements.

Between 1987 and 2010, the ACR developed and implemented a total of nine modality specific accreditation programs in:

- Mammography;
- Radiation oncology;
- MRI;
- Ultrasound;
- Stereotactic breast biopsy;
- Breast ultrasound and biopsy;
- Nuclear Medicine and PET;
- Radiography/Fluoroscopy[1]; and
- CT.

---

[1] Discontinued in 2005 due to lack of participation.

Elements of ACR Accreditation

As with all accreditation programs in healthcare the final objective is an assessment of the quality of care provided. In radiology, quality of care is a result of the integration of all of the elements that go into producing the image or therapy. Thus, there are standards set for personnel qualifications including the physician, the radiologic technologist and the medical physicist as well as equipment specifications, quality control and quality assurance procedures. For the diagnostic imaging modalities, there is an assessment of the clinical image quality and phantom image quality and dose (where appropriate). For radiation oncology, there is an assessment of the treatment record of patients.

Personnel requirements include their training and experience for initial accreditation. At the time of renewal there are requirements for continuing education and ongoing experience. Equipment specifications vary across the modalities and may be specific to the type of imaging being performed. Quality control procedures are defined. An annual evaluation by a medical physicist is required for all equipment including dose assessment for modalities that use ionizing radiation. Additionally, there is more frequent testing that is done by the technologist. There must be documentation of specific policies and procedures relative to patient education, infection control and safety for patients and personnel. All physicians are required to participate in a peer-review program that systematically evaluates the accuracy of interpretation and the appropriateness of the procedure.

Clinical images are evaluated to assure they are adequate for interpretation. The specific scoring criteria are defined as appropriate to the modality. In general, the categories of assessment include anatomic coverage, technique parameters, artifacts, procedure identification and filming technique (for hard copy film only).

As with the clinical images, the criteria for scoring phantoms are quite specific to each modality. The phantom assessment provides an opportunity to evaluate the capabilities of the equipment without the variables introduced by patients. For the modalities that use ionizing radiation a dose assessment is also performed using a phantom.

Due to the size of the U.S. and the large number of imaging providers, the ACR diagnostic modality accreditation programs were originally conducted by mail. Today the majority of the documentation is submitted via the web. However, the ACR recognizes that this might allow some providers to provide misleading information to indicate compliance when, in fact, they are non-compliant. To address this issue every site agrees that they may be subjected to a random site survey sometime during the 3-year accreditation cycle. The site survey is performed by a team that includes a radiologist, a medical physicist and a radiologic technologist. They do a thorough assessment of compliance with all standards but also use the opportunity to provide education on ongoing quality improvement.

Federal Government and Accreditation

In 1992 the US Congress passed the Mammography Quality Standards Act (MQSA), which requires all providers of mammography in the US to be accredited and certified [23]. In fact, this law made it illegal to perform any mammography without accreditation and certification. The legislation tasked the Food and Drug Administration (FDA) with oversight and implementation of regulations to support the legislation. The ACR Mammography Accreditation is the only program approved by the FDA to accredit mammography facilities on a national level although some states have been approved to accredit within their own states.

Initially, the FDA adopted most of the standards of the ACR's voluntary accreditation program. However, they also added regulations that ACR was then required to incorporate these into its requirements. When the final regulations became effective in 1999, additional regulations for equipment and patient reporting were added and again ACR was required to make changes to its accreditation program in order to maintain its approval by the FDA [25]. In fact, once the initial regulations were issued, the ACR no longer had the authority to make any modifications to its accreditation criteria without prior approval by the FDA.

In addition to accreditation and certification, the MQSA also requires the FDA to conduct annual inspections of all accredited providers. In most states the FDA has contracted with the state department of radiation control to perform these inspections. During the inspection there is significant amount of documentation and paperwork that the site must provide in order to demonstrate compliance.

In 2008 Congress passed the Medicare Improvements for Patients and Providers Act (MIPPA). One section of this legislation requires all providers of Advanced Diagnostic Imaging Service (ADIS) to be accredited by an accreditation organization approved by the Centers for Medicare and Medicaid Services (CMS) by January 1, 2012 [24]. ADIS is defined as MRI, CT, Nuclear Medicine and PET. Because of specific wording in the law this mandate applies only to imaging centers not associated with a hospital. However, it does apply to any in-office imaging such as that done by cardiologists, orthopedists and other non-radiologists.

In contrast to the regulations for MQSA, CMS issued only very general regulations [1]. CMS allowed the accrediting organizations to establish their own standards for measuring compliance with the law and the regulations. However, any substantive changes to the accreditation standards going forward must be approved by CMS.

CMS approved three accrediting organization: the ACR, the Intersocietal Accreditation Council (IAC) and The Joint Commission (TJC). Each organization has different standards and assesses provider compliance with those standards in different ways. It remains to be seen how this mandate will affect the quality of imaging care in the US.

3.2 Challenges and Opportunities

When accreditation is voluntary there are a number of challenges. First is uptake: If there is no mandate by the government or payers there is frequently little incentive to participate in the rigors of an accreditation evaluation. For modalities where there are no mandates the volume of accredited sites remains low. However, a more disturbing consequence of voluntary accreditation is that providers who do not pass the evaluation may decide not to implement corrective action and re-apply. Thus, a site that is providing inadequate care may choose to withdraw from the accreditation process but continue to offer imaging services. In fact, when the mammography accreditation program was voluntary 31.3 % of those that did not pass accreditation withdrew without correcting their deficiencies [4].

Mandatory accreditation seems to encourage facilities to maintain the necessary level of quality over time. After the implementation of MQSA and sites became more familiar with the criteria the initial pass rate rose from 70 % to 88.3 % and the rate on second attempt reached 88.7 % for an overall pass rate of 98.2 % [4]. With the new mandate for accreditation in MRI, CT, Nuclear Medicine and PET the ACR will monitor the pass/fail rates for these modalities to assess whether a similar effect is seen.

4 Clinical Audit Versus Accreditation

Clinical audit and the audits used in accreditation have much in common, in the sense that both type of audits aim at improving or ensuring the good quality of practices (Table 17.2). The main difference between clinical audit and accreditation can be seen in the freedom or versatility of their practical implementation and the different weights in the use of the concept.

Clinical audit is very flexible and can take several forms, ranging from internal to external, partial to comprehensive, and single to multi-level assessments. It is based on the standards of good practice, which are of various levels and a facility can select one or more standards to be used in the audit. It is generally a useful tool of quality management and aims at ensuring that the best available clinical procedures are in use. It is not so much intended to prove the quality to any outside stakeholders and there is usually no penalty served to those who are unable to meet the criteria of good practice.

Accreditation is more formal, always external, and aims at giving a public proof that the facility audited is competent to undertake the given procedures in accordance to the given standards. The procedures and processes must comply with these requirements. The systematic assessment by peers will determine if competency and requirements are met and if accreditation could be granted to or maintained by the facility.

Table 17.2 Key clinical audit and accreditation elements

	Clinical audit	Accreditation
Aim	To ensure and improve good quality care, the provision and organization of services	
	An internal quality management tool, not usually for a public demonstration of competency	Public demonstration of competency to provide certain service or perform certain procedure
Assessment	Systematic assessments by clinical experts competent in the field of practice to be audited	Systematic assessment by peers
Agent	Internal or external	External: public or private
Process	Variety of forms	Formal
	Flexible	Structured
	Internal or external	External
Site visit	Vary, assessment can be on-site or off-site	Compulsory on-site peer visit, or
		Off-site by declaration + random site visit
Failed assessment	Provision of recommendations to enable practice improvement	Provision of recommendations to enable practice improvement
		Possible denial of accreditation
Standards	Agreed standards for good medical practice	
Scope	Variable, can assess one or more elements of a service	Fixed
	Comprehensive or partial	Comprehensive, covering all aspects of a service
Priorities	Structure	Professional
	Processes	Equipment
	Outcome	Quality processes, including review of images, dose records, treatment records etc.

5   Conclusion

The continuing increase use of medical radiological procedures gives rise to increased radiation dose to the population and calls for better dose management, justification and optimization. As tools of quality management, both clinical audit and practice accreditation are gaining more and more importance. Both concepts provide a systematic review of the procedures in order to improve the quality of practice and the outcome of patient care. While clinical audits and accreditation have much in common, they are different and complementary procedures and cannot replace each other. Both concepts have been widely applied but there are still a number of challenging issues in their implementation. One of the key factors to ensure their on-going success is an appropriate follow-up and the publication of the results to prove the expected benefits to the stakeholders.

References

1. 42 CFR, Parts 410,411,414, 415 and 485 [CMS-1413-P] (2009) Department of Health and Human Services, Centers for Medicare & Medicaid Services, Washington, DC
2. Almen A, Leitz W, Richter S (2009) 2009:03 National survey on justification of CT-examinations in Sweden. SSM, Stockholm. http://www.stralsakerhetsmyndigheten.se/Publikationer/Rapport/Stralskydd/2009/200903/. Accessed 16 Mar 2012
3. Barter S, Drinkwater K, Remedios D (2009) National audit of provision of MRI services 2006/07. Clin Radiol 64(3):284–290
4. Destouet JM, Bassett LW, Yaffe MJ et al (2005) The ACR's mammography accreditation program: ten years of experience since MQSA. J Am Coll Radiol 2(7):585–594
5. European Commission (1997) Council Directive 97/43/Euratom of 30 June 1997 on health protection of individuals against the dangers of ionizing radiation in relation to medical exposure, and repealing Directive 84/466/Euratom. Official J Eur Commun L 180:22–27
6. European Commission (2009) Radiation Protection No 159: European Commission guidelines on clinical audit for medical radiological practices (diagnostic radiology, nuclear medicine and radiotherapy). Publication Office of the European Union, Luxembourg
7. Ferreira IH, Dutreix A, Bridier A et al (2000) The ESTRO-QUALity assurance network (EQUAL). Radiother Oncol 55(3):273–284
8. Agency International Atomic Energy (2007) Comprehensive audits of radiotherapy practices: a tool for quality improvement. Quality Assurance Team for Radiation Oncology (QUATRO). IAEA, Vienna
9. International Atomic Energy Agency (2010) Comprehensive clinical audits of diagnostic radiology practices: a tool for quality improvement, IAEA human health series no. 4. IAEA, Vienna
10. International Standards Organisation (1995) BS EN ISO 8402:1995 Quality management and quality assurance. Vocabulary. British Standards Institution, London
11. Izewska J, Svensson H, Ibbott G (2003) Worldwide quality assurance network for radiotherapy dosimetry. In: Standards and codes of practice in medical radiation dosimetry, vol 2. IAEA, Vienna, pp 139–155
12. Jacobs JA (2004) Certification and accreditation law handbook. The American Society of Association Executives, Washington, DC
13. Järvinen H (2011) Clinical auditing and quality assurance. A refresher course lecture. In: Proceedings of the third European IRPA congress 2010. http://www.irpa2010europe.com/proceedings.htm. Accessed 16 Mar 2012
14. Järvinen H, Alanen A, Ahonen A et al (2011) Guidance on internal audits and self-assessments: support to external clinical audits. In: Proceedings of the conference of the Nordic Society of Radiation Protection (NSFS). http://nsfs.org/NSFS-2011/documents/session-07/S7-O1.pdf. Accessed 16 Mar 2012
15. Lau LS (2007) The design and implementation of the RANZCR/NATA accreditation program for Australian radiology practices. J Am Coll Radiol 4(10):730–738
16. Malone J, Guleria R, Craven C et al (2011) Justification of diagnostic medical exposures, some practical issues: report of an International Atomic Energy Agency Consultation. Br J Radiol. doi:10.1259/bjr/42893576
17. Oikarinen H, Meriläinen S, Pääkkö E et al (2009) Unjustified CT examinations in young patients. Eur Radiol 19(5):1161–1165
18. Rekstad BL, Levernes S, Heikkilä IE et al (2008) Norwegian experience with workshop as a clinical audit tool for radiotherapy of specific cancer diagnoses. In: Abstract book international workshop on clinical audit 2008. http://www.clinicalaudit.net/Book%20of%20abstracts_Final_HJ.pdf. Accessed 16 Mar 2012
19. The Royal College of Radiologists (2009) AuditLive. http://www.rcr.ac.uk/audittemplate.aspx?PageID=1016. Accessed 16 Mar 2012

20. Soimakallio S (2003) A nationwide organization for clinical audit in Finland. In: Proceedings of the international symposium on practical implementation of clinical audit for exposure to radiation in medical practices 2003. http://www.clinicalaudit.net/img/Proceedings2003.pdf. Accessed 16 Mar 2012

21. Soimakallio S, Alanen A, Järvinen H et al (2011) Clinical audit: development of the criteria of good practices. Radiat Prot Dosimetry 147(1–2):30–33

22. United Nations Scientific Committee on the Effects of Atomic Radiation (2010) Sources and effects of ionizing radiation. In: UNSCEAR 2008 report to the General Assembly with scientific annexes, vol I. United Nations, New York

23. United States of America (1992) Mammography quality standards act of 1992. House Resolution 6182, Public Law 102–539. US Congress

24. United States of America (2008) Medicare improvements for patients and providers act of 2008. House Resolution 6331, Public Law 110–275, Section 135. US Congress

25. United States of America FDA (1997) Code of Federal Regulation Title 21. http://www.accessdata.fda.gov/scripts/cdrh/cfdocs/cfcfr/cfrsearch.cfm?cfrpart=11. Accessed 3 Feb 2012

Chapter 18
The International Basic Safety Standards and Their Application in Medical Imaging

John Le Heron and Cari Borrás

Abstract In 2011, several international organizations, led by the International Atomic Energy Agency, completed the latest revision of the document commonly known as the Basic Safety Standards (BSS), and during the course of 2012 its formal approval by each of the cosponsoring organizations was completed. The BSS is used worldwide as the basis for radiation protection legislation and regulations. As in the previous versions, this updated BSS sets out the radiation protection requirements for workers, members of the public and patients. Particular attention is given to the use of radiation in medical applications, the potential benefits of which are increasing, especially with the developments of new imaging technologies and wider use of image-guided interventions. The provision of radiological procedures involve a multidisciplinary team led by a physician who often is not the licensee of an authorized facility, thus the responsibilities in medical exposures are shared by several individuals. The BSS establishes the education, training and competence requirements and lists the functions of the radiological medical practitioners, referring medical practitioners, medical radiation technologists and medical physicists. Procedure justification is to be shared between the referrer and the medical radiological practitioner, who is also responsible for optimization of protection, aided by the medical radiation technologist who chooses the appropriate patient techniques, and by the medical physicist, who calibrates sources, assesses image quality and patient dose, and is responsible for the physical aspects of the quality assurance programme, including equipment acceptance testing and commissioning. They all participate in preventing, detecting and assessing unintended

J. Le Heron (✉)
Radiation Protection of Patients Unit, Division of Radiation, Transport and Waste
Safety, International Atomic Energy Agency, A-1400, Vienna, Austria
e-mail: J.Le.Heron@iaea.org

C. Borrás
Radiological Physics and Health Services Consultant, 1501 44th St. NW,
Washington, DC 20007, USA
e-mail: cariborras@starpower.net

L. Lau and K.-H. Ng (eds.), *Radiological Safety and Quality: Paradigms
in Leadership and Innovation*, DOI 10.1007/978-94-007-7256-4_18,
© Springer Science+Business Media Dordrecht 2014

and accidental medical exposures. The practical implementation of these BSS requirements should ensure the radiation protection and improve the quality of practice in a radiological facility.

Keywords Human imaging for non-medical purposes • Justification • Medical exposure • Occupational exposure • Optimization • Public exposure • Radiation protection • Safety standards • Unintended and accidental medical exposures

1   Introduction

The use of radiation in medicine continues to grow throughout the world, bringing immense benefit to patients and to society. The growth is due to several factors. At one level there are simply more machines, increasing the ease of access to radiological procedures, especially in the developing world. Secondly there continue to be developments in technology and techniques that have changed how radiological procedures are performed and what they can achieve – for example, digital technologies are replacing analogue systems, MDCT is replacing SDCT, image-guided interventional procedures are utilised in many areas of medicine replacing surgical or other procedures, and virtual procedures are becoming a reality. Due to the increasing capabilities of radiological procedures, the role of imaging in particular is changing – on the one hand, radiology is becoming the first "port of call" in determining what is wrong with the symptomatic patient and, on the other, it is increasingly taking on a role in the early detection of disease, often in asymptomatic individuals.

The potential benefits arising from radiation in medicine are increasing. But radiation can also cause harm, with deterministic, carcinogenic and hereditary effects. The responsible use of radiation requires a regulatory framework that gives a formal structure to ways of ensuring that the benefits can be utilised and, at the same time, keeping the harm to an acceptable level. Regulatory requirements for radiation protection should be essentially the same throughout the world and this naturally leads to the role of international safety standards in providing a consistent approach to ensuring radiation protection.

The development of safety standards is a statutory function of the International Atomic Energy Agency (IAEA). In particular, the IAEA statute (Article III.A.6) expressly authorizes the Agency "to establish standards of safety for protection of health" and "to provide for the application of these Standards". The IAEA published the first set of basic safety standards in June 1962 as Safety Series No. 9 [4], following the guidance of their Board of Governors in 1960 that "The Agency's basic safety standards ... will be based, to the extent possible, on the recommendations of the International Commission on Radiological Protection (ICRP)". A revised version was published in 1967 and a third revision in 1982. This edition, titled the 1982 Edition of Safety Series No. 9 [5], was jointly sponsored by the IAEA, the International Labour Organisation (ILO), the Nuclear

Energy Agency of the Organisation for Economic Co-operation and Development (OECD/NEA), and the World Health Organization (WHO). The revision of these Standards started in 1991 with the incorporation in the process of the Food and Agriculture Organization of the United Nations (FAO), which had just published jointly with WHO the Codex Alimentarius, and the Pan American Health Organization (PAHO), which had had a radiological health programme since 1960 with strong emphasis on patient protection [3]. The new Standards, with the title International Basic Safety Standards for the Protection against Ionizing Radiation and for the Safety of Radiation Sources (known as the BSS) was published in 1996 [2], following the approval or endorsement of the governing bodies of all the cosponsoring organizations, a consensus of 192 countries. Since then, the BSS, promoted by the cosponsoring organizations, has been used as the basis for radiation protection legislation/regulations worldwide.

A review of the 1996 BSS, carried out in 2006 by the then cosponsoring organizations, concluded that, while there was no single major reason for a revision, a number of factors – including the then imminent publication of the new ICRP recommendations – justified preparing a new edition. The revision process started in 2007, with the incorporation of the European Commission and the United Nations Environment Program as potential cosponsors, and culminated in IAEA Board approval in 2011 for publication of an interim edition of the new document [6], pending final publication as a jointly sponsored standard when also approved by the co-sponsoring organizations. During the course of 2012, all the cosponsoring organizations completed their respective approval processes, clearing the way for the final publication in 2013. In the remainder of this chapter, the term BSS refers to the 2011 publication [6].

The BSS, in keeping with all the IAEA Safety Standards, uses as its starting point the scientific data of United Nations Scientific Committee on the Effects of Atomic Radiation (UNSCEAR) and the recommendations of the ICRP, but it also draws upon extensive research and development work by national and international scientific and engineering organizations on the health effects of radiation and on techniques for the safe design and operation of sources.

The BSS has no particular mandatory status, but is available for countries to adopt/adapt if they wish. However, if a country wishes to receive technical assistance from the IAEA or from a co-sponsor of the BSS, then compliance with the BSS becomes a condition of that assistance. This has given the BSS a very important status in the developing world as the "bible" of radiation protection.

2   BSS Structure

Table 18.1 presents the structure of the BSS. Within this structure, the BSS consists of a series of requirements – statements that specify what must be done and by whom. They are grouped into requirements applicable for all exposure situations with additional separate requirements for planned exposure situations (situations of

Table 18.1 Structure of the International Basic Safety Standards – section and subsection headings

---

1. Introduction
   Background
   Scope
   Structure

2. General Requirements for Protection and Safety
   Definitions
   Interpretation
   Resolution of conflicts
   Entry into force
   Application of the principles of radiation protection
   Responsibilities of the government
   Responsibilities of the regulatory body
   Responsibilities for protection and safety
   Management requirements

3. Planned Exposure Situations
   Scope
   Generic requirements
   Occupational exposure
   Public exposure
   Medical exposure

4. Emergency Exposure Situations
   Scope
   Generic requirements
   Public exposure
   Exposure of emergency workers
   Transition from an emergency exposure situation to an existing exposure situation

5. Existing Exposure Situations
   Scope
   Generic requirements
   Public exposure
   Occupational exposure

   Schedule I: Exemption and clearance
   Schedule II: Categories for sealed sources used in common practices
   Schedule III: Dose limits for planned exposure situations
   Schedule IV: Criteria for use in emergency preparedness and response
   References
   Annex: Generic criteria for protective actions and other response actions in emergency exposure
      situations to reduce the risk of stochastic effects
   Definitions

---

exposure that arise from activities that typically require authorization), emergency exposure situations (situations of exposure that arise from accidents or events and require prompt action in order to avoid or reduce adverse consequences) and existing exposure situations (situations of exposure that already exist when decisions on the need for control need to be taken). For each of the three types of exposure situation, the requirements are further grouped into requirements for occupational exposure, public exposure and (for planned exposure situations) medical exposure. They apply to all situations involving radiation exposure that are amenable to control, and this clearly includes all uses of radiation in medicine, and hence in radiology.

In a medical facility in which radiation generators or radioactive sources are used (called here a radiology facility for simplicity), consideration needs to be given to the patient, the personnel involved in performing the radiological procedures and to members of the public that may be around. In addition, there may be carers and comforters of patients undergoing procedures, and persons who may be undergoing a radiological procedure as part of a biomedical research project. This chapter discusses the requirements of the BSS, in particular those for medical exposure resulting from medical imaging procedures, and how such requirements when applied practically in a radiology facility should ensure the twin goals of quality radiology and radiation protection. Clearly this chapter cannot purport to be a comprehensive or exhaustive summary of the BSS, and the reader must refer to the actual BSS for full details.

3   Medical Exposure

The BSS defines medical exposure as exposure incurred by patients for the purposes of medical or dental diagnosis or treatment; by carers and comforters; and by volunteers subject to exposure as part of a programme of biomedical research.

3.1   Responsibilities

Because radiological procedures involve a multidisciplinary team led by a physician who often is not the registrant or licensee of an authorized facility, responsibilities in medical exposures are shared by several individuals. At the higher level, there are requirements for the government, the regulatory body (for radiation protection) and, in the case of medical exposures, the health authority and professional bodies. At the individual level, in a radiology facility there are three key roles that are crucial to radiation protection for medical exposures – those of the radiological medical practitioner, the medical radiation technologist and the medical physicist. The radiological medical practitioner is the generic term that the BSS uses to refer to a health professional with specialist education and training in medical uses of radiation, who is competent to perform independently or to oversee procedures involving medical exposure in a given specialty. Clearly the radiologist is a radiological medical practitioner, but many other specialists may also take on that role, including cardiologists, orthopaedic surgeons, dentists, to name just a few. Medical radiation technologist is the generic term used to cover the variety of terms that are used throughout the world such as the radiographer and the radiologic technologist, again to name just a few. The term medical physicist (medical physics is now classified by the ILO as a profession in the International Standard Classification of Occupations-08 [7]) is used in the BSS, replacing the expressions in the 1996 version of "expert" in "radiotherapy physics", "radiodiagnostic physics" and/or "nuclear medicine physics". In the BSS, a medical physicist is a health professional with specialist education and training in the concepts and techniques of applying physics in medicine, and competent to practise independently in one or more of the subfields (specialties) of medical physics, (e.g. diagnostic radiology, radiation therapy, nuclear medicine). Other individual responsibilities in medical exposures are assigned to the registrant or licensee; the manufacturers, suppliers of sources, equipment, or software; the workers; the referring medical practitioners; and the ethics committees.

Because the key persons acting in these roles have a significant impact on radiation protection, it is very important to ensure that only appropriate persons are permitted to act in these roles – it cannot be left to good luck. It therefore falls on the radiation protection regulatory body to ensure that the conditions of the authorization for a radiology facility establish that any individual seeking to act as a radiological medical practitioner, a medical radiation technologist or a medical physicist is specialized in the appropriate area, and meets the respective education, training and competence requirements in radiation protection. Typically a specialization would be recognized through a national system of registration, accreditation or certification, and appropriate area refers to diagnostic radiology and image guided interventional procedures in the first instance, but in many cases would be narrower, such as cardiology or urology, to give examples. Clearly the health authority and professional bodies are involved in this recognition process.

The general medical and health care of the patient is, of course, the responsibility of the individual physician treating the patient; however, when the patient presents in the radiology facility, the radiological medical practitioner has the particular responsibility for the overall radiation protection of the patient. This means the responsibility for the justification of the given radiological procedure for the patient in conjunction with the referring medical practitioner, and responsibility for ensuring the optimization of protection in the performance of the examination.

The role of the medical radiation technologist is crucial as his/her skill and care in the choice of techniques and parameters determines to a large extent the practical realization of the optimization of a given patient's exposure in many modalities.

The medical physicist provides specialist expertise with respect to radiation protection of the patient. The medical physicist has responsibilities in the implementation of the optimization of radiation protection in medical exposures, including source calibration, image quality and patient dose assessment, and physical aspects of the quality assurance programme, including medical radiological equipment acceptance and commissioning. In radiotherapy, these functions are to be conducted "by or under the supervision of a medical physicist". For diagnostic radiological procedures and image-guided interventional procedures, they are to be performed "by or under the oversight of or with the documented advice of a medical physicist, whose degree of involvement is determined by the complexity of the radiological procedures and the associated radiation risks".

The medical physicist is also likely to have responsibilities in providing radiation protection training for medical and health personnel. In addition, he/she may also perform the role of the radiation protection officer (RPO), whose responsibilities are primarily in occupational and public radiation protection.

For a radiology facility, the radiation protection responsibilities described above for the radiological medical practitioner, the medical radiation technologist and the medical physicist will be assigned through an authorization (or other regulatory means) issued by the radiation protection Regulatory Body in that country or state.

The rights of the patient are being increasingly recognized in medical care, and this is extending to radiological procedures. The BSS requires that the licensee for the radiology facility ensures that the patient be informed, as appropriate, of both the potential benefit of the radiological procedure and the radiation risks.

3.2   Justification

The requirements in the BSS embrace the three ICRP principles of radiation protection – justification of the practice, optimization of the protection and dose limitation. Only the first two apply to medical exposure, i.e. dose limitation does not apply to medical exposure.

The concept of justification as it applies to medical exposure is reasonably well established, with the current 3-level approach having been introduced with the ICRP Publication 73 [8] and more recently reiterated in ICRP Publications 103 and 105 [9, 10]. However the transfer into day-to-day practice has proven more difficult, especially with respect to "level 3" justification for an individual patient undergoing a given radiological procedure. The evidence for the ineffective application of the principle has been gaining greater publicity in recent years, with many media reports appearing on inappropriate, unnecessary or unjustified radiological procedures. The process of revising the BSS provided an opportunity to improve requirements in this area.

Responsibility for both the generic justification (level 2) of a radiological procedure and the justification of radiological procedures performed as part of a health screening programme for asymptomatic populations falls on a country's health authority, in consultation with appropriate professional bodies. Ethics committees take on the responsibility for justification of medical exposures that occur as part of a programme of biomedical research.

In the BSS, justification for an individual patient (level 3) is a joint responsibility, performed by consultation between the radiological medical practitioner and the referring medical practitioner. The BSS also states what needs to be taken into account when performing the justification: appropriateness of the request; use of relevant national or international guidelines; urgency of the procedure; characteristics of the exposure; characteristics of the individual patient; and last but not least, relevant information from previous radiological procedures. In some countries professional bodies have produced guidance on imaging procedures, including when they should be performed, when they should not be performed, advantages, limitations and radiation risk. Such so-called referral guidelines or appropriateness criteria [1, 12] are a useful bridge between the radiological medical practitioner and the referring medical practitioner in their joint responsibility.

A particular issue addressed in the BSS is that of justification of radiological procedures performed on asymptomatic individuals, intended for the early detection of disease but not as part of an approved health-screening programme. The situation might arise through "entrepreneurial" medicine, often with self-presenting patients, or it may be a "grey area" of medicine where the procedure has not yet become widely accepted medical practice. The BSS places joint responsibility on the referring medical practitioner and the radiological medical practitioner, but also puts the onus on relevant professional bodies or the Health Authority to develop guidelines. Further, the individual must be informed in advance of the expected benefits, risks and limitations of the procedure.

3.3 Optimization

Optimization of protection and safety is the second principle of radiation protection, and it comes into action once the justification has taken place. Protection and safety must be optimized for each and every medical exposure. What this means is described briefly below. As noted above, the radiological medical practitioner has prime responsibility for optimization, but the medical radiation technologist and the medical physicist also have very important roles to play.

The first step in the optimization process is to ensure that any medical radiological equipment or any software that could influence the delivery of medical exposure is used only if it conforms to established national or international standards.

The performance of a given radiological procedure typically involves many decisions over technique and technical parameters, most of which will have some impact on the image quality and the patient dose. Achieving the right balance is aim of optimization. The BSS requires that, for diagnostic radiological procedures and image guided interventional procedures, the radiological medical practitioner, in cooperation with the medical radiation technologist and the medical physicist, must ensure that the following are used:

• Appropriate medical radiological equipment and software; and
• Appropriate techniques and parameters to deliver a patient exposure that is the minimum necessary to fulfil the clinical purpose of the procedure, taking into account of the:

  • Relevant norms of acceptable image quality; and
  • Relevant diagnostic reference levels.

In addition to this general requirement applicable to all exposures in radiology, there is a further requirement for special consideration in the following situations in radiology:

- Paediatric patients subject to medical exposure;
- Individuals subject to medical exposure as part of a health screening programme;
- Volunteers subject to medical exposure as part of a programme of biomedical research;
- Relatively high doses to the patient, such as CT and image guided interventional procedures;
- Exposure of the embryo or foetus, in particular for radiological procedures in which the abdomen or pelvis of the pregnant woman is exposed to the useful radiation beam or could otherwise receive a significant dose;
- Exposure of a breast-fed infant as a result of a female patient undergoing a radiological procedure with radiopharmaceuticals.

A prerequisite for being able to perform any radiological procedure successfully is that the equipment about to be used is going to function correctly, and its radiation output is reproducible and predictable. For these reasons, the BSS has requirements for all X-ray systems to be calibrated and their performance to be monitored through a programme of quality assurance. Responsibility for calibration falls on the radiology medical physicist, and calibrations must be in terms of appropriate quantities using nationally or internationally accepted protocols. They must take place with new equipment at commissioning prior to clinical use, after any maintenance that could affect the dosimetry and at intervals approved by the Regulatory Body. A team approach is needed for the quality assurance programme, with the active participation of medical physicists, medical radiation technologists and radiological medical practitioners, and taking into account principles established by WHO, PAHO and relevant professional bodies. Regular and independent audits must be a feature of the quality assurance programme with audit frequencies in accordance with the complexity of the radiological procedures being performed at the radiology facility and the associated risks.

Every radiology facility must know what levels of radiation dose they are typically using when performing common radiological procedures that result in acceptable image quality, and this is another requirement of the BSS. Responsibility for performing the measurements falls on the medical physicist, using calibrated dosimeters and following nationally or internationally accepted protocols. But in addition, the results of the patient dose assessments must be compared with the nationally or regionally established diagnostic reference levels (DRLs). If the comparison shows that the typical dose for a given radiological procedure exceeds the relevant DRL, or if the typical dose falls substantially below the relevant DRL and the exposures do not provide useful diagnostic information, then a review must be conducted to determine whether the optimization of protection and safety for patients is adequate or whether corrective action is required.

The establishment of DRLs is the responsibility of government, as a result of consultation between the health authority, professional bodies and the regulatory body. The use of DRLs as briefly outlined above is a very important aspect of optimization of radiation protection. Their use is in effect a "litmus test", identifying situations where the optimization might be outside what would be considered acceptable practice – identifying outliers. The setting of a value for a DRL for a given procedure necessarily is based on assessment of current practice in achieving a particular clinical goal. The DRL value is then applied prospectively. However, review of DRLs must take place periodically, especially as new techniques and technologies are introduced into practice. The prospective application of the then new DRL values starts a new cycle, and in this way, over time, the use of DRLs will influence current practice in the direction of quality radiology coupled with appropriate patient radiation protection.

3.4   Unintended and Accidental Medical Exposures

With all the above requirements in place in a radiology facility, the desired outcomes of quality radiology and patient radiation protection should be ensured. However, there is always the potential for things to go wrong or awry, and the BSS require that steps are taken to minimize the likelihood of unintended or accidental medical exposures arising from flaws in design and operational failures of medical radiological equipment, from failures of and errors in software, or as a result of human error.

The events of concern in radiology are:

- Any diagnostic radiological procedure or image guided interventional procedure in which the wrong individual or the wrong tissue of the patient has been subject to exposure;
- Any exposure for diagnostic purposes that is substantially greater than was intended;
- Any exposure arising from an image-guided interventional procedure that is substantially greater than was intended;
- Any inadvertent exposure of the embryo or foetus in the course of performing a radiological procedure; and
- Any failure of medical radiological equipment, software failure or system failure, or accident, error, mishap or other unusual occurrence with the potential for subjecting the patient to a medical exposure that is substantially different from what was intended.

Should any of these occur, then prompt investigation is required to determine the doses involved, and to identify and implement corrective actions needed to prevent a recurrence. Records of the investigation need to be kept, and the referring medical practitioner and the patient need to be informed.

3.5   Reviews and Records

The BSS requires periodical radiological reviews – an investigation and critical review of the current practical application of the radiation protection principles of justification and optimization for the radiological procedures that are performed in the medical radiation facility. Responsibility for the radiological review falls on the radiological medical practitioners in the radiology facility, in cooperation with the medical radiation technologists and the medical physicists.

The final requirements in the BSS for medical exposures relate to record keeping. There are requirements for records in three broad areas – personnel records, including delegations of radiation protection responsibilities and training of personnel in radiation protection; records of calibration, dosimetry and quality assurance, including results of the calibrations and periodic checks of the relevant physical and clinical parameters selected during treatment of patients, dosimetry of patients, local assessments and reviews made with regard to DRLs, and records associated with the quality assurance programme; and records of medical exposures – for diagnostic radiology and image guided interventional procedures, information necessary for the retrospective assessment of doses; exposure records for volunteers in biomedical research; and reports on investigations of unintended and accidental medical exposures.

4   Occupational Exposure

The BSS defines occupational exposure as exposure of workers in the course of their work. In a radiology facility, many persons will be subject to occupational exposure. However, the level of occupational exposure associated with radiology is highly variable and ranges from potentially negligible in the case of simple chest X-rays, to significant for complex interventional procedures. In most cases, the main determinant for occupational exposure is proximity of personnel to the patient when exposures are being made and the length of these procedures. Interventionists are a particular concern as they typically receive the highest occupational exposure in medicine.

Occupational radiation protection is achieved by application of the three ICRP principles of justification, optimization and dose limitation. Table 18.2 lists the dose limits for workers given in the BSS. Following the latest recommendation of the ICRP [11], the equivalent dose limit to the lens of the eye in a year is now 20 mSv – in the 1996 BSS it was 150 mSv. In practice, to afford radiation protection for medical personnel in radiology, it is the application of optimization and dose limitation that is arguably the most important. However, it should be recognized that the lack of rigorous application of the justification principle is resulting in the performance of significant numbers of unnecessary imaging examinations, and these add to occupational exposure.

Table 18.2 Dose limits[a] for occupational exposure, from the BSS [6]

Workers > 18 year of ages	Apprentices and students[b], 16–18 years of age
An effective dose of 20 mSv per year averaged over 5 consecutive years (100 mSv in 5 years), and of 50 mSv in any single year	An effective dose of 6 mSv in a year
An equivalent dose to the lens of the eye of 20 mSv per year averaged over 5 consecutive years (100 mSv in 5 years) and of 50 mSv in any single year	An equivalent dose to the lens of the eye of 20 mSv in a year
An equivalent dose to the extremities (hands and feet) or the skin[c] of 500 mSv in a year	An equivalent dose to the extremities (hands and feet) or the skin[c] of 150 mSv in a year

[a]The dose limits are for planned exposure situations
[b]Apprentices who are being trained for employment involving radiation and students who use sources in the course of their studies
[c]The equivalent dose limits for the skin apply to the average dose over 1 cm^2 of the most highly irradiated area of the skin. The dose to the skin also contributes to the effective dose, this contribution being the average dose to the entire skin multiplied by the tissue weighting factor for the skin

In the BSS, the employer and licensee have joint responsibility for the protection of workers against occupational exposure, and that includes ensuring that:

- Protection and safety is optimized and that dose limits for occupational exposure are not exceeded;
- A radiation protection programme is established and maintained, including classification of areas, establishment of local rules and procedures, and provision of personal protective equipment;
- Arrangements are in place for the assessment of occupational exposure through a personnel monitoring programme; and
- Adequate information, instruction and training on radiation protection and safety are provided.

Personnel also have responsibilities with respect to occupational radiation protection, including that they must follow the rules and procedures and they make proper use of the monitoring equipment and personal protective equipment provided.

The reader is referred to the BSS for full details on requirements for occupational radiation protection.

5   Public Exposure

Public exposure refers to exposure incurred by members of the public excluding any occupational exposure or medical exposure. In a radiology facility this means personnel whose duties do not involve the use of radiation and any visitors to the facility, including patients except when they are undergoing their own radiological procedure.

Table 18.3 Dose limits[a] for public exposure, from the BSS [6]

An effective dose of 1 mSv in a year[b]
An equivalent dose to the lens of the eye of 15 mSv in a year
An equivalent dose to the skin of 50 mSv in a year

[a]The dose limits are for planned exposure situations
[b]In special circumstances, a higher value of effective dose in a single year could apply, provided that the average effective dose over 5 consecutive years does not exceed 1 mSv per year

The BSS sets out requirements needed to ensure adequate protection for members of the public. For a radiology facility, this usually means ensuring appropriate shielding for areas where radiation is used and controlling access to areas where radiation is used. Responsibility for this lies with the licensee. Public dose limits are listed in Table 18.3.

6   Human Imaging Using Radiation for Non-Medical Purposes

Not all human imaging takes place for medical purposes. Global events and social considerations have led to an increasing worldwide usage of human imaging for various aspects of security screening. This typically takes place in public places, such as airports, border control, prisons and court houses. For these applications, specially designed inspection imaging devices are used. The BSS sets out radiation protection requirements for such activities.

There are however some other situations where again human imaging takes place for non-medical purposes, but the procedure is conducted by medical personnel, using medical clinical radiological equipment. Such situations are employment related, or for legal or health insurance purposes, performed without reference to clinical indications. In these cases, the optimization requirements of the medical exposure section of the BSS apply.

7   Impact of the BSS

The impact of the BSS in a given country will depend, in the first instance, on the degree to which the country's legislation and regulations adopt or incorporate the requirements of the BSS. This process depends on governments, and is influenced by politics, resources and priorities. The health authority and the professional bodies, particularly medical imaging (i.e. radiology), radiation technology and medical physics societies, all have important advisory and advocacy roles to play in this process.

But what is critical is not what the law says but the practical implementation of the BSS in radiology facilities. Awareness about the BSS and its role needs to be promoted. If and when required, education and training resources should be provided and infrastructure should be strengthened. In a given radiology facility, the health professionals not only need to assume their respective radiation protection responsibilities described above, but they also need to demonstrate to management that improving radiation safety culture yields additional dividends, i.e. better working conditions, better patient care and, as a consequence of greater patient satisfaction, potentially more financial revenues. The implementation of the BSS will be facilitated if it is seen as adding value to the process of providing radiology services, and not as yet another burden.

8 Conclusions

The BSS is an excellent example of successful collaboration among leading global stakeholders in radiation protection. It provides a set of requirements that form the basis for ensuring acceptable levels of radiation safety and protection for patients, personnel and members of the public, and at the same time leads to the practice of radiology being performed in a manner that results in quality patient care.

References

1. American College of Radiology (2013) ACR appropriateness criteria. Available from http://www.acr.org/Quality-Safety/Appropriateness-Criteria. Accessed 1 Oct 2013
2. Food and Agriculture Organization of the United Nations, International Atomic Energy Agency, International Labour Organisation, OECD Nuclear Energy Agency, Pan American Health Organization, World Health Organization (1996) International Basic Safety Standards for Protection against Ionizing Radiation and for the Safety of Radiation Sources. Safety Series No. 115. IAEA, Vienna
3. Hanson H, Borrás C, Jiménez P (2006) History of the radiological health program of the Pan American Health Organization. Pan Am J Public Health 20(2/3):87–98
4. International Atomic Energy Agency (1962) Basic Safety Standards for Radiation Protection, Safety Series No. 9. IAEA, Vienna
5. International Atomic Energy Agency (1982) Basic Safety Standards for Radiation Protection (1982 edition), Safety Series No. 9. IAEA, Vienna
6. International Atomic Energy Agency (2011) Radiation Protection and Safety of Radiation Sources: International Basic Safety Standards – interim edition, IAEA Safety Standards Series GSR Part 3 (Interim). IAEA, Vienna
7. International Labour Organisation (2008) International standard classification of occupations. ISCO-08 Group definitions – final draft (word), Page 57. Available from http://www.ilo.org/public/english/bureau/stat/isco/isco08/index.htm. Accessed 19 Feb 2012

8. International Commission on Radiological Protection (1996) Radiological Protection and Safety in Medicine. ICRP Publication 73, Ann ICRP 26(2):1–47

9. International Commission on Radiological Protection (2007a) The 2007 Recommendations of the International Commission on Radiological Protection. ICRP publication 103, Ann ICRP 37 (2–4):1–47

10. International Commission on Radiological Protection (2007b) Radiological Protection in Medicine. ICRP publication 105, Ann ICRP 37(6)

11. International Commission on Radiological Protection (2011) Statement on tissue reactions, 21 Apr 2011. http://www.icrp.org/page.asp?id=123. Accessed 19 Dec 2011

12. Royal College of Radiologists (2011) iRefer: making the best use of clinical radiology services, 7th edn. RCR, London, 2011. Available from http://www.irefer.org.uk/index.php/about-irefer. Accessed 17 Feb 2012

Chapter 19
International Clinical Teleradiology: Potential Risks and Safeguards

Adrian K. Dixon and Arl Van Moore Jr.

Abstract   Teleradiology, in conjunction with Picture Archiving and Communications Systems (PACS), has brought about dramatic changes and opportunities in the way in which radiology can be provided and practised, with networking links integrating a wide range of both national and international centers. This chapter addresses these issues from the point of view of the radiology department, the radiologist, the individual patient, the radiology manager and the radiology and medical community as a whole.

Keywords   International teleradiology • Picture archive communication system • Regulation • Teleradiology • Teleradiology standards

1   Introduction

Teleradiology based on both the continued advances as well as the reduced costs in computing power, improved data storage efficiencies and networking bandwidth and transmission speed, in its broadest sense, has been one of the great innovations in the practice of medicine of the last two decades. To be able to view radiological images at any appropriate site inside or outside a medical facility (Fig. 19.1) is a huge boon to timely interpretation and medical diagnosis, all to the benefit of the treatment and management of our patients (Table 19.1) [11]. Specialized consultation about

A.K. Dixon (✉)
University of Cambridge, Addenbrookes Hospital Box 218, Hills Rd,
Cambridge CB2 2QQ, UK
e-mail: akd15@radiol.cam.ac.uk

A. Van Moore Jr.
Department of Radiology, Carolinas Medical Center Charlotte Radiology,
1701 East Blvd, Charlotte, NC 28203, USA
e-mail: Van.Moore@charlotteradiology.com

L. Lau and K.-H. Ng (eds.), *Radiological Safety and Quality: Paradigms in Leadership and Innovation*, DOI 10.1007/978-94-007-7256-4_19,
© Springer Science+Business Media Dordrecht 2014

Fig. 19.1 Teleradiology
viewing and reporting
options: intra-facility, inter-
facility, home, regional,
national and international

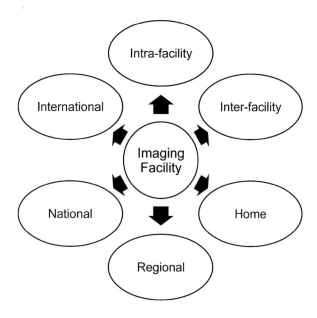

Table 19.1 Teleradiology
applications and benefits for
key stakeholders. Impact of
teleradiology use:
+ + = good improvement;
+ = some improvement;
+/− = little/no improvement

	Consumers	Providers	System
Clinical services			
Local access	+ +	+/−	+ +
Workload balancing	+/−	+ +	+
Second opinion	+ +	+ +	+ +
Subspecialty reporting	+ +	+ +	+ +
After-hour cover	+ +	+ +	+ +
Telemedicine	+ +	+	+ +
Education and training	+/−	+ +	+ +
Peer review	+/−	+	+
QA	+/−	+	+
Research	+/−	+ +	+/−

complex cases (e.g. unusual neuroradiological problems) is another benefit that
developments such as Picture Archiving and Communications Systems (PACS)
have brought about (Fig. 19.2). Indeed, many centers now have one senior neuroradiologist available to review all neuroradiological examinations from a wide population catchment area 24 h a day, 7 days a week. In taking advantage of the different
time zones may well involve overseas/international discussion and diagnosis. There
are clear benefits for a radiologist and the referring clinician to be able to view and
discuss an image simultaneously at different locations. Those locations can be in the
same room with the radiologist and referring clinician in the same room or the two
individuals may be half a world apart. The referring clinician might be on the West
Coast in the United States at 2 a.m. and the diagnostic radiologist might be in
Switzerland to cite an example. Quite apart from neuroradiology, such discussion

Fig. 19.2 Teleradiology applications: clinical services, i.e. primary interpretation, workload balancing, second opinion, subspecialty reporting, after-hour cover, telemedicine for clinicians; education and training; peer review; quality assurance; and research

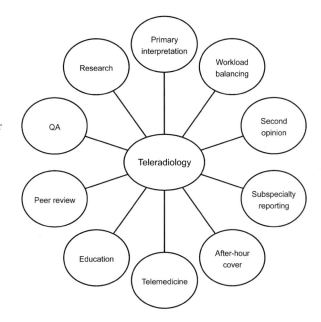

with other experts can have great benefit in other specialized fields – non-accidental injuries in pediatric care and rare bone tumors in musculoskeletal radiology spring to mind.

The opportunities provided by this helpful dissemination of the diagnostic images may brings problems that pose threats to the established practice of radiology as we know it because of this powerful changing dynamic (Table 19.2). This chapter will discuss these pros and cons of such external and international involvement in what is hoped is a balanced and measured way and will offer some final overall personal insights and perspectives as seen through the eyes of two authors who have experienced the changes introduced in their profession by the rapid introduction of teleradiology during the last two decades.

2 The Radiology Department

Radiology departments have evolved into large multidisciplinary organizations embracing the skills of radiologists, radiographers, medical physicists, physician assistants, nurses, nursing assistants, IT professionals, administrators, secretaries, and many other support staff. All of these personnel are encouraged and stimulated to rally round as a team that strives to cover problems, handle enquiries and elevate standards. The successful introduction of teleradiology to a radiology department hinges on good IT integration with the existing picture archiving and communication, radiology information, hospital information and voice recognition systems, whether

Table 19.2 Teleradiology and its impact on practice performance

Practice performance	Positive	Negative
Access	Enables timely access to radiology reports, irrespective of location, time and subspecialty	
Integrated care	Supports telemedicine	Fragments clinical and imaging teams
		Reduces access to past medical and imaging records, consultation and discussion
Appropriateness	Alerts timely referral to tertiary centres, if and when appropriate	Eliminates referrer and patient contact
		Hinders consent and justification
Safety		Reduces supervision and technique optimization
		Limits risk and adverse event management
Responsiveness		Reduces awareness of the patient's needs
Report	Improves turn-around time, 24/7 cover	Reduces vetting of radiologists
		Opens to telecommunication risks
Capability	Promotes sub-specialization	Deskills and isolates general radiologists
	Offers opportunities for training, audit and research	Varying radiologist's training, language and local knowledge
Quality assurance	Utilizes as a QA tool	Reduces team-based quality improvement actions
Efficiency	Opens to competition, rationalization Tailors workflow	Could lead to unsafe workload
Effectiveness	Enables local care of condition Reduces social disruption	Could result in ineffective management of imaging facilities
Sustainability	May temporarily address workforce shortage by transferring to another country	Has unclear medico-legal issues Has hidden cost
	Reduces morbidity and mortality by subspecialty reporting	

outsourcing or not. If radiology image interpretations are to be provided by radiologists based outside the department in the same metropolitan area or afar, there are many potential barriers because of communication or cultural issues, which can potentially reduce the quality of service and benefit to the patient. Likewise, because there is no close "rubbing of shoulders" between radiologists, physicists and administrators on site, problems may arise such as protocols in obtaining images, equipment replacement, etc. The remote radiologist, whether next-door or in another country, may not feel so responsible for, nor be aware of radiation dose or examination protocol issues relating to the images they are interpreting. Although he or she may be "directing" the examination either under the letter of the 1997 European Union Council Directive 97/43/Euratom or the ACR-ACPM-SIIM Technical Standard for Electronic Practice of Medical Imaging [1, 3], this "direction" becomes tenuous in the extreme with respect to outsourced radiological supervision.

When considering both the national and international perspectives, the individual radiology department will have to consider the statutory and legal requirements pertaining to the local, states, or countries in question. One question that should be asked is "Can an entire radiology department be 'managed' from another country whether a neighbor or on another continent?" There are now several examples within the UK where multinational providers have set up reporting centers outside the conventional normal National Health Service axis and offer provision of various services (reporting, mobile MR/CT/US services, staffing, etc.). And in the UK, recent legislation is making it even easier for promotion of outsourced services including the management of an entire radiology department.

Whilst in the short term managers may appreciate being able to offload some of the problems of running a department such as purchasing, maintenance along with managing complex equipment and staffing issues prove economically advantageous, in the long and intermediate term there are inevitably good and bad points. On the plus side, the large multinational outsourced provider assumes the responsibilities for the problems of ensuring that reporting radiologists are available and registered in the country (no trivial task) so that reports are issued in a timely fashion. The large provider also carries the responsibility of ensuring the quality and the clarity of the language of the reports. For example, when outsourced mobile MR with overseas reporting was first introduced in the UK to overcome lengthy waiting list problems, a clinical guardian had to be appointed to deal with initial complaints about the quality and lack of clarity in the language and wording of the overseas-outsourced reports. This led to intensive quality control and audit of performance culminating in a series of over 250,000 MR reports being double read; major discrepancies were reviewed by the clinical guardian [4]. Such safeguarding – introduced somewhat belatedly – means that outsourced reporting may now be subject to closer scrutiny than routine reporting.

Against these potential advantages comes the accusation that financially 'for profit' driven companies will 'cherry-pick' the simple radiological workload. There is perceived profit in performing and reporting high volumes of simple procedures such as outpatient (ambulant) knee, spine and shoulder MR examinations and reporting these off shore in a country where radiologist salaries may be lower. This leaves the radiologists in the base hospital with an even greater proportion of complex procedures, which may have a lower reimbursement per unit time. Examples may be complex imaging in cancer restaging where several examinations may need to be compared with the current examination or the cases where a radiologist must be on site and providing active supervision such as paediatric cases under sedation/anesthesia; these on-site radiologists (and their associated radiologists in training) are also disadvantaged in that they will also lose the valuable on-going experience provided by a high volume of straightforward and less complex work.

3   The Radiologist

One of the upsides of the implementation of teleradiology across national boundaries is that it inevitably raised questions about the overall standard of reporting. In the United States, the use of different time zones for afterhours preliminary reports was begun at the very beginning of the twenty-first century with the efforts of NightHawk® and other enterprises that emulated this model. US radiologists in Australia read images obtained in the US. For many years these preliminary reports have been over read by the attending radiologists in the respective facilities that originated the images. Quality assurance was kept at a high level by means of this double reading of the examinations with feedback mechanisms established to provide feedback when discrepancies were noted.

When MRI reporting services in the UK were first out-sourced beyond national borders in 2007, there was an outcry that the reports would be issued by less qualified radiologists and in the beginning this was partly true. However, when full audit and Quality Assurance Review, was introduced and the quality of the reports returned to the clinician was analyzed there were few overall differences between the reports coming from the distant teleradiology providers than in most district general hospitals in the UK. Occasionally there were problems with respect to the fact that the outsourced radiologist did not have full access to the previous imaging studies and other data about the patient. Nor could the outsourced radiologist readily discuss the patient with the referring clinician on site (and vice versa), although this issue could be partly solved by using phone or video consultation.

This in turn has led to an overall improved audit of radiological procedures in all branches of UK practice, which has been a very good thing for patient care. It has led to much better report quality assurance and some purchasers of radiological services now insist that 10 % of all reports are double-read to monitor reporting discrepancies within all radiological practices and such practice is slowly being encouraged throughout Europe. In response, some teleradiology providers incorporate QA into the reporting session and radiologists are asked to double read some cases as part of the reporting work list. This inevitably leads to specialization and in some cases hyper-subspecialization where some radiologists relinquish the reporting/interpretation of certain procedures and concentrate their reporting/interpretation on areas where they have designated subspecialty skills. This more targeted reporting/interpretation in some settings means that radiologists will have to gain extensive subspecialty skills. Radiology practices then have to adapt to become a complex network of subspecialty medical practice.

While in some areas the net effect, created in part by teleradiology, increasingly imposes problems for coverage in small and rural communities – if the consultant radiologist is not seeing enough mammograms to keep his or her radiological eye in and maintain their respective diagnostic skills, it will mean the mammograms are no longer reported/interpreted by that individual. Ultimately, it could be that all neuroradiological procedures (e.g. a conventional cranial CT) will have to be reported by a card-carrying, designated, accredited neuroradiologist. Thus, one can envisage that over time the smaller radiological practices becoming much more closely aligned with the larger central hub on a "hub and spoke" model covered by a large PACS network and this is increasingly happening in certain areas of the UK.

In the US for example, there are now several large practices where the radiologist is either only on site a limited number of days during the week or in some cases not at all. All of the imaging, however, is provided to the facility by high functioning sub-specialized radiologists 24 h a day, 7 days a week. In essence the highly functioning radiologist is brought to the patient bedside through the magic of teleradiology and the patient is receiving the best level of care available heretofore only provided to patients directly within the walls of a major medical centre. In the small hospital in Wadesboro, NC the patients of that facility have access to imaging interpretations provided by the Carolinas Medical Centre subspecialty radiologists Charlotte, NC. But of course this does mean that there is no radiologist immediately on hand to perform a simple interventional radiological procedure such as CT guided drainage of an abdominal abscess.

In other countries around the globe such hub and spoke models involve overseas links so that reports can be issued in real time during the night with the staff radiologist reviewing the reports and formally signing them off the next morning. An interesting discussion in the American Journal of Roentgenology in 2011 [2] between that doyen of the medico-legal world, Leonard Berlin, and a reader about quality assurance, where Dr Berlin was asked whether there is an industry standard for acceptable discrepancy rates between preliminary radiology reports and final reports. Much depends on the level of discrepancy. What are the acceptable levels of discrepancy? What are considered trivial discrepancies and what are egregious discrepancies? One outcome of asking these questions is that within the UK all practices must now be audited [5] and the use of teleradiology has highlighted the need for this.

In the UK NHS audit of 386,567 outsourced MR examinations [8], examinations and reports were graded on a system of 1–5; audit showed gradual improvement over 3 years in the quality of the examinations, the language of the report and the clinical opinion offered. Most centers introduced an English-speaking language checker for the final report; some of the original reports were not easily understood by UK clinicians due to radiologists using correct but somewhat old fashioned anatomical terms. Happily the overall number of serious life-threatening discrepancies was very low, but this is perhaps not surprising as the cases selected were mainly ambulant patients with traumatic or degenerative conditions. With regards to the numerous articles in the radiology literature, which have measured discrepancy rates in different situations, they range mainly in the 2–7 % for discrepancies of intermediate importance [4]. Some teleradiology providers lead by turning these case discrepancies into learning opportunities for the radiologists as part of a regular internet-based group learning process (webinar). A similar learning and team-building model could be applied to improve image optimization and other QA actions between the outsourcing and reporting facilities.

There is a range of day-to-day practical issues, which could impact on radiologist performance and human error when working with examinations originating from different institutions. This is due to significant variations in radiology information system, system security, imaging protocol, image presentation, image dispatch, referral format, worksheet layout and illegible handwriting. Such variations could increase the time taken to extract information and human error due to overlooking or missing data. However, these potential risks could be resolved by standardization if and when possible.

For the local radiologist, it may be frustrating to have to review reports coming in from international centers that he/she did not report or review. In addition to waiting for the images to be assimilated onto the local PACS and then having to discuss them in detail at subsequent clinico-radiological or multidisciplinary team meetings the findings will need to be accurately communicated to all of the physicians involved in the patient care. On the other hand the inconvenience of these added responsibilities may be offset by not having to be disturbed so much by calls in the middle of the night.

For the radiologist working at the remote imaging interpretation site, it may at times be depressing to have so little if any immediate communication and feedback from fellow radiologists, especially when difficult cases are involved. The lack of discussion with clinical colleagues (and indeed the patient) has been described by some radiologists as working in a vacuum. The difficulty of comparing the transmitted images with previous radiological examinations can also pose a major problem; this can lead to inappropriate and occasionally inaccurate reports. We all occasionally like to hear of our clever pick-ups in routine reporting. And, sadly but profitably, we all learn from our errors at discrepancy meetings or Quality Assurance conferences.

4   The Patient

The patient will, in all likelihood, be blissfully unaware of all the controversial aspects concerning the alternative models whereby radiological services can be provided. All he or she usually cares about is getting the procedure performed quickly and cheaply by someone who is competent. The more inquisitive patient may consider factors such as the quality of the examination (e.g. up to date imaging technology and low radiation dose) and the qualifications of the individual generating the report (e.g. training and QA factors such as the radiologist's history of false negatives/positives) important factors. But Sir Raymond Hoffenberg, long ago pointed out the difficulties of providing health care that is good, cheap and timely; two of these factors often come with a knock on deleterious effect on the third. With the value equation, where value is determined by quality per unit cost, there will need to be limits on minimal levels of training and expertise as well as competency in the workplace. This is a struggle that the patient rarely is exposed to in their quest for medical care, especially in imaging. They rely on others to examine and enforce these parameters.

Timeliness of the examination is paramount in the patient's mind. All patients have a level of anxiety when told they should undergo any examination. Thus an electronic booking system that immediately offers a (choice of) time and date for any given procedure is attractive. For the patient to be told 'we will send you an appointment/we will contact you' is not reassuring – the thoughts 'will they', 'has it got lost in the post' inevitably occur. Furthermore the patient cannot plan other events when 'hanging on' for an appointment. Thus, from the patient's perspective, a provider (perhaps a multinational company or other out of system provider) who supplies an examination at an alternative site in a more timely fashion may be preferable to a less prompt service from the home base.

Cost is also a major factor for the patient. In a nationalized system the patient may not directly bear the cost. But, in the private marketplace, there can be wide variation in costs according to prestige of the medical centre to which the patient is referred to; different radiologists/radiological practices may also have widely differing reporting fees. There may be different charges for an urgent procedure; conversely charges may be cheaper at less popular time-slots. Even within a nationalized system, a patient may 'crack' and choose to pay to have a private CT examination performed (e.g. by an international company) the next day rather than wait to have their turn for their nationalized one in say approximately 10 days time. The irony is that, for something like an oncological referral, such a privately performed CT examination often ends up being re-reported by a radiologist in the nationalized system following PACS transfer of the patient's images. The overall cost to the health care system is greater because of duplication in both radiologist efforts and system costs.

Quality of the total radiological experience (diagnostic accuracy, booking, comfort etc.) is hardly something discussed by patients. One does not usually overhear gossip such as 'I always go to Dr X; she/he has such good results' about radiologists, as rumored to occur for gynecologists, etc.! Few patients realize that it was clever Dr Y in the radiology department who spotted the atypical lesion in the colon – which turned out to be a malignant lesion – which was removed at an early stage by laparoscopic surgery – thereby not only making an early diagnosis and saving the patient's life but important in effecting treatment in a much earlier stage, reducing the overall treatment cost to the system and likely prolonging the patient's life. Subsequent praise will still often go to the surgeon or oncologist.

Lay understanding of the importance of radiological services and expertise in treatment at all levels in the community still has a long way to go! From a legal point of view, the patient in some venues may need to be asked whether they accept that the report is to be provided by a radiologist from another country. This does not necessarily pertain within Europe so long as the radiologist in the other country is listed on the national specialist register of the country where the examination is performed. Likewise, within the US the radiologist may be licensed, credentialed in good standing within another state. In Europe, for a report to be generated outside the European Economic Area, the patient's permission and consent must be obtained – even though the radiologist in question is on the patient's national specialist register or is an accepted provider within the patient's healthcare network. In the US the ACR Teleradiology Task Force recommendation is that any time that a patient requests information regarding the location of the radiologist interpretation that the information be provided to the patient.

In summary, the patient's experience is greatly assisted by teleradiology and other electronic aspects of care. The booking should be timely and the images, once obtained, should be instantly available throughout the medical enterprise. The report can be viewed and discussed. Teleradiology also facilitates previous imaging availability to allow comparison at case conferences and follow-up radiological examinations.

5   The Radiology Manager

Managers may be very impressed that their 'out of hours' radiology can be sent to another provider whether in the building next door or to a country halfway around the world for a rapid turnaround time provisional report. This may be seen as a quick fix to providing coverage outside of the "routine" working hours. Legislation in different countries, states or provinces determines whether or not a radiology report issued by a radiologist in another country can be viewed as definitive. In the USA, almost all images are still reviewed by a board certified staff radiologist the next day who issues the final authenticated report following a provisional report generated in another state or another country. One obvious solution is for remote reporting services to hire radiologists with approved credentials in the state/province or country where the care is being provided – but that drives costs up – which negates one of the economic attractions of outsourced reporting and negatively impacts the quality equation. One creative solution which has been successfully pioneered in a few situations is where a radiology department encourages/persuades a staff member to live in another country where time shifting can be used advantageously in order to handle the bread and butter night time reporting from afar during the normal working hours in the adopted country; an annual return back to the home country is often needed to maintain credentialing and to meet citizenship and licensure requirements.

With the globalization of health care, many big medical centers now have satellite centers around the world; for example a major USA centre may run a satellite hospital in the Far East; their own specialist radiologists in the Far East can cover the US urgent and emergent night calls and vice versa. This solution may be preferable in maintain and controlling quality while being a more cost effective alternative to hiring radiology reporting services from another organization. Yet another solution which is becoming quite widely used in the UK is to train emergency radiographers (technicians) to go one stage further than the original 'red dot' scheme (which highlighted unsuspected findings) and to issue a provisional report; indeed in some centers the radiographer who has undergone designated additional training and is well supervised by radiological colleagues may issue the definitive report which is timely and cost-effective. From a managerial perspective, such local solutions may turn out cheaper than teleradiology solutions and carry the great advantage of keeping the radiological department intact. While answering the value equation in reducing costs there is a trade off in keeping quality high since these provisional reports will be under a high level of scrutiny in the provision of contemporaneous care.

In many countries the increased mortality over weekends is a cause for concern; much has been made of the lack of availability of experienced physicians and surgeons; just as important is the timely input of radiological expertise. If there is no 24/7 provision of expert diagnostic and interventional skills at a local level, the patient's diagnosis and treatment may be delayed with severe consequences. Much hangs on what expertise can be provided locally. For example it seems essential that every medical institution with emergency admissions should have expert radiologists available on a 24/7 basis to provide abdominal ultrasound and CT and that those radiologists or their colleagues should be able proceed to drain an offending empyema or abdominal abscess or to perform the emergency embolization. Every such department should be offering around the clock cranial CT service. However not every medical centre will be providing neurosurgical facilities and this is exactly where teleradiology come into its own; the images can be discussed with remote expert neuroradiologists and neurosurgeons at the tertiary centre and the all-important decision whether or not to transfer the seriously ill patient (by ambulance or even air) can be made by teleconference.

On the other side of the argument come the advantages portrayed by a multinational integrated radiological services provider. At best, a single organization covers all the purchasing of radiological equipment, subsequent servicing, managing replacement staffing and reporting for a large group of hospitals or a large population. In this way resources can be distributed evenly to the geographical points where they are most needed; CT and MR units in close proximity to trauma and radiotherapy centers; ultrasound machines with appropriately supervised trained sonographers working out in the community. As stated above, appropriate rosters can be constructed so that a single expert neuroradiologist or group of neuroradiologists can cover the neuroradiological referrals from a wide catchment area. Likewise, interventional radiological services can be concentrated at appropriate sites.

Nevertheless, it is extremely important that such developments are under the control of experts in the field and a good model would be that in Boston where many hospitals fall under the overall umbrella of the Massachusetts General Hospital as an academic centre example or Carolinas Healthcare System as a large not for profit community institution. An integrated PACS service is the critical bedrock on which such radiological services depend. In a nationalized service such as the UK National Health Service, this is slowly being further deployed across regional boundaries. In the US this is also becoming the norm for large hospital systems. Freely transmittable and rapid availability of data from other centers can then be embedded in the local patient electronic record; but a national electronic record still seems a long way off in many countries! The cost of the systems is currently staggering. Of course the radiological equipment manufacturers and other service providers are all keen to establish long term relationships with guaranteed annual fees – such a guaranteed long term source of revenue is becoming almost more important than the costs of individual pieces of expensive imaging equipment.

6 The Radiology Community

Any change in the market place is bound to be viewed with apprehension by those intimately involved in the process. Indeed many local and professional organizations have tried to obstruct such developments. But if patients are to gain the maximum advantages of the drive to more integrated healthcare where protocols, best practice and guidelines can be shared across countries, change is inevitable and the leaders in medicine and more specifically radiology must facilitate this change. In order to monitor such change, various statements and opinions on this topic have emanated from professional organizations such as the American College of Radiology [12, 13]. The European Society of Radiology (ESR) has extensively discussed and deliberated on teleradiology as well as its implications at legal, organizational, technological and scientific levels; it has produced an ESR White Paper "Teleradiology in the European Union" [6] and an ESR Response to the Council Common Position concerning the cross-border healthcare directive [7]; it also has a dedicated subcommittee on e-health. The ESR strongly advocates that patients need to give informed consent before teleradiology is performed and that they should receive full information that their health data will be transferred to another country and that their images will be reported or consulted by individuals who have had no direct contact with the patient.

Focusing on the primacy of patients and providing guidance to good teleradiology practice the International Radiology Quality Network with the assistance and support from its members has published "Principles of International Clinical Teleradiology" [9]. These principles outline the requirements to safeguard good patient care and minimize potential risks by addressing the various aspects of teleradiology service: (1) technical issues e.g. image quality and quality assurance; (2) professional issues e.g. continuing education for radiologists and radiographers; and (3) administrative issues e.g. communication, documentation, privacy, security and ethical considerations. Some professional organizations, e.g. The Royal Australian and New Zealand College of Radiologists (RANZCR) have adopted these principles as position statement for the practice of teleradiology [10]. Under the Quality Use of Diagnostic Imaging Program, the RANZCR has conducted research into the background, practice and standards for teleradiology [11]. However, more work is needed to gain a better understanding of the legal and ethical implications, to explore its use as an education and training tool, to develop web-based education and QA tools for teleradiology practices and to advocate the adoption of practice standards.

It is important to bridge the gap between the publication of standards and their use in practice. In this regards, UN agencies can play an important role by advocating the adoption of practice standards and their implementation into national policy. Despite limited resources and infrastructure in developing countries, there is an opportunity for UN agencies to lead and facilitate the trial of innovative and affordable teleradiology solutions in these settings by using low cost (less resolution) digitization systems, standard communication, workstation and free software. This action would provide specialist opinion to a remote community, which was not previously possible.

7 Conclusion

The increasing use of teleradiology is driven by clinical needs, consumer expectation for prompt turn-around-time, in and after hour cover, workforce shortage, increase in workload both in volume and complexity, budget challenges, rationalization and outsourcing. Advances in information and telecommunication technology further facilitate this trend, so is the availability of faster and more secure private networks. In a recent survey 67 % of Australian radiologists use teleradiology in their daily work [10]. As in the use of radiation in medicine, the ultimate goal for teleradiology is to maximize its benefits and minimize its risks.

The impact of teleradiology on practice performance was previously reported [11] and this is summarized in Table 19.2. Teleradiology is like a "two-edged sword" that requires careful consideration and balancing, needing uniform standards to guide quality care while ensuring patient safety. Specialist radiologists can make a pivotal contribution to clinical decision-making and management. Using radiologists and radiology wisely could reduce the burden on the healthcare system by not only improving diagnosis and management but also by reducing unnecessary and repeated radiation exposure, thus optimizing overall patient care. Therefore, by providing leadership and implementing innovative actions, the radiology stakeholders will ensure safeguards are in place to maximize the benefits for international clinical teleradiology and to minimize its potential risks, thus leading to better patient-focused care.

References

1. ACR–AAPM–SIIM (2012) Technical standard for electronic practice of medical imaging, Revised April 2012
2. Berlin L (2011) Quality assurance and nighthawk radiology. AJR Am J Roentgenol 197(5): W963
3. Council of the European Union (1997) Council Directive 97/43/Euratom of 30 June 1997 on health protection of individuals against the dangers of ionizing radiation in relation to medical exposure, and repealing Directive 84/466/EURATOM. OJEC 1997/L 180/22
4. Dixon AK, Fitzgerald R (2008) Outsourcing and teleradiology: potential benefits, risks and solutions from a UK/European perspective. J Am Coll Radiol 5(1):12–18

5. European Commission (2009) Radiation Protection 159. European guidelines on clinical audit for medical radiological practices (Diagnostic Radiology, Nuclear Medicine and Radiotherapy). EU Publications Office. http://ec.europa.eu/energy/nuclear/radiation_protection/doc/publication/159.pdf. Accessed 2 Apr 2012
6. European Society of Radiology (2006) Teleradiology in the European Union, white paper. http://www.myesr.org/html/img/pool/1_ESR_2006_VII_Telerad_Summary_Web.pdf. Accessed 2 Apr 2102
7. European Society of Radiology (2010) ESR position on the proposal for a directive of the European Parliament and of the council on the application of patients' rights in cross-border healthcare (Spanish Presidency compromise paper). http://www.myesr.org/html/img/pool/20100713_ESR_Position_Cross_border_healthcare_directive_Final.pdf. Accessed 2 Apr 2012
8. Golding SJ, Webster P, Dixon AK (2010) Impact of a clinical governance programme on reporting standards in MRI: findings of the national service in outsourced MRI. In: Proceedings of UK Radiological Congress 2010, BJR Congress Series. http://bjr.birjournals.org/site/misc/Proceed_2010.pdf. Accessed 16 June 2012
9. International Radiology Quality Network (2009) Principles of international clinical teleradiology. Available at http://www.irqn.org/work/teleradiology.htm. Accessed 11 June 2012
10. Kenny LM, Lau LSW (2008) Editorial: clinical teleradiology – the purpose of principles. Med J Aust 188(4):197–198
11. Lau LSW (2006) Clinical teleradiology in Australia: practice and standards. J Am Coll Radiol 3(5):377–381
12. Van Moore A, Allen B Jr, Campbell SC et al (2005) Report of the ACR task force on international teleradiology. J Am Coll Radiol 2(2):121–125
13. Van Moore A Jr (2009) ACR presidential address: with change inevitable, can we survive? J Am Coll Radiol 6(11):749–755

Part IV
Global, Regional and National Challenges, Actions and Opportunities

Chapter 20
Leadership and Innovations to Improve Quality Imaging and Radiation Safety in Africa

Michael G. Kawooya, Azza Hammou, Hassen A. Gharbi, and Lawrence Lau

Abstract Most of the African communities are rural and have poor access to general healthcare and diagnostic imaging. There is a severe shortage of healthcare professionals, including radiologists, ultrasonographers, radiographers and medical physicists, especially in the rural areas. Other challenges for the region include work volume, population exposure, equipment, imaging budget and awareness in radiation safety and protection.

To address these challenges and to improve quality radiology and radiation safety in Africa, the regional leaders initiated a range of innovative actions. These actions are discussed, including detailed examples focusing on the capacity building of imaging professionals by education and training and the strengthening of region-wide radiation protection regulatory framework by collaboration. The authors aim to share this experience so that actions could be replicated in other regions if appropriate.

The stakeholders work together by collaboration, advocacy, experience sharing, mutual assistance, and maximizing resources to improve awareness; education and

M.G. Kawooya (✉)
Department of Radiology, Ernest Cook Ultrasound Research and Education Institute (ECUREI), Mengo Hospital, Kampala, Uganda
e-mail: kawooyagm@yahoo.co.uk

A. Hammou
Tunisia Medical School, Tunisian National Center of Radiation Protection (CNRP), Hôpital d'Enfants, Tunis, Tunisia
e-mail: aza.hammou@rns.tn

H.A. Gharbi
Radiology and Medical Biophysics, African Society of Radiology, World Federation for Ultrasound in Medicine and Biology, Tunis, Tunisia
e-mail: hassen.agharbi@planet.tn

L. Lau
International Radiology Quality Network, Reston, VA, USA
e-mail: lslau@bigpond.net.au

L. Lau and K.-H. Ng (eds.), *Radiological Safety and Quality: Paradigms in Leadership and Innovation*, DOI 10.1007/978-94-007-7256-4_20,
© Springer Science+Business Media Dordrecht 2014

training; provision of tools and guidance; radiation safety and radiation protection; system infrastructure and the implementation of effective policies. The leaders from the professional organizations, institutions and regulatory authorities play important roles to improve the access to quality imaging and radiation safety in Africa.

Keywords Leadership • Radiation protection • Radiation safety • Radiology education • Ultrasound education

1 Introduction

Africa is the second most populous continent with a population of over one billion. Among all the continents, Africa has the highest number of countries totaling 54. The continent is geographically grouped as Northern, Western, Central, Eastern and Southern Africa. Despite the variations in population density, socioeconomic status and healthcare access, there is general acceptance of the beneficial use and pivotal role of diagnostic imaging and ionizing radiation in healthcare. The challenge for all is to ensure whenever and wherever diagnostic imaging or radiation medicine procedure is available, it is provided by qualified personnel and all equipment is properly installed and well maintained. For equipment employing ionizing radiation, it should comply with international safety requirements and is supported by a radiation safety program.

This paper presents some of the issues and challenges for diagnostic imaging, radiation safety and implementation of radiation safety standards in Africa; leadership actions by individuals, organizations and authorities; workforce capacity building by strengthening education and training; innovations to strengthen national radiation protection infrastructure; and evaluation of the effectiveness and outcome of some of these interventions.

2 Challenges

2.1 Geography and Infrastructure

There is a growing access to modern diagnostic imaging and therapeutic technologies in African countries. However, the access to and the sophistication of diagnostic and therapeutic procedures vary within and between countries. More than 80 % of the population in most Sub-Saharan African countries is rural. While many urban communities have access to modern imaging facilities almost comparable to those in developed countries, there is very limited or no imaging services in the rural communities. This uneven distribution of resources between urban and

Table 20.1 Radiological equipment distribution in different regions of Tunisia, expressed in the number of units (N) (Centre National de Radio Protection, Tunisia national register 2011). Abbreviations: *NE* North East, *NW* North West, *C* Central East, *CE* Central West, *SE* South East, *SW* South West, *mammo* mammography, *OPG* orthopantomography, *BMD* Bone densitometry

Regions	Population '000	X-ray units	Dental X-ray units	CT units	Mammo units	OPG units	BMD units	Total X-ray units
Tunis	2,399	493	657	59	62	45	13	1,329
NE	1,465	105	171	15	16	17	6	330
NW	1,225	55	72	11	4	3	2	147
CE	2,422	292	371	35	29	26	17	770
CW	1,398	45	52	8	4	6	2	117
SE	961	86	78	14	8	10	2	198
SW	586	37	35	4	2	0	0	78
Total	10,458	1,113	1,436	146	125	107	42	2,969

Table 20.2 Radiological equipment distribution in different regions of Tunisia, expressed in the number of units per 100,000 inhabitants (N/I) (Centre National de Radio Protection, Tunisia national register 2011)

Regions	Population '000	X-ray units/ 100,000	Dental X-ray units/ 100,000	CT units/ 100,000	Mammo units/ 100,000	OPG units/ 100,000	BMD units/ 100,000	Total X-ray units/ 100,000
Tunis	2,399	20.50	27.38	2.45	2.50	1.87	0.54	55.39
NE	1,465	7.16	11.67	1.02	1.09	1.16	0.41	22.52
NW	1,225	4.49	5.87	0.89	0.32	0.24	0.16	12.00
CE	2,422	12.00	15.31	1.44	1.19	1.07	0.70	31.79
CW	1,398	3.22	3.72	0.57	0.28	0.42	0.14	8.37
SE	961	8.94	8.11	1.45	0.83	1.04	0.21	20.59
SW	586	6.28	5.94	0.67	0.33	0.00	0.00	13.25
Total	10,458	10.64	13.70	1.39	1.19	1.02	0.40	28.38

rural communities is accompanied by a corresponding lower radiation protection and radiation safety infrastructure in rural settings (Tables 20.1 and 20.2).

Services, which exist, face numerous interrelated challenges. For example, the hospital-based utilization in Uganda is 125/1,000 in a major urban hospital and 8/1,000 in a rural hospital. Only 10 % of those patients that should have imaging in the rural setting received imaging compared to 56 % in a major urban hospital.

2.2 Workforce Shortage

Workforce capacity varies between the African countries, e.g. Uganda has a physician to population ratio of 4.7 per 100,000, compared to Kenya's 14.7, and South Africa's 15.0 [17, 18]. Human resource stands out as a key limiting factor towards a more equitable access to diagnostic imaging and to attain the Millennium Development Goals (MDGs) in Africa [16, 18]. The High Level Forum for the United Nations Millennium Development Goals estimated that a work force of 2.5 healthcare workers per 1,000 is required to meet the MDGs. For Uganda, this figure is 1.5, which means there is a need for 26,739,955 [6].

Within a country, there is significant variation in the workforce between urban and rural areas. For example, in Ugandan public hospitals, radiologists are available only in the Regional Referral and Tertiary levels hospitals. The country has 40 radiologists but more than 30 of these works in the capital city of Kampala. There is no radiologist to interpret or perform procedures in the rural areas. This misdistribution of work-force is similar for radiographers, sonographers and medical physicists, i.e. 50 % (total = 200) of radiographers, 95 % (total = 50) sonographers and 100 % (total = 4) medical physicists work Kampala. The established posts for radiographers are only found in the General Hospitals, Regional Referral and Tertiary levels. There is no radiographer at primary healthcare level. In remote or rural areas, many equipment operators may have little or no training and even if trained, often have no access to continuing medical education. The supervision and support are minimal.

2.3   Workload

The radiologists are overworked. For example, a Ugandan radiologist performs 20,000 procedures per year, works more than 8 h per day and almost 300 days in a year, with a much lower remuneration in comparison to radiologists in developed countries. As a result of this workforce shortage, less than 10 % of all X-ray procedures were reported and only at tertiary and regional referral levels. To overcome this challenge, the adoption of a more pragmatic and innovative approach by the training of other healthcare personnel, e.g. radiographers in plain film interpretation is suggested [20].

2.4   Population Exposure

The risks for radiation use in medicine are well documented. Radiation safety is becoming a public healthcare concern, especially with certain diagnostic radiology, nuclear medicine and radiotherapy procedures. In recent years there is a significant increase in population exposure from non-natural sources, most of which arising from computed tomography, fluoroscopy and interventional radiology procedures [2, 14, 21].

2.5   Workplace Issues

The equipment is often old, outdated, hardly calibrated and functions poorly. Radiation protection measures are often not enforced. Breakdowns are frequent and there is virtually no maintenance or quality assurance system in place [8]. Spare parts are scanty and often unaffordable. The electricity supply if available is erratic and unregulated with dangerous voltage surges.

Patients present late and referrals to higher-level healthcare facilities are hampered by poor transport and infrastructure. Alternative non-ionizing imaging modalities like ultrasound are hardly available and the tendency is to rely solely on radiography. Referral guidelines for diagnostic imaging are not available. The image quality and imaging data are often substandard and there is often no qualified person to provide interpretation. Practice standards for diagnostic imaging have been developed but not yet implemented.

2.6   Budget

Health budgets in Africa are often insufficient, but the African governments are trying to step up the allocation. The 2001 Abuja declaration under the African Union committed African governments to spend 15 % of their national budget on health programs. This is as yet to be achieved by many governments [13]. For example, only 10 % of Uganda's 2007 recurrent budget was spent on health, which reflects the prevalent under-spending possibly responsible for the inequitable access to healthcare and imaging. To aggravate the situation, some governments like Uganda run a capped health budget [11, 12]. Since imaging is not a priority, it is implicit that capital and recurrent budget financing for imaging are low, yet imaging departments are capital intensive. Making the existing challenging situation worse, rural imaging departments receive less funding than their urban counterparts [8].

2.7   Radiation Safety Awareness

The existing radiation safety measures in many African countries do not comply with international radiation safety standards. The awareness of radiation protection principles among medical staff was poor. This was evident in a survey conducted by the National Radiation Protection Centre of Tunisia (NCRP). The aim was to assess the awareness, knowledge, skill and practice of radiation protection measures in a group of orthopedic surgical theater personnel in Tunisia (Figs. 20.1 and 20.2). The NRPC monitors and regulates the use of ionizing radiation in the medical and non-medical sectors. An effective regulatory framework with adequate capacity

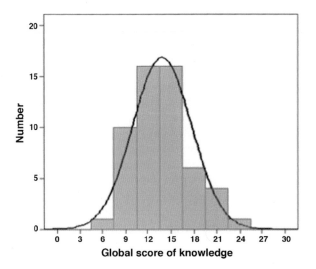

Fig. 20.1 Distribution of knowledge scores for radiation protection. The number of personnel (y-axis) is plotted against global score of knowledge (x-axis). Sixty-five orthopaedic theatre staff were surveyed and 54 responded, age between 23 and 51 and a mean of 32. The M:F ratio was 1:1. Duration of working with radiation was from 1 to 29 years and a median of 5 years. The group consisted of 18 % surgeons, 17 % trainee surgeons, 32 % nurses, 20 % technicians and 13 % service workers. In this survey 53 % had no knowledge of ionizing radiation and X-ray, 48 % were unaware of occupational exposure risk, 52 % failed to apply basic radiation safety and protection measures, 96 % ignored annual dose limits and 67 % did not use a monitoring dosimeter. Awareness was scored with a total of 30 points. It ranged from 4.60 to 25.20 with a mean of 13.85 points

plays an important role to ensure compliance to standards for all authorized facilities.

A self-administered questionnaire was conducted in Tunisia and other African countries to evaluate the effectiveness of radiation protection programs and conformance to radiation safety regulations. This study showed that the weakness of a regulatory system was directly linked to the absence of a radiation safety culture and a respect of radiological risk control.

3 Strengthening Education and Training

3.1 *ECURIE*

In 2002, to address the acute shortage of trained sonographers, nine radiologists and one medical physicist founded an ultrasound training institution – the Ernest Cook Ultrasound Research and Education Institute (ECUREI). ECUREI is a private institution of higher learning in Uganda and is registered by the National Council

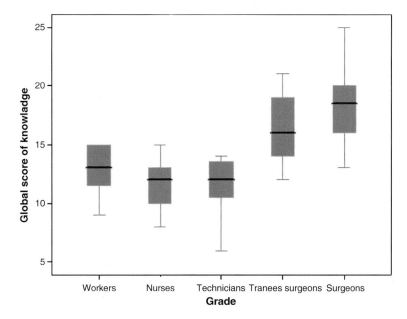

Fig. 20.2 Global knowledge scores for different grades. The global knowledge scores for radiation protection (y-axis) differ between theatre personnel (x-axis). The awareness of radiation protection global score is significantly correlated to the grade ($p = 0.001$) and the seniority ($p = 0.005$) of staff

of Higher Education. It is the only institution in Sub-Saharan Africa apart from the Republic of South Africa, which offers specialized courses in ultrasound at different levels. ECUREI is recognized as a Centre of Excellence by the World Federation of Ultrasound in Medicine and Biology [9].

3.2 Tailored Training

ECUREI started by offering training to healthcare workers to a diploma level in ultrasound in affiliation with Jefferson Ultrasound Research and Education Institute (JUREI) of the Thomas Jefferson University to improve workforce capacity and to address the issue of procedures performed by untrained personnel. Subsequently, responding to the need in other areas, the institute launched programs (Tables 20.3, 20.4, and 20.5) in radiography, biomedical engineering, X-ray pattern recognition; and postgraduate courses to train teachers for the upcoming teaching institution in Africa. ECUREI collaborated with Fontys University of Applied Sciences of the Netherlands and MEDUPROF-S and launched a free online version of its X-ray film interpretation diploma course.

Table 20.3 Ultrasound courses offered by ECUREI

Courses	Duration	Objective	Content	Competency
Certificate in basic obstetrical US	3 months	Impact skills in basic obstetrical US for rural facilities	US physics, cross-sectional anatomy and obstetrical US	Perform and interpret basic obstetrical US to detect pointers for high risk pregnancy
Diploma in diagnostic US	6–12 months	Impact skills in obstetrical, gynecological and general abdominal US	US physics, cross-sectional anatomy, obstetrical, gynecological and general abdominal US	Perform and interpret obstetrical, gynecological and general abdominal US
Bachelor in diagnostic US	2–3 years	Impact of skills in obstetrical, gynecological, general abdominal, small parts, and musculoskeletal US; management and research skills	Basic health sciences, US physics, cross-sectional anatomy, obstetrical, gynecological, general abdominal small parts, vascular, and musculoskeletal US; management and research	Perform and interpret obstetrical, gynecological, general abdominal, small parts, and musculoskeletal US, and manage an US department
Master in diagnostic US	2 years	Impact of skills in obstetrical, gynecological, general abdominal, small parts, and musculoskeletal, interventional US, echocardiography, management and research	Obstetrical, gynecological, general abdominal small parts, vascular, and musculoskeletal interventional US, echocardiography; management; research; and evidence-based medicine	Perform and interpret obstetrical, gynecological, general abdominal, small parts and musculoskeletal US, echocardiography; perform US guided interventions; set up and manage an US department; and conduct operational research

Table 20.4 Radiography courses offered by ECUREI

Courses	Duration	Objective	Content	Competency
Diploma in X-ray pattern recognition	4 years	Impart plain-X-ray interpretation skills to non-radiologists for service provision to under-served areas	Physics, radiation protection, quality assurance, radiographic anatomy, and plain-film interpretation	Interpret plain X-ray examinations
Bachelor in diagnostic imaging technology	4 years	Impart skills for conventional radiography, US and other imaging modalities; impact skills for X-ray interpretation and management skills	Physics, radiation protection, quality assurance, radiographic anatomy, basic health sciences, radiographic techniques and plain-film interpretation; obstetrical, gynecological and general abdominal US	Perform and interpret plain radiography, obstetrical, gynecological and general abdominal US; perform CT, mammography and MRI examinations
Master in diagnostic imaging technology	2 years	Impart knowledge and skill for general radiography, mammography, CT, MRI and nuclear medicine; skills for X-ray interpretation; skills for the teaching of radiography, the setting up and management of a radiology department and radiation protection; and skills to conduct operational research	Physics, radiation protection, quality assurance, radiographic anatomy, basic health sciences, radiographic techniques and plain-film interpretation; obstetrical, gynecological and general abdominal, small parts, MSK, echocardiography, interventional US; management; entrepreneurship; research; and evidence-based medicine	Perform and interpret plain radiography obstetrical, gynecological and general abdominal US; perform CT, mammography and MRI examinations; set up and manage a radiology department; and conduct operational research

The institution provides training to other healthcare workers, mainly nurses and midwives, in basic obstetrical ultrasound to address the high maternal mortality and morbidity rates attributed to mothers not attending antenatal care and not delivering in facilities where there are trained obstetrical personnel [9].

Since 2002, over 700 professionals from many different countries have graduated after completing ultrasound, radiography and biomedical engineering courses (Table 20.6). Figures 20.3 and 20.4 show the distribution of the ECUREI ultrasound trainees in Uganda and Africa respectively.

Table 20.5 Biomedical engineering course offered by ECUREI

Courses	Duration	Objective	Content	Competency
Diploma in biomedical engineering	4 years	Impart skills for the management and maintenance of imaging and other medical equipment throughout the entire medical technology lifecycle	Basic engineering sciences and patient care; electrical, mechanical, electronic and hospital engineering principles and practice; electronic prototype manufacturing; biomedical devices technology; quality assurance and systems; software engineering; and management principles	Troubleshoot and repair of common medical equipment, and to manage the entire equipment life cycle; quality assurance, calibration and safety checks; and manage a maintenance workshop

3.3   Outreach Innovations

Portable Maternal Ultrasound Unit (PMU) Project

ECUREI and the University of Washington with support from General Electric USA have trained 22 midwives and supplied 7 ultrasound machines to the Isingiro district in the Western Region of Uganda. The midwives were trained in basic obstetrical ultrasound for 3 months before returning to the healthcare centers in their districts. Obstetrical ultrasound was then integrated into routine antenatal care with the aim of identifying those mothers at risk, i.e. twin pregnancy, placenta preavia and mal-presentation. As a result, antenatal clinic attendance and delivery outcome have improved significantly.

Midwives Antenatal Ultrasound Project (MAUP)

ECUREI and the University of Washington, with sponsorship from USAID and General Electric USA have trained 26 midwives and supplied 7 ultrasound machines to the Mpigi district in the Central region of Uganda. The midwives were trained in basic obstetrical ultrasound for 3 months before returning to the healthcare centers in their districts. Obstetrical ultrasound was then integrated into routine antenatal care with the aim of identifying those mothers at risk, i.e. twin pregnancy, placenta preavia and mal-presentation. As a result, the proportion of pregnant women attending all recommended four antenatal visits (ANC) has increased from 26 % to 65 % and the proportion of those delivering in health facilities has increased from 41 % to over 50 % within 1 year of project

Table 20.6 Total number of students enrolled in ECUREI programs from 2002 to 2011. The enrolments under "other countries" were from Cuba, India, the Netherlands and United Kingdom

Programs	Total enrolled	Uganda	Kenya	Tanzania	Zambia	Sudan	D.R. Congo	Rwanda	Burundi	Nigeria	Somalia	Others countries
Ultrasound												
Diploma	421	382	11	5	10	4	2	1		1	3	2
Bachelor	37	14	3	8	10						1	1
Masters	10	5	1	3								1
Radiography												
Diploma	55	54	1									
Bachelor	156	101	3	19	15	2		14		2		
Master	14	10	1					1		2		
Biomedical engineering												
Diploma	30	30										
Total	723	596	20	35	35	6	2	16		5	4	4

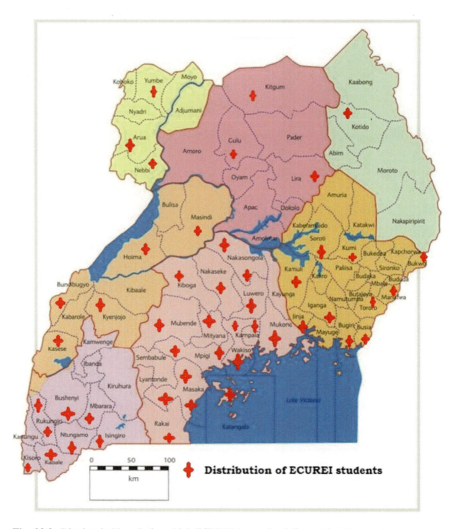

Fig. 20.3 Districts in Uganda for which ECUREI has trained diagnostic ultrasound users

implementation. It should be noted that prior to this intervention, the majority of women delivered at home or with untrained birth attendance, and both these delivery modes are associated with high maternal mortality.

Equipment Support Project

ECUREI with USAID provided support to the repair of medical equipment in Luwero and Mpigi by training of healthcare workers in these hospitals to undertake first line equipment maintenance. The project is in its infancy stage but so far, the stipulated project indicators are met. Among the indicators is the number of machines repaired, calibrated and restored to normal function.

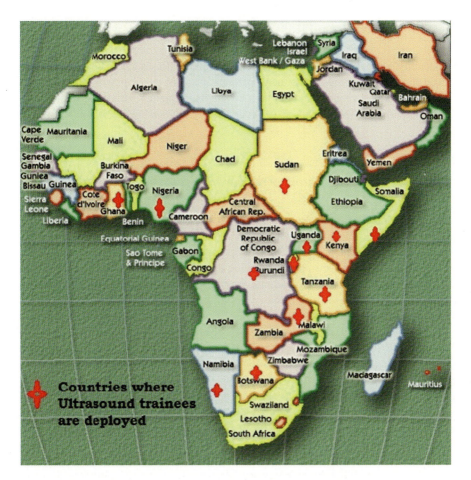

Fig. 20.4 Countries in Africa for which ECUREI has trained ultrasound users

3.4 Impact Assessment

Before the ECUREI interventions, there were hardly any users of ultrasound equipment at the primary healthcare and general hospital level. Trained sonographers were available only at tertiary and regional referral levels. This was a bleak status since one primary healthcare care facility serves up to 200,000 people. Furthermore, there was only one trained imaging machine maintenance technician for all public hospitals.

The institution is working with the University of Washington to assess the impact of its training programs and the effectiveness of its trainees and graduates on rural healthcare service delivery, focusing on maternal health; and on equipment repair, maintenance and quality assurance in rural settings. The initial results are encouraging.

4   Strengthen Regulatory Infrastructure

A country's radiation protection and radiation safety infrastructure could be strengthened by on-going needs assessment, practice standard development, policy implementation, training support and system evaluation. Together with these steps, closer stakeholders collaboration and experience sharing will lead to better outcome.

4.1   National Regulatory Framework

The International Atomic Energy Agency's Integrated Review of the Regulatory System (IRRS) compares a country's nuclear and radiation regulatory infrastructure to international standards and where appropriate, to good practice. The IRRS recognizes that the organizational structure and regulatory framework vary between countries, pending on legal and administrative systems, nuclear and radiation protection programs and available resources. The objective is to assess existing status, identify weaknesses and provide recommendations to the State the means to strengthen the effectiveness of its nuclear and radiation regulatory body and infrastructure.

Under IRRS, a pilot was conducted in Tunisia, followed by assessments in other participating countries. The methodology included the completion of questionnaires by self-declaration, verification of data by observation and subsequent analysis, aiming to determine the effectiveness of a facility's radiation protection program and its compliance to radiation safety regulations. A range of facilities using ionizing radiation was assessed, including hospitals, education institutions, manufacturing industries and others.

Based on the findings, it was established that questionnaire by self-assessment was an efficient and effective approach. The weakness of a regulatory system was directly linked to the absence of a radiation safety culture and a respect for radiological risk control. System weaknesses and issues were identified, solutions were developed and recommendations were provided to the regulatory authority to improve the effectiveness of its organization and infrastructure.

4.2   Radiation Safety training

To complement the e-learning resource from the International Atomic Energy Agency [7], the stakeholders have developed a training program with the goal to provide healthcare professionals with the skills necessary: to implement a radiation protection programs in hospitals; to understand and demonstrate compliance to national and international regulations including the International Basic Safety Standards for Protection against Ionizing Radiation and for the Safety of Radiation Sources (BSS); and to handle compliance issues.

4.3 *Radiation Safety Officer*

Even though it may not be a national requirement, each of the facilities providing radiation medicine procedures is encouraged to appoint a qualified Radiation Safety Officer (RSO) to monitor the use and disposal of radionuclide or sealed sources. This person must have the training, qualifications and experience specified in the licence issued by the regulatory body. This program provides a regulatory framework for qualified professionals to work as a RSO.

In the RSO training program, the radiation safety topics cover: medical health physics, hazard control, basic effects of ionizing radiation for ionizing radiation and dosimetry in medicine. Other subjects are: X-ray radiation and radioactive source management, safety and security, radioactive materials transportation, radioactive waste management, quality management and emergency preparedness and response. The practical areas included are: managing the risk of exposure to ionizing radiation resulting from non-compliance of radiation protection, radiation protection measures for workers, hands on and practical training in radiation protection, detectors and measurement tools manipulation and calibration.

5 Leadership and Collaboration

5.1 *Individual Leaders*

Drs. Ernest Cook, Albert Cook and Jack Cook played pioneering multi-specialist roles as internists, surgeons, radiologists and teachers in Uganda. Under Dr. Ernest Cook's leadership, the first medical school in East Africa opened at the Mengo Hospital in 1916. The basics of radiography and plain X-ray interpretation were taught in the Mengo Medical School [1, 4, 5]. Other key leaders and pioneers contributing to the early development of diagnostic imaging in Africa were Drs. L.R. Whittaker, H. Middlemiss, P. Palmer and P. Cockshott. Dr. Whittaker founded the Department of Radiology and Nairobi University in Kenya and the Black Lion Teaching Hospital in Addis Abba. Drs. Middlemiss, Palmer and Cockshott have written numerous articles and books on radiology in the tropics detailing the radiological appearances of many tropical diseases in Africa.

In recent time Professors Barry B Goldberg, Hassen Gharbi and Harald Lutz, working through the World Federation of Ultrasound in Medicine and Biology, are instrumental in the diffusion and training of ultrasound in many African countries. Other individuals who have founded teaching Departments of Radiology in Sub-Saharan Africa include: Professor Henry Kasozi (Makerere University College of Health Sciences, Uganda), Professor Helmut Diefenthal (Kilimanjaro Christian Medical College, Tanzania), Dr. Rhamadan Kazema (University of Dar es Salaam, Tanzania), and Dr. Elias Onditi (Moi University) and Professor S.B. Lagundoye (University College Hospital and University of Ibadan, Nigeria).

5.2 Professional Organizations

The regional and continental imaging societies in Africa (Table 20.7) include: African Society of Radiology (ASR), Mediterranean and African Society of Ultrasound (MASU), and Pan African Congress of Radiology and Imaging (PACORI). Professional organizations play leadership and key roles in promoting quality radiology practice by continuing medical education, research, networking and advocacy. They focus on the challenges and develop solutions for Africa, which are unique given the resource-limited settings.

5.3 Regulatory Authorities

The Forum of National Regulatory Bodies in Africa (FNRBA) is a network created to enhance radiation safety in the region and is similar in concept to other regional networks in Europe, Asia and Latin America. The Member States share the same objectives, language, geographic location and/or other factors. The main objective of FNRBA is to create a platform for the exchange of experience in radiation protection and radiation safety in different fields between African countries, to provide an opportunity for mutual support and the coordination of regional initiatives, to leverage the development and optimization of resources, and to contribute to the strengthening of the effectiveness and sustainability of national regulatory infrastructure.

Working in different thematic groups enables members from various States to better define their needs, issues, strengths and weaknesses, and to develop solutions in cooperation with other stakeholders, including international organizations and agencies, e.g. IAEA and WHO. Networking is an effective means to enable countries to fulfill the requirements of the BSS towards the protection of patients, the public and workers.

5.4 Collaboration and Mutual Assistance

Aiming to strengthen of the effectiveness and sustainability of national regulatory infrastructure, the FNRBA conducted a multi-national survey to assess and compare the radiation safety infrastructure, experience and expertise in the participating States and to identify means to provide mutual assistance.

The FNRBA's thematic group on medical exposure used self-assessment by questionnaires to evaluate the independence of national regulatory authorities and the effectiveness of their policies in radiation safety in medicine with a focus on radiotherapy covering patients, public and workers.

Table 20.7 African regional and continental imaging societies include: African Society of Radiology (*ASR*), Mediterranean and African Society of Ultrasound (*MASU*), and Pan African Congress of Radiology and Imaging (*PACORI*)

Society	Date founded and founders	Membership	Region	Aims and objectives	Major achievements
MASU	1986 F Cicero and H Gharbi	All health personnel engaged in the use of US	Africa and Mediterranean	Promote the diffusion and appropriate use of US	Formation of national US societies and WFUMB Centers of Excellence, US training and professional development
PACORI	2001 M Kawooya, H Kasozi, and J Wambani	All health personnel engaged in diagnostic imaging and radiotherapy	Africa, especially sub-Saharan	Promote safe and rational use of radiology and imaging and training in imaging	Mobilization, advocacy, networking, and professional development
ASR	2007 H Gharbi and J Labuscagne	Radiologists	Africa	Encourage and promote the development of radiology in Africa	Mobilization, networking, and professional development

The questionnaires examined the justification of procedures; optimization measures and dose minimization; equipment calibration, maintenance and QA program; qualification and training of professionals. Other areas included: documentation of procedures; establishment and use of reference levels; use of guidance references; and continuing professional development program for medical specialists, radiation protection officers and medical physicists. Other issues explored were: the availability and effectiveness of radiation protection education from professional organizations and academic institutions; the existence of ethical research committees; the availability of dosimetry and calibration services; and the qualification requirements for medical physicists in charge of radiation protection and for radiation protection officers.

The answer to each question was graded from one to five and each country was given a global score based on the compliance to the radiation protection requirements laid down by the BSS [3]. Based on these scores, the participating countries were grouped under five classes: excellent, satisfactory, insufficient needing improvement, seriously insufficient, completely insufficient or inexistent.

Countries in Class 1 and Class 2 can provide assistance to those in Class 4 and Class 5. The countries in Class 3 can become providers of assistance after receiving assistance and implementing improvements. The countries able to provide full assistance were: Egypt, Ghana, Morocco and Tunisia. The countries with potential to provide assistance after improving their services were: Ethiopia, Kenya, Madagascar, Namibia, Nigeria, Tanzania, Zambia and Zimbabwe. The countries needing improvement were: Cameroon, Gabon, Mali, Uganda and Seychelles. The countries needing assistance were: Niger, Mauritania, Cote-Devoir, Sierra Leone and Senegal. Centers of Excellence can be established in the Class 1 countries to train the trainers and dispatch experts to other countries to strengthen their safety programs and develop a safety culture.

5.5 Experience Sharing

The radiation protection stakeholders from the 10 member countries meet to share experience and lessons learnt, facilitated by continental professional bodies, e.g. ASR, PANCORI; national diagnostic imaging societies; and national medical physics societies. This sharing of best practice covers the different aspects of radiation protection, radiotherapy, nuclear medicine and medical physics. At this stage, there are medical physics societies in 13 African countries.

6 Other Opportunities

In additional to the above, a range of other actions is used to improve a more equitable access to diagnostic imaging and radiation safety in Africa by focusing on: resources, i.e. financial and equipment as well as human; system and procedures

Interplay of Factors for Increasing Imaging Access in Africa

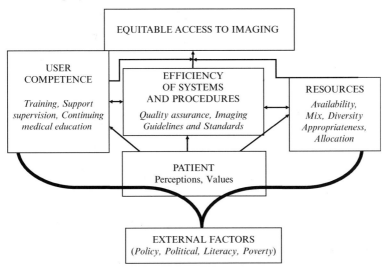

Fig. 20.5 Factors leading to more equitable diagnostic imaging in Africa

infrastructure, such as quality assurance, standards and guidelines; patient factors; and other issues, such as policy, politics, literacy, poverty etc. (Fig. 20.5).

6.1 Human Resource

The strategy to increase the workforce is a priority for the African governments in order to achieve the three health-related MDGs [19]. Increasing the number of healthcare workers for imaging can only be achieved by scaling up formal training. ECURIE has adopted this approach. To maintain the competency of these trained personnel, regular continuing medical education (CME) is essential. ECUREI as the training centre for the WFUMB Center of Excellence conducts regular ultrasound CME courses. National and continental societies such as the ASR, PACORI and MASU play a key role in continuing professional development. To determine the relevance and effectiveness of any training program, an assessment of its impact on healthcare delivery is necessary.

6.2 Financial and Physical Resources

Financial and physical resources should be available in the right diversity and mix, and their allocation appropriate and equitable. Rural areas are likely more under-resourced than urban areas because of political and socio-economic reasons [15]. The only solution is a favorable health policy aiming at more equitable access. It is important that the health policy makers apply evidence to resource allocation. For example, the utilization levels in Ugandan hospitals should be at 50 % but a figure of 5–20 % is still being used for planning purpose [10].

6.3 Guidelines and Guidance

Referrers in Africa do not have or use referral guidelines for diagnostic imaging. Locally suitable referral guidelines have to be developed, piloted and disseminated and their impact assessed. Professional societies are instrumental in the preparation of these guidelines [8]. Quality assurance program for imaging equipment should be given a higher priority since it impacts on diagnostic accuracy and radiation safety.

6.4 Patient Awareness

Patients have their perceptions, conceptions and misconceptions [10]. Insufficient education and awareness in health issues may be the cause of patients presenting late for diagnosis and treatment. It is important to educate patients and make them understand the benefits of an early diagnosis. They should be aware of the risks of ionizing radiation and the benefits of alternative non-ionizing imaging procedures like ultrasound. Governments, professional societies and the media have the responsibility of educating patients.

6.5 Effective Advocacy

Issues such as policy implementation, literacy and poverty etc. [8] are largely influenced and controlled by governments and are best approached by advocacy. Advocacy should address the plight of the rural communities focusing on a more equitable access to vital services like obstetrical ultrasound and plain film imaging. Professional societies play important role as lead advocates by pointing out these deficiencies and referring to the research evidence from academic institutions.

7 Conclusion

Diagnostic imaging, radiation medicine and radiation safety are important elements to underpin quality healthcare service delivery. Urban communities in Africa have access to imaging services similar to those in developed countries. However, remote communities, which comprise more than 80 % of Africa's population, have limited access. Human resource is a key limiting factor. Although governments have a mandate for education and training; collective efforts by technical experts, training institutions and professional societies are critical in the strengthening of this common goal, as well as the implementation and improvement of training interventions based on evaluation and feedback.

The deficiencies in radiation safety infrastructure were successfully tackled by leadership and innovative solutions through the formation of the FNRBA and the networking of the national regulatory bodies and stakeholders; collaboration and provision of mutual assistance; experience sharing of best practices; and the training of radiation safety officers. It is intended that those countries that are more experienced and with expertise would assist others by facilitating education and training and more Training Centers of Excellence for radiation safety be progressively established.

The actions described in the building of workforce capacity by education and training (ECURIE) and in the strengthening of national radiation safety infrastructure by regulatory framework (FNRBA) are excellent examples initiated by the regional leaders. Such innovative approaches could be replicated in other settings. The summation of these and other collaborative actions will lead to more equitable access to quality imaging and safer practice of radiation medicine throughout Africa.

Acknowledgement The authors acknowledge the assistance from the following individuals towards the preparation of this manuscript: Robert Nathan, M.D., MPH Acting Assistant Professor of Radiology, Harborview Medical Center; Edith Namulema MBChB, M.Sc. (Epid), Racheal Ankunda MPH, Peter Dungu and Fred Kirumira, ECURI; Hager Kammoun M.D., Head, Medical and Occupational Exposure Protection and Chadha Hammou Chouchane M.D., Head, National Register of Radiologic Sources, Occupational Exposure Protection and Emergency Preparedness, NCRP.

References

1. Billington WR (1970) Albert Cook 1870–1951: Uganda pioneer. Br Med J 4(5737):738–740
2. Dawson P (2004) Patient dose in multislice CT: why is it increasing and does it matter? Br J Radiol 77(Spec No 1):S10–S13
3. Food and Agriculture Organization of the United Nations, International Atomic Energy Agency, International Labour Organisation, OECD Nuclear Energy Agency, Pan American Health Organization, World Health Organization (1996) International Basic Safety Standards for protection against ionizing radiation and for the safety of radiation sources, Safety series no. 115. IAEA, Vienna
4. Foster WD (1974) Makerere Medical School: 50th anniversary. Br Med J 3(5932):675–678
5. Goodchild RT (1947) A medical jubilee in Central Africa. Br Med J 2(4521):342–344
6. High-Level Forum on the Health Millennium Development Goals (2004) Addressing Africa's health workforce crisis: an avenue for action. High Level Forum For MDGs, Abuja Nigeria
7. International Atomic Energy Agency (2012) Radiation protection of patients. http://rpop.iaea.org/RPOP/RPoP/Content/index.htm. Accessed 12 Mar 2012
8. Kawooya MG, Pariyo G, Malwadde EK et al (2012) Assessing the performance of imaging health systems in five selected hospitals in Uganda. J Clin Imaging Sci 2:12

9. Kawooya MG, Goldberg BB, De Groot W et al (2010) Evaluation of US training for the past 6 years at ECUREI, the World Federation for Ultrasound in Medicine and Biology (WFUMB) Centre of Excellence, Kampala. Uganda Acad Radiol 17(3):392–398
10. Kawooya MG, Pariyo G, Malwadde KE et al (2011) Assessing the diagnostic imaging needs for five selected hospitals in Uganda. J Clin Imaging Sci 1:53
11. Odaga J, Lochoro P (2006) Budget ceilings and health in Uganda. Caritas, Uganda
12. Okuonzi SA (2004) Dying for economic growth? Evidence of a flawed economic policy in Uganda. Lancet 364(9445):1632–1637
13. Ostrowski CM, Clarke G (2009) Improving ministry of health and ministry of finance relationships for increased health funding. Woodrow Wilson International Center for Scholars, Washington, DC
14. Rehani MM, Ortiz-Lopez P (2006) Radiation effects in fluoroscopically guided cardiac interventions – keeping them under control. Int J Cardiol 109(2):147–151
15. Suri T, Tschirley D, Irungu C et al (2009) Rural incomes, inequality and poverty dynamics in Kenya. Tegemeo Institute of Agricultural Policy and Development, Nairobi
16. Tsegaye M (2010) Health workers needed: poor left without care in Africa's rural areas. The World Bank, Washington, DC
17. World Health Organization (2000) The World Health report 2000: health systems, improving performance. WHO, Geneva
18. World Health Organization (2006) The World Health Report 2006: working together for health. WHO, Geneva
19. World Health Organization (2009) Increasing access to health workers in remote and rural areas through improved retention – Plan of action. WHO, Geneva
20. Yawn B, Krein S, Christianson J et al (1997) Rural radiology: who is producing images and who is reading them? J Rural Health 13(2):136–144
21. Yilmaz M, Albayram S, Yasar D et al (2007) Female breast radiation exposure during thorax multidetector computed tomography and the effectiveness of bismuth breast shield to reduce breast radiation dose. J Comput Assist Tomogr 31(1):138–142

Chapter 21
Leadership and Innovations to Improve Radiology Quality and Radiation Safety in China: Regional, National and Global Perspectives

Bin Song, Wentao Wu, Xiaoyuan Feng, and Lawrence Lau

Abstract This chapter begins with a description of the current use of and resources for radiation medicine in China, covering diagnostic and interventional radiology, nuclear medicine and radiation therapy procedures. This is followed by a discussion of the workforce, workload, equipment and infrastructure issues and their impact on quality control and radiation safety. Examples of the relevant national laws and regulations are listed and their objectives are outlined. The solutions used to tackle these challenges are presented. Education and training are important and key strategies. Regional Radiation Quality Control Centers are being established to spearhead local implementation of actions. Other actions include research, increasing awareness, and building competent team in radiological quality control and radiation safety within medical facilities. The initial focus is on radiologists but these efforts will be extended to other clinicians and the public. As there are regional variations in economic development and healthcare infrastructure, local adaptation of overarching principles together with a consideration of the available resources will lead to better local solutions. From a global perspective, there are useful and freely available resources in radiation protection, and programs that will add value to local radiation protection efforts. Effective communication and collaboration with other stakeholders will result in better outcome. Lead by the authorities and professional bodies, in collaboration with other stakeholders, a range of innovative actions will be progressively introduced in China resulting in

B. Song (✉) • W. Wu
Department of Radiology, West China Hospital, Sichuan University, 37 Guo Xue Xiang,
Chengdu, Sichuan 610041, People's Republic of China
e-mail: cjr.songbin@vip.163.com

X. Feng
Department of Radiology, Huashan Hospital, Fudan University, Shanghai,
People's Republic of China

L. Lau
International Radiology Quality Network, Reston VA 201911-4326, USA

L. Lau and K.-H. Ng (eds.), *Radiological Safety and Quality: Paradigms in Leadership and Innovation*, DOI 10.1007/978-94-007-7256-4_21,
© Springer Science+Business Media Dordrecht 2014

nation-wide strengthening of system infrastructure and improvements in radiation safety and quality in the radiation medicine facilities.

Keywords Quality control • Radiation medicine • Radiation protection in China • Radiation protection laws • Radiation safety

1 Introduction

With a population of more than 1.3 billion, China is the world's most populous country and its medical imaging and radiation therapy workload is heavy. However, this volume of work varies considerably among the provinces. Sichuan is the most populated province and its medical imaging and radiation therapy workload varies significantly between the regions and cities, due to differences in economic development and healthcare infrastructure.

In Sichuan Province with a population of more than 80 million, the annual utilization of radiation medicine is approximately 15.8 million procedures, among which radiological procedures account for 14.654 million, diagnostic and therapeutic nuclear medicine 1.058 million and 38,000 respectively, and radiation therapy 56,000 (Fig. 21.1). Meanwhile, the workforce is far from enough to match this volume of work. Currently, there are 9,038 radiation medicine practitioners and supporting staff in the entire province, consisting of 6,889 radiologists and medical imaging technicians, 793 interventional physicians, 308 nuclear medicine staff, 559 radiation therapy staff, 88 engineers and 401 other supporting staff (Fig. 21.2). This is equivalent to an average workload of 11 patients per day per radiologist in this province.

To illustrate variation in workload, the West China Hospital is used as an example. The hospital is a Regional Medical Center (RMC) for Sichuan and its department of radiology has established a Regional Radiation Quality Control Center (RRQCC) for the province. The department performs approximately 229,000 CT and 299,000 x-ray procedures per annum. However, the 44 radiologists and 49 medical imaging technicians in this department handled all these procedures, which are equivalent to 33 patients per day per radiologist, or three times the province's average.

Education, training and other actions are being used to improve quality control for radiation medicine, radiation safety, and radiation protection in China. Government departments, agencies and authorities, including the Ministry of Health [8], have developed policies, laws, and regulations to guide radiation safety, use of medical radiation, and radiation quality control. Some examples include: 'Protection law of radioactive contamination of the People's Republic of China', 'Radiology licenses of the People's Republic of China' etc. [9–11]. These laws and regulations are followed by the hospitals and copies are filed in the management reference library.

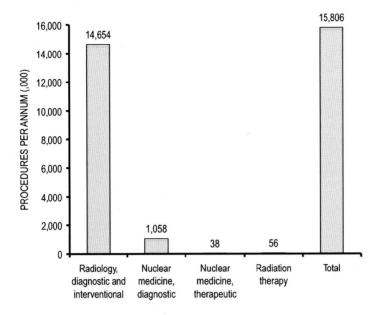

Fig. 21.1 Annual numbers of radiological procedures performed in Sichuan Province

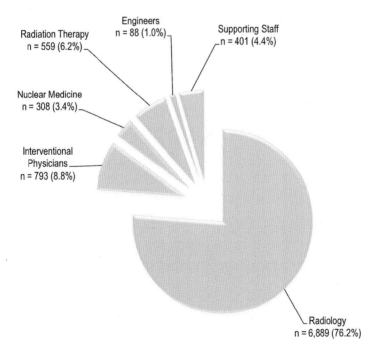

Fig. 21.2 Radiological practitioners and supporting staff in Sichuan Province

The health departments in the provinces are developing supporting local laws and regulations, to facilitate smooth implementation of quality control, radiation protection, and training actions. At the same time, RRQCCs are being established to standardize and supervise radiation safety and quality control in the region. The objective is to incorporate quality control procedures into daily practice and radiation safety as an integral part of a facility's safety culture.

2 Issues and Challenges

The stakeholders are working hard on actions to improve radiation safety and quality control in China. There is a range of quality control and radiation safety issues requiring attention. One of these is workload: both clinical and administrative, as a result of an increasing demand for medical imaging and radiation therapy procedures. The other is the resource required to identify, develop, prioritize, and implement solutions. The experience in Sichuan is used to illustrate these issues and challenges.

2.1 Organization and Record Keeping

Among the 2,696 medical facilities in Sichuan, 2,412 maintain health records of radiation medicine practitioners; 2,298 use a system to maintain and monitor radiation exposure of radiation medicine practitioners; 2,319 have radiation protection supervising groups; and 1,539 have radiation emergency rescue teams (Fig. 21.3). Therefore, about 10 % and 15 % respectively of these facilities do not have a system to monitor the health or radiation exposure of radiation medicine practitioners, and those facilities in which record keeping is incomplete are not counted.

2.2 Practitioner Licensing, Training and Monitoring

Among the 9,038 radiation medicine practitioners in Sichuan, 7,396 have working license; 3,035 participate in annual training; 1,604 have operating license for main medical equipment; 7,859 have health monitoring; and 7,857 use a personal dose monitor (Fig. 21.4). Based on these figures, only one third of practitioners receive annual training on quality control and radiation protection and about 15 % do not have health surveillance or radiation exposure monitoring.

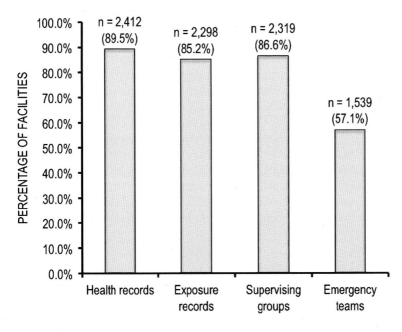

Fig. 21.3 Organization and record keeping in 2,696 medical facilities in Sichuan Province. The x-axis describes the infrastructure elements and the y-axis the percentage of facilities having these elements

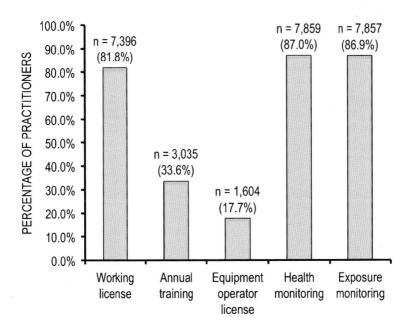

Fig. 21.4 Profile of the 9,038 radiation medicine practitioners in Sichuan Province. The x-axis shows the licensing, monitoring and training features and the y-axis the percentage of practitioners having these features

2.3 Quality Control and Radiation Protection Resources

Thirty-eight radiation therapy units are available in the Sichuan province. The supporting equipment includes 33 analog positioning devices, 54 treatment planning systems, 49 radiation dosimeters; one afterloading machine dedicated activity meter, nine radiation dose scanning devices, 13 intensity modulated radiation therapy (IMRT) verification devices, and 12 water calibration phantoms. This means 55 % of the facilities are equipped with analog positioning devices and 20 % with water calibration phantoms.

Among the nuclear medicine facilities in the province, there are 36 nuclear medicine activity meters, 34 surface contamination detection and measurement devices, 36 environmental radiation monitors, 68 fixed radiation alarms, 221 personal dose radiation alarms, 32 image quality control units, and 24 stability-testing units.

There are 6,395 protection shields available for the radiation protection of patients. These include 1,754 thyroid shields, 1,698 gonadal shields, 1,551 breast shields and 1,392 leaded eyewear.

2.4 Equipment Profile

There are 3,695 pieces of equipment for radiation medicine in Sichuan, consisting of 3,568 units for diagnostic radiology, 79 units for interventional radiology, 10 units for nuclear medicine, and 38 units for radiation therapy (Fig. 21.5). Out of these 3,695 units, 2,598 are licensed, 1,395 have equipment performance tested, 2,533 have radiation protection monitored, 412 use individual dose monitoring, 842 have undergone occupational hazard assessment and 1,065 have performed quality control evaluation (Fig. 21.6). Therefore, the licensing of equipment is lower at 70.3 % when compared to the licensing of practitioners at 81.8 %. A significant percentage of units are aging, e.g. 745 (20.2 %) units are aged over 10 years and 1,027 (27.8 %) units between 5 and 10 years. There are ionizing radiation warning signs, warning lights, patient radiation safety notifications, and relevant operating instructions in 2,600 procedural rooms (70.4 %).

3 Radiation Protection Laws and Regulations

To strengthen the quality and management of radiation medicine practice and the use of radioactive material, to establish standards guiding good medical practice, and to ensure the radiation protection and safety of patients and practitioners; laws, legislations, administrative regulations, department regulations and practice standards are developed documenting the requirements for practitioners, facilities, and environment. Some examples are listed below. The most powerful and important radiation protection law is the 'Protection law of radioactive contamination of the People's Republic of China' [9], which underpins other related laws and regulations.

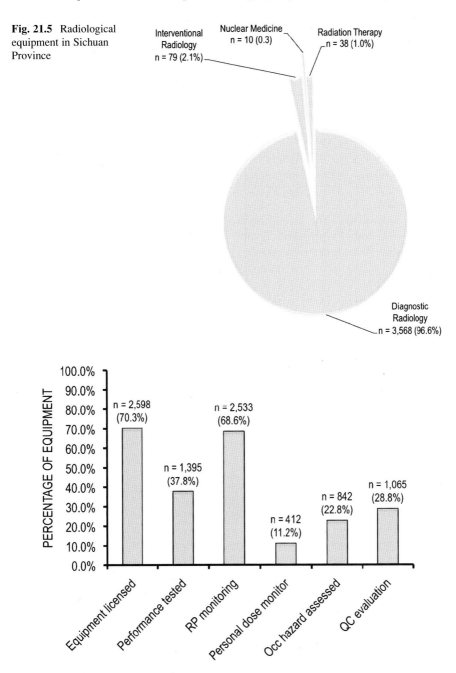

Fig. 21.5 Radiological equipment in Sichuan Province

Fig. 21.6 Profile of the 3,695 pieces of radiation medicine equipment in Sichuan Province. The x-axis indicates the equipment features and the y-axis the percentage of equipment having these features

3.1 Radiation Medicine Facilities

These laws and regulations are developed to document the standards and requirements for day-to-day administrative and clinical working procedures, aiming to decrease radiation risk and harm for patients and practitioners. Some of these examples are shown below.

- Protection law of radioactive contamination of the People's Republic of China;
- Radioisotopes and radiation device security and protection regulations;
- Radioisotopes and radiation device security license management approach;
- Radioisotopes and radiation device security and protection management approach;
- City radioactive waste management approach;
- Classification of x-ray devices; and
- Radioactive sources encoding rules.

3.2 Environmental Protection

These laws and regulations are developed to document the standards and requirements for environmental protection and emergent pollution management, aiming to decrease the radiation risk and harm to the public and personnel involved in the use of radioactive material. Some of these examples are shown below.

- Protection law of radioactive contamination of the People's Republic of China;
- Classification of radioactive sources;
- Radioactive material transportation security license management approach;
- Safety of radioactive waste management ordinances; and
- Radioactive material transportation safety regulations.

3.3 Practitioner Protection

These laws and regulations are developed to document the standards and requirements for the radiation protection of practitioners and the emergent management of radiation-induced injuries, aiming to decrease the radiation risk and injury for practitioners involved in radiation medicine procedures [10]. Some of these examples are shown below.

- Protection law of radioactive contamination of the People's Republic of China;
- Protection law of occupational diseases of the People's Republic of China;
- Notice on further strengthening the management of individual dose monitoring of radiation staff;
- Safety requirements for gamma ray facility operators; and
- Radiation safety requirements of gamma ray detection devices.

4 Improvement Actions

To achieve tangible improvements in radiation protection, radiation safety and quality control in radiation medicine, these laws and regulations must be applied in practice. The radiation medicine facilities and practitioners should adopt a radiation safety culture and integrate these recommendations into daily working procedure. This approach is equally applicable to the protection of the environment and to the minimization of radioactive waste contamination.

4.1 Strengthening Infrastructure

Regional Radiation Quality Control Center

The administrative framework for radiation safety and radiation control in China is evolving and the infrastructure is being established. In some regions, RRQCCs have been established by the RMCs under the supervision of MOH and regional department of health to standardize and guide radiation safety and quality control procedures. Under the direct supervision of the MOH, a RMC is a region's peak medical facility and is in charge of the region's RQCC. RRQCCs play administrative and supervisory roles.

The formation of RRQCCs requires the establishment of infrastructure, building of a radiation quality control team, training of personnel, development of administrative processes and quality standards, application of laws and regulations, development and implementation of standardized procedures including periodic radiation equipment safety checks. Among all these components, infrastructure establishment and radiation quality control team building and training are paramount, because the former reflects a centre's philosophy and the latter its capability and competency.

The stakeholders are working hard to improve team building in different settings. Radiation protection should be adapted to suit the available resources in the different regions. The aim is to apply arrangements that are most appropriate to a particular setting.

The National Institute for Radiological Protection, Chinese Center for Disease Control and Prevention (NIRP, China CDC), also named Chinese Center for Medical Response to Radiation Emergency (CCMRRE), founded in 1965, is a specialized national technical institution for radiological medicine and radiation protection, affiliated to the Ministry of Health and the Chinese Center for Disease Control and Prevention, and a professional guiding centre in this field nationwide. The Institute fulfils and assumes the responsibilities and tasks of technical support to the state and technical guidance and training to the local professionals [13].

Standard Setting and Other Functions

Supported by the technical expertise and experience of the RMCs, the RRQCC develop standards and working protocols based on the laws and regulations promulgated by the MOH [11] and the regional departments of health. The ultimate aim is to incorporate quality procedures into daily practice and radiation safety as an integral part of a facility's safety culture.

In addition to developing standards, the RRQCCs conduct periodic meetings, organize education and training sessions, and undertake assessments relating to radiation safety and radiation quality control thus ensuring more effective implementation of the requirements.

Evaluation of Radiation Quality Control

Two methods are currently used to evaluate the impact and effectiveness of quality control in radiation medicine and radiation protection. The MOH and the regional departments of health conduct regular and random checks of the RRQCCs. In addition the RRQCCs are under the supervision of the regional departments of health. In the future, the MOH will supervise quality control in radiation medicine and radiation protection.

4.2   Education and Training

Objectives

The education and training programs in radiation safety and quality control will initially focus on the radiation medicine practitioners. The aim is to inform them the goals, fundamental tasks and missions; to improve their understanding and awareness of quality control and radiation protection; to encourage change in practice; and to adopt a quality and safety culture. The fundamental missions are to ensure the radiation safety of patients and practitioners including their families, and to protect the environment. The role of and need for radiation research is stressed. Education and training will improve knowledge, thus preventing and reducing the probability and extent of radiation-induced health effects.

Practitioners

The Chinese Society of Radiology [2] has developed a range of education and training programs focusing on radiation safety and quality control procedures catering for radiation medicine practitioners, including radiologists, interventional physicians, radiation therapists, radiographers, nurses and engineers. The contents include laws and regulations, standard operation procedures, case-based education, environment protection and emergency management topics in relation to radiation safety and quality control.

The society hosts annual Chinese Congress of Radiology (CCR), during which relevant radiation protection and quality control researches are reviewed, selected and presented, in addition to other clinical topics. Combining the research findings and incorporating the current legal requirements, the CSR develops, promotes and distributes guidelines and guidance handbooks throughout the country. These useful resources assist the RRQCCs and RMCs in improving the efficiency and effectiveness of radiation safety and quality control actions, which are appropriate to and meeting the needs of the local setting.

Undergraduates

Another component of the CSR education program deals with undergraduate education to improve their awareness in radiation safety and quality. It is encouraging to note the increasing number of medical students participating in activities to learn more about and to promote radiation safety and quality control. These activities provide students with relevant information, increase their awareness in radiation safety and quality control at an early stage, and cultivate good habits; therefore laying sound foundation towards more appropriate use of radiation medicine and good practice in the future.

4.3 Procedure Justification

The awareness in technological advances, new clinical techniques and applications, radiation safety, and quality control among many referring clinicians and radiologists is limited and will benefit from periodic updates focusing on the appropriate use of procedures and radiation safety. This CSR initiative provides training material and conducts classes to inform radiologists and other clinicians of these advances and to guide them in choosing an appropriate procedure for a given condition. Experienced radiologists and clinicians prepare the training material and document their recommendations for the most appropriate imaging or treatment for the different conditions.

4.4 Optimization of Protection

In many provinces, RRQCCs are responsible for the development of specific courses for equipment operations, training classes on radiation quality control and safety, and in-service training to meet the national requirements for the licensing of practitioners.

4.5 Team Building and Safety Culture

Quality control in radiation medicine and radiation protection requires collective efforts. Team building in facilities will facilitate the implementation of lessons and recommendations. The development of a radiation quality and safety culture enables sustainable change in practice of the radiation medicine practitioners, so that the radiation protection measures are progressively becoming part of daily working procedure.

4.6 Advocacy and Awareness

The CSR is initiating a public education campaign for consumers and the community to promote awareness in radiation safety and basic understanding of radiation medicine procedures; and to participate in radiation protection actions as guided by the practitioners. Through the provision of useful information, the campaign aims to reduce unnecessary and unintended exposure, radiation-induced injury and psychological stress.

4.7 Research in Radiation Protection

The CSR, under the supervision of the Chinese Medical Association; leading Chinese radiologists; and other stakeholders are conducting research in radiation safety and quality control. The Chinese Journal of Radiological Medicine and Protection was established in 1981 [12], with the responsible institution of the China Association for Science and Technology, as well as the sponsor of Chinese Medical Association. It publishes articles on the advances and researches in radiological medicine and protection in China covering biological effects of ionizing radiation, clinical studies, radiotherapy, radiation hygiene, radiation protection and management, environmental radioactivity monitoring, medical emergency response in nuclear accident, and radiation dosimetry. The research findings contribute to improvements in these areas. The current research direction falls into three main streams.

Optimization of Exposure and Contrast Media

These research projects focus on popular CT procedures, e.g. 3-D CT coronary angiography, vertebral body imaging etc. The aim is to optimize the use of contrast media and radiation exposure, and to improve image processing and diagnostic data. The findings will contribute to a reduction in radiation exposure and contrast media, while ensuring image quality, diagnostic accuracy and therapy are not compromised.

Use of Procedures Without Radiation

There are collaborations underway with equipment manufacturers to explore new and to improve existing ultrasound applications and MRI imaging sequences, which could potentially replace CT and x-ray procedures in suitable conditions. There are research projects underway focusing on the substitution of modalities involving ionizing radiation with non-ionizing alternatives to better protect the patients and practitioners.

Radiation Safety Management and Education

These research projects focus on the development and implementation of practice standards, daily management procedures, and education and training for radiation protection, radiation safety, quality control, waste disposal etc. The findings will be implemented by the local facilities, aiming to balance among efficiency, accuracy, quality, and safety.

4.8 Collaboration and Sharing Responsibility

According to the 'National notification about the safety supervision of radioactive sources', the MOH should administer and evaluate the occupational diseases and harm related to radioactive sources; manage an appropriate use of radiation in diagnosis and treatment and supervise radiation medicine facilities; and manage radiation accidents and environmental pollution due to radioactivity. Under the national healthcare framework, the MOH is the leading central agency. RRQCCs are being established in the provinces under the supervision of regional health departments.

The CSR is responsible for conducting academic researches, organizing annual conferences, designing and hosting education and training courses, drafting and updating of practice guidelines, sharing experience and knowledge in the use of procedures, radiation safety and radiation protection, promoting awareness in new developments and trends, advocating for more appropriate use of radiation medicine, and evaluating these actions on a regular basis etc. The CSR assists the MOH in the establishment of radiation safety requirements and quality control teams.

5 Global Perspective

Collaboration to improve quality control and radiation protection extends beyond professional and national boundaries, by involving stakeholders from other disciplines, sectors, and regions. For example, there are issues and solutions common to the providers of nuclear energy and radiation medicine. In the medical use of radiation, the key stakeholders include the healthcare systems, practitioners, patients, payers, equipment manufacturers, medical indemnity insurers etc. The stakeholders leading improvement actions are individual experts, and local, regional, national and international institutions, organizations, authorities and agencies.

Globalization and advances in telecommunication and information technologies open many opportunities, which could be used to improve quality control and radiation protection in different settings, including China. For example, useful resources are freely available which could be adopted or adapted to suit local needs, catering for the patients, practitioners and healthcare systems. There are programs where collaboration adds value and saves time. Healthcare resources are limited worldwide and a joint approach by information exchange and experience sharing will ensure the common objectives are attained with less time and cost. Some examples of these possibilities are outlined.

5.1 Education Resources

A range of valuable education resources on radiation protection and radiation safety for practitioners, patients, and authorities are *freely* available. An example is the International Atomic Energy Agency's (IAEA) Radiation Protection of Patients website [5]. The scope includes diagnostic radiology, interventional radiology, nuclear medicine, and radiation therapy etc. There are helpful hints on technique optimization and exposure reduction in radiological procedures, posters on radiation protection and safety, and a large range of lectures on the various topics on radiation safety and radiation protection.

5.2 Guidance and Recommendations

The International Commission on Radiological Protection (ICRP) publishes recommendations on the various aspects of radiation protection e.g. procedure justification, optimization of protection, and the use of Diagnostic Reference Levels (DRLs) etc. to improve radiation protection of patients and practitioners [6]. Together with other international agencies and organizations the IAEA revised the International Basic Safety Standards for the Protection Against Ionizing Radiation and for the Safety of Radiation Sources in 2011 [4]. This document provides valuable guidance to maintain and improve radiation protection in healthcare systems and facilities.

The Alliance for Radiation Safety in Pediatric Imaging – Image Gently, is a coalition of healthcare organizations dedicated to provide safe and high quality pediatric imaging nationwide [1]. The primary objective is to raise awareness in the imaging community of the need to adjust radiation dose when imaging children. The ultimate goal of the Alliance is to change practice. It has developed useful information for patients and practitioners; some of this is available in translation.

5.3 Exposure Survey and Safety Initiative

The United Nations Scientific Committee on the Effects of Atomic Radiation (UNSCEAR) conducts regular global surveys on the frequencies of procedure and levels of exposure, equipment and staffing levels and to monitor the evolving trends [14]. China participates in these surveys and contributes to the global database. The involvement of more countries in surveys improves the understanding of the emerging utilization trends and the regional and healthcare system variations. The sharing of data facilitates the development and implementation of evidence-based radiation safety recommendations, guidance tools and policies worldwide.

The WHO launched a Global Initiative on Radiation Safety in Health Care Settings [15] to mobilize the health sector towards a safer and more effective use of radiation in health. This innovation offers an inclusive global framework to bring all the stakeholders together in concerted efforts to improve the implementation of radiation safety actions in medical settings. One of the actions under this initiative is the development, trial and implementation of a set of global referral guidelines for medical imaging and radiology [7, 16]. This is a joint collaboration between the WHO and International Radiology Quality Network (IRQN) together with over 30 other organizations and agencies. The CSR contributes to the action as a member of the IRQN. The document is a useful resource, which can be adapted or adopted to suit the local setting.

5.4 Evidence-Informed Policy Network

An example of evidence-based healthcare policy development is the Evidence-informed Policy Network model [3]. EVIPNet comprises networks that bring together national teams, which are coordinated at both regional and global levels. In Asia, there are EVIPNet national teams in China, Laos, Malaysia, the Philippines, and Vietnam. The WHO created the network in 2005 to 'establish mechanisms to transfer knowledge in support of evidence-based public health and healthcare delivery systems, and evidence-based health-related policies'. The network promotes a systematic use of health research evidence in policy-making and facilitates partnerships at country level between policy-makers, researchers and community to develop and implement policies through the use of the best scientific evidence. The EVIPNet model could be explored as a possible mechanism to advance evidence-based policies for quality control and radiation protection.

6 Conclusion

In recent years, there is a significant increase in the use of radiation medicine procedures in China as in many other parts of the world. There are challenges requiring specific strategies and tailored actions for particular settings. These include issues relating to equipment, practitioners and system infrastructure. The awareness is low and infrastructure suboptimal for radiation protection and radiation quality control. Radiation protection devices and radiation protection tools may not be available or used in facilities. Procedure justification, optimization of protection, and quality assurance for radiation therapy are among some other issues. However, the regulatory authorities and professional organizations are taking steps to strengthen system infrastructure and implement actions to improve quality control and radiation protection. Some examples include education and training, establishment of RRQCCs, research, advocacy and awareness raising etc.

The drive to improve quality control, radiation safety and radiation protection in China is led by the Ministry of Health, regional departments of health, the Chinese Society of Radiology and other stakeholders through the RRQCCs and radiation medicine facilities. Leadership from the professional organizations such as the CSR contributes by providing scientific and evidence-based recommendations and from the authorities by developing laws and regulations. This joint collaborative approach will improve quality patient care and safety in practice.

China is a large country with many provinces, regions and cities. These localities differ in economic development, healthcare infrastructure and radiation safety resources. To achieve global improvement, a single panacea is not applicable. The adaptation of overarching principles together with a consideration of the available resources will lead to solutions that are more suitable and appropriate to meet the local needs.

Similar to the experience in other systems, effective communication and collaboration with other stakeholders will add value to the radiation protection and radiation safety actions in China and ensure their better implementation. The collaborators include local, regional and international academic and research institutions, professional organizations, and agencies. Interdisciplinary and intersectorial collaboration with other stakeholders, i.e. radiographers, medical physicists, other user of radiation medicine, referring physicians, and patients is very important. The ultimate decider influencing success is participation of the radiation medicine practitioners, without whose support these recommendations will not be used in practice.

References

1. Alliance for Radiation Safety in Pediatric Imaging – Image Gently (2013) Global resources. Available at http://www.pedrad.org/associations/5364/ig/index.cfm?page=542. Accessed 4 Jan 2013

2. Chinese Society of Radiology (2013) Available at http://chinaradiology.org/csr/en/. Accessed 4 Jan 2013
3. Evidence-informed policy Network (2013) Evidence-informed policy-making. EVIPNet Asia. Available at http://www.who.int/evidence/resources/country_reports/asia/en/index1.html. Accessed 4 Jan 2013
4. International Atomic Energy Agency (2011) Radiation protection and safety of radiation sources: international basic safety standards – interim edition, IAEA safety standards series GSR part 3 (Interim). IAEA, Vienna
5. International Atomic Energy Agency (2013) Radiation protection of patients. Available at https://rpop.iaea.org/RPoP/RPoP/Content/index.htm. Accessed 4 Jan 2013
6. International Commission on Radiological Protection (2007) ICRP publication 105. Radiation protection in medicine. Ann ICRP 37(6):1–63
7. International Radiology Quality Network (2013) Referral guidelines. Available at http://www.irqn.org/work/referral-guidelines.htm. Accessed 4 Jan 2013
8. Ministry of Health of the People's Republic of China (2013a) Ministry of Health. Available at http://www.moh.gov.cn. Accessed 9 Feb 2013
9. Ministry of Health of the People's Republic of China (2013b) Protection law of radioactive contamination of the People's Republic of China. Available at http://www.gov.cn/flfg/2005-06/27/content_9911.htm. Accessed 26 Jan 2013
10. Ministry of Health of the People's Republic of China (2013c) Protection law of occupational diseases of the People's Republic of China. Available at http://61.49.18.65/zwgkzt/pfl/201203/54444.shtml. Accessed 26 Jan 2013
11. Ministry of Health of the People's Republic of China (2013) The health industry standard of the People's Republic of China: basic requirements for licensing of radiodiagnosis and radiotherapy. MOH People's Republic of China, Beijing
12. National Institute for Radiological Protection (2013a) Chinese journal of radiological medicine and protection. Available at http://www.nirp.cn/en/index.php?option=com_content&view=article&id=41&Itemid=25. Accessed 26 Jan 2013
13. National Institute for Radiological Protection (2013b) National Institute for Radiological Protection, Chinese Center for Disease Control and Prevention (NIRP, China CDC). Available at http://www.nirp.cn/en/index.php?option=com_content&view=article&id=44&Itemid=6. Accessed 26 Jan 2013
14. United Nations Scientific Committee on the Effects of Atomic Radiation (2010) Sources and effects of ionizing radiation. Volume I: sources. UNSCEAR 2008 report to the General Assembly with scientific annexes A and B. United Nations, New York
15. World Health Organization (2008) WHO global initiative on radiation safety in health care settings technical meeting report. WHO, Geneva
16. World Health Organization (2013) Medical imaging specialists call for global referral guidelines. Available at http://www.who.int/ionizing_radiation/about/med_exposure/en/index2.html. Accessed 4 Jan 2013

Chapter 22
Status of Medical Diagnostic Radiation Safety in Korea

Dong-Wook Sung and Byung Ihn Choi

Abstract The regulation of medical use of radiation is dualistic in Korea. Nuclear medicine and therapeutic radiation are under the supervision of the Ministry of Education, Science and Technology while diagnostic radiation is managed by the Ministry of Health and Welfare. In some countries, the regulatory framework is more unified with specialists addressing each field and working under one system. With the Korean system, if the International Atomic Energy Agency (IAEA) were to recommend a policy in diagnostic radiation, this would have to be communicated through the Ministry of Education, Science and Technology and might not reach those societies in diagnostic radiology, thus posing a barrier to its implementation. Considering the Korean circumstances, there are numerous obstacles preventing this dualistic regulatory system to be unified as one. Therefore, the government should strengthen its efforts to directly communicate issues, discussions and policies in diagnostic radiation through the Ministry of Health and Welfare to the relevant societies and organizations. Despite this dualistic system, topics in diagnostic radiation are actively researched, taught, and promoted by the academic societies, including the Korean Society of Radiology (KSR). This chapter will further explain the current advances in diagnostic radiation through the Korean government and academic societies.

Keywords Optimization of protection • Procedure justification • Radiation protection • Radiation safety

D.-W. Sung (✉)
Department of Radiology, Kyung Hee University Hospital, Seoul, South Korea

Radiation Safety Committee, Korean Society of Radiology, Seoul, South Korea
e-mail: sungdw@khmc.or.kr

B.I. Choi
Department of Radiology, Seoul National University, Seoul, South Korea

Asian Oceanian Society of Radiology, Seoul, South Korea
e-mail: bichoi@snu.ac.kr

L. Lau and K.-H. Ng (eds.), *Radiological Safety and Quality: Paradigms in Leadership and Innovation*, DOI 10.1007/978-94-007-7256-4_22,
© Springer Science+Business Media Dordrecht 2014

379

1   Issues for a Dualistic Regulatory System

1.1   Communication of Radiation Risk

After the Fukushima nuclear power plant accident in March 2011, the international community has shown an increased interest in radiation exposure and its adverse effects. Although radiation from a power plant accident and that used in medicine differ greatly, the public is often offered confusing and inappropriate explanations, such as by comparing the exposure from radiological procedures to a nuclear power plant accident. There is further concern and confusion about radiation exposure from computed tomography (CT) through the Internet and media. Statements comparing radioactivity exposure to diagnostic radiation and claiming that it is relatively harmless are the result of a dualistic regulatory system such as the one being used in Korea. The degree of radioactive contamination is repeatedly compared to the number of chest x-rays or a CT scan. This problem is in part related to the management of diagnostic radiation matters solely by the Atomic Energy Act. Diagnostic radiation issues have been communicated through individual contacts instead of the Ministry of Health and Welfare, Korea Food and Drug Administration, or associated academic societies. Such issues are partially resolved with the increased efforts and interest from specialists and medical diagnostic radiation academic societies. However, the government or international organizations seem to have not taken notice of the reality. The fear of radioactive contamination has lead to the general fear of radiation from diagnostic x-rays and some people are under the false belief that radiation from x-rays will remain within the human body and later develop into cancer [17].

1.2   Korean Alliance for Radiation Safety and Culture in Medicine

In order to overcome these issues, medical diagnostic radiation academic societies gathered in 2011 to establish the Korean Alliance for Radiation Safety and Culture in Medicine (KARSM) to educate the public and personnel working in medical radiation and to promote the safety of radiation through cultural events, educational campaigns, and the use of websites such as Image Gently and Image Wisely. The KARSM's website is scheduled to open in late 2012. Promotions, education, and information will be provided to facilitate the reduction of radiation dose in diagnostic radiology for the patients, doctors, radiological technologists, and medical physicists alike. Not only is the KARSM working to rectify the misunderstanding about radiation and radioactivity amongst the general public, the KSR is participating with education, events, and campaigns. There should be a more unified participation from the Ministry of Health and Welfare, Korea Food and Drug Administration (KFDA), and KSR with the IAEA in establishing policies in radiation safety and radiation protection.

2   Issues for Radiation workers

2.1   *Occupational Exposure*

Korea established 'Regulations for Management of Safety in Diagnostic Radiation' in 1995 and has implemented regular exposure monitoring of diagnostic radiation facilities and personnel since.

Radiation workers are those personnel working in the fields related to diagnostic radiation including radiological technologists, doctors, dentists, dental hygienists, radiologists, nurses, nursing assistants and other health care workers employed in a health facility. There are 55,614 radiation workers in 2010 of which 77.7 % are radiological technologists, dentists, and doctors [12]. The average annual radiation exposure in 97.2 % of radiation workers is below 5 mSv while 0.2 % showed a level above 20 mSv in 2010. The mean effective dose of all radiation workers decreased from 0.77 mSv in 2006 (Fig. 22.1) to 0.58 mSv in 2010 [12].

The average annual dose in 2010 of radiation workers in different occupational groups and radiologists from different institutions are summarized in Tables 22.1 and 22.2 respectively. Radiologists from general hospitals and dental hospitals received higher annual dose and are in need of appropriate education and training.

In Korea, only one person is legally required to be trained as the supervisor of radiation safety. The above data suggests that the radiation workers in general hospitals and dental hospitals are in need of additional education and training. The government is making efforts to develop education programs and regulations for radiation workers.

According to 'Regulation for Management of Safety in Diagnostic Radiation', those radiation workers receiving radiation dose above 5 mSv over a 3 month period will receive a warning and those above 20 mSv will be investigated to determine if this is due to actual radiation exposure or a dosimetry error. In 2010, 1.6 % of radiation workers received a warning for exceeding 5 mSv over a 3-month period.

Considering radiation workers are under strict regulations to protect themselves against radiation, it is notable that there are no regulations for a dose limit for patients or the general population. This is because each patient has different medical needs and requires different management. Therefore, it is not practical to specify a standardized dose limit for the patients.

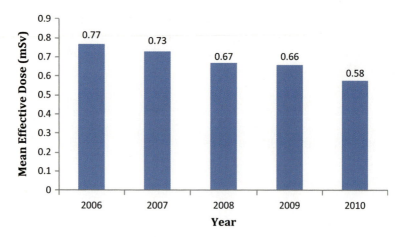

Fig. 22.1 Mean effective dose for all radiation workers. The mean annual dose has decreased from 0.77 mSv in 2006 to 0.58 mSv in 2010

Table 22.1 Average annual dose for different radiation worker groups in Korea in 2010

Occupation	Average annual dose (mSv)
Radiation technologist	1.21
Physicians	0.34
Dentist	0.16
Dental hygienist	0.13
Diagnostic radiologist	0.41
Nurse	0.40
Nursing assistant	0.30
Medical assistant	0.30
Other radiation workers	0.47

Table 22.2 Average annual dose for radiologists from different institution types in Korea in 2010

Medical Institution	Average annual dose (mSv)
General hospital	0.61
Hospital	0.21
Clinic	0.22
Dental hospital	0.59
Dental clinic	0.09
Public health center	0.37
Testing agency/Personal dosimetry service	0.00
Others	0.06

2.2 Procedure Utilization

There is no statistical data on population exposure in Korea. However, according to the United Nations Scientific Committee on the Effects of Atomic Radiation

(UNSCEAR) in 2008, the number of CT procedures per million people in Korea was the same as in the USA at 32.2 and was the third highest amongst the OECD nations [23]. The number of CT procedures according to the Health Insurance Review Agency has increased two folds from approximately 2.26 million in 2005 [15] to 4.8 million in 2009. The number of CT procedures performed in Korea is similar to the USA and this is increasing each year. It can be assumed that the amount of radiation dose per capita in Korea is increasing rapidly.

2.3   Awareness in Radiation Safety and Radiation Protection

In a survey conducted among a group of radiation workers on medical diagnostic radiation exposure, most workers gave commonsensical answers when responding to questions from patients about radiation dose. Twenty percent of the surveyed showed even less interest in radiation exposure than the patients or the general population. Less than half of the workers were aware of the deterministic or stochastic effects of radiation and only half of the group knew the difference between radiation and radioactivity. Twenty-four percent of radiation workers believed two to three CT procedures would have a carcinogenic effect while 75 % recognized the need for education in radiation dose. In a different study, 76 % of radiologists, 73 % of emergency medicine doctors, and 100 % of patients were underestimating the radiation dose [14]. Such reports are found in other international papers as well. There is an imminent need for active education, campaigning, and appropriate communication between the radiation workers and patients; a better understanding amongst the general population; and the inclusion of radiation protection topics in medical school curricula [2, 13, 19].

There are efforts being made in Korea to improve radiation safety and radiation protection and to reduce radiation dose, e.g. by developing regulations, providing guidelines, and using diagnostic reference levels (DRL) as guidance tools. Recently, there are actions to develop a diagnostic reference level in children through national surveys [9].

3   Efforts to Reduce Exposure

The most important aspect of medical use of radiation is that it should be justified by medical need. Any use of radiation in patients need to be formally justified by ensuring the benefits outweigh the risks. All patients need to be informed of alternative diagnostic options. The referrers and providers should ensure the risk is small relative to the benefit and information obtained. Once justified, the next step is to minimize the dose, while obtaining adequate diagnostic data. This includes the employment of standardized protocols and tailoring the procedure to the individual. CT procedures, which usually involve higher radiation doses, can be

Table 22.3 Diagnostic reference levels (*DRL*) for radiographic procedures in Korea

Radiographic procedure	ESD (mGy)	Radiographic procedure	ESD (mGy)
Skull AP	2.23	L-spine AP	4.08
Skull lateral	2.97	L-spine lateral	10.53
Chest PA	0.34	L-spine oblique	6.35
Chest AP	1.63	Clavicle AP	1.82
Chest lateral	2.80	Shoulder AP	1.73
Abdomen AP	2.77	Humerus AP	0.64
Pelvis AP	3.42	Elbow AP	0.35
C-spine AP	1.86	Wrist AP	0.28
C-spine lateral	1.03	Hip AP	3.42
T-spine AP	3.79	Knee AP	0.61
T-spine lateral	8.15	Ankle AP	0.28

ESD entrance surface dose

made less hazardous by optimization. In summary, justification and optimization of radiological procedures require participation from the referrers and providers to ensure that patients are not subjected to unnecessary exposure. The patients should be assured that the procedures could be optimized through individually adapted protocols and the use of correct protective gears.

3.1 Dose Reduction for Radiography

KFDA's efforts to document and assess radiation dose in Korea started in 2007 with a nation-wide survey on the radiation doses delivered to the patients for routine chest x-rays. In 2009, a similar survey was conducted among the pediatric patients. A survey on 22 radiographic procedures was carried out from 2010 to 2011 (Table 22.3). Based on these surveys, diagnostic reference levels (DRLs) were established. In Korea, the recommended DRL for a chest x-ray is 0.34 mSv for adults [10] and 100 μGy for children [11]. Although the results of the 2010–2011 survey have not been officially published, the DRLs of these 22 procedures, with the exception of chest x-ray, appear to be in agreement with those used in the United Kingdom, a nation renowned for her strict management of radiation dose to the patients [22].

The single most important factor to reduce radiation dose is by using accurate and standardized protocols. Generally, the image quality is proportional to the radiation used. Therefore the referrers and providers, i.e. radiologists, are unconsciously driving high radiation dose by preferring extra-high image quality. However, extra-high quality images are not always needed for a diagnosis to be made

Table 22.4 Diagnostic reference levels for CT procedures in Korea

Procedure	CTDIvol (mGy)	DLP (mGy·cm)
Brain	60	1,000
Abdomen	20	700

[18]. Therefore, efforts to minimize radiation dose while attaining images of adequate quality with sufficient imaging data for a diagnosis are most important. The guidelines to reduce radiation dose in radiography are as follows:

- Provide education on the relationship between radiation dose and image quality;
- Develop individualized protocol to attain appropriate image with minimal radiation dose;
- Optimize exposure parameters (mAs, kVp); and
- Minimize radiation field

3.2   Dose Reduction for CT

The introduction of CT probably is the single most important advance in diagnostic radiology. However, when compared to radiography, a CT procedure is associated with much higher radiation dose, resulting in greater radiation burden for the general population. In Korea, the number of CT units was 1,724 in 2009, which ranks third among the OECD countries. The introduction of CT into clinical practice has been followed by an exponential increase in the number of CT procedures performed. According to Health Insurance Review and Assessment Service, the number of CT procedures in 2003 was 1.7 million and increased to 4.8 million in 2009, increasing at a rate of 20 % per year (NHIC) [15]. Although a large part of this increase in procedures is from CT-based screening programs for the asymptomatic patients, the Canadian Association of Radiologists reported that 30 % of CT procedures carry little diagnostic values [26]. To address unnecessary exposure, KFDA conducted a nationwide survey in 2008 to evaluate the utilization of CT in 2008. Based on this survey, the DRLs for CT procedures were been established (Table 22.4). For cranial CT procedures, the recommended CTDIvol is 60 mGy and DLP 1,000 mGy; and for abdominal CT procedures, the recommended CTDIvol is 20 mGy and DLP 700 mGy [9, 20].

The referrer and provider significantly influence the radiation dose delivered to a patient, they must ensure the CT procedure is clinically justified and that the associated risk is small relative to the diagnostic information. Further, they should consider replacing CT, when practical, with other options, such as ultrasonography and MRI. Therefore, they should be well informed of current practice trends and ensure that the potential benefit from an indicated CT procedure is far greater than the potential risk. This is called justification, the first general principle in radiologic protection [20, 25]. Inappropriate use of radiation leads to unnecessary exposure, usually arises from (1) repeating the same procedure, (2) using procedures of little diagnostic value, (3) requesting a follow up procedure too frequently, even before disease remission, (4) choosing inappropriate procedures, i.e. with the rapid advances in radiology, guidance from radiologists in the selection of the most suitable procedure has become imperative, (5) poor communication between the referrers and providers (radiologists) regarding the clinical information and the reasons for the procedure, and (6) referrers' over dependency on imaging procedures [20].

Besides justification, optimization is the second general principle in radiation protection. Optimization means that imaging should be performed using doses that are as low as reasonably achievable (ALARA), consistent with the diagnostic task. The first step to optimization is patient preparation through accurate communication between the examiner and examinee, by minimizing patient anxiety and providing patient information. The most important aspect of radiation protection is the development of appropriate and effective protocol for a specific clinical situation. Encouraging patients to wear radiation protective gears could reduce radiation dose [3]. Minimizing radiation dose during procedure by using automatic exposure control (AEC) technique are recommended for dose optimization. Unnecessarily large radiation field should be avoided and contrast-enhanced CT should be used only when indicated. The exposure and dose information for a procedure are displayed on the CT console. However, 30 % of the CT units in Korea do not provide such display. The Korean government, together with KSR, is currently addressing this issue.

3.3   Dose Reduction for Interventional Radiology and Fluoroscopy

During the past decades, there has been a substantial increase in the number, type and complexity of interventional radiology (IR) procedures. Interventional procedures are characterized by long fluoroscopy time and sometimes require radiographic images to be acquired. As a consequence, interventional radiology procedures impose a greater radiation threat than CT, and the developments in this field have profound radiation protection implications for both patients and staff. Total fluoroscopy time correlates closely to dose in IR. The expertise of the interventionalist determines the dose. For the patient, IR procedures may lead to potential side effects, e.g. skin burns, as documented in several reports [1]. The ICRP has devoted significant effort over the recent years to improve radiation safety in IR, and in 2001 released 'Publication 85: Avoidance of radiation injuries from medical interventional procedures', which included recommendations addressing different aspects of IR [24].

The exposure delivered during fluoroscopic procedures is determined by entrance surface dose and dose-area product. The maximum entrance skin dose depends on the automatic dose-rate control setting, patient size, focus-skin distance

Table 22.5 Interventional radiation dose metrics based on a 2008 Korean survey

Procedure	Fluoroscopic time (min)	Number of images	ESD (mGy)	DAP (Gy·cm^2)	Cumulative dose (mGy)
TACE	16.6	108.0	511.7	210.0	512.0
AVF	18.8	55.3	31.8	27.7	31.8
PTBD	4.2	2.9	58.6	18.5	58.6
Cerebral angiography	9.3	345.6	373.9	226.0	405.0
GDC coil embolization	51.1	272.0	2,264.0		2,264.0

All values shown are mean levels. *TACE* transcatheter arterial chemoembolization, *AVF* arteriovenous fistula, *PTBD* percutaneous transhepatic drainage, *GDC* Guglielmi detachable coils, *ESD* entrance skin dose, *DAP* dose-area product

Table 22.6 Radiation dose for fluoroscopic procedures, based on a 2010–2011 Korean survey

Procedure	DAP (Gy·cm^2)
Barium enema	38.1
Upper GI series	22.0
Barium swallow	9.9
Small bowel series	31.4
IVP	8.0
ERCP (Diagnostic)	53.5
Hysterosalphingography	19.6
Cerebral angiography (Diagnostic)	66.3
Coronary angiography (Diagnostic)	21.2
Femoral angiography (Diagnostic)	48.3

All values are mean levels. *IVP* intravenous pyelography, *ERCP* endoscopic retrograde cholangiopancreatography

and the exposure time. In Korea, an evaluation of the current trends in IR was carried out in 2007. The results are presented in Table 22.5 [8]. For example, although the average entrance skin dose is 511.7 mGy for TACE, the maximum entrance skin dose is 4.3 Gy. Therefore despite patient protection, possible deterministic effects in the skin are inevitable, especially in difficult cases. A survey on the exposures for fluoroscopic procedures was carried out from 2010 to 2011, and the results are presented in Table 22.6. The variation in exposures for the same procedure is due to the difference in complexity and levels of the procedure. Many interventionalists are not aware of the potential radiation risks for interventional procedures or the simple methods to decrease exposure and the incidence of side effects, by using dose control strategies. Such lack of knowledge calls for adequate education and training in both clinical techniques and radiation protection [6]. The ICRP Publication 85 contains recommendations for both patients and staff, and the IAEA suggests employing technological methods to reduce radiation dose [4, 5].

The lack of information on radiation dose for fluoroscopy and fluoroscopically guided interventions has failed to generate as much concern as CT procedures have. However, as IR is already an established part of mainstream medicine and it is expected to expand with on-going developments, it is important that the radiation protection implications to both patients and staff are not overlooked in the zealous drive to improve health care.

3.4   Dose Reduction for Children

In pediatric imaging, particular attention should be given to reduce radiation dose as younger children have higher potential radiation risk. Depending on the reports, that rate of children being exposed unnecessarily ranges from 10 % to 30 % [21]. As in adults, pediatric health care professionals should seek ways to decrease radiation dose by adhering to radiation protection principles. There is an international accepted agreement that the benefits of an indicated procedure should far outweigh the risks. Because children are more prone to radiation risk compared to adults, radiologists should always be consulted when developing pediatric imaging strategies and creating specific protocols with techniques optimized for pediatric patients. The ALARA principle should be strongly endorsed, unnecessary procedures and repeats avoided, and utilization of single-phase CT procedures encouraged [7].

In 2010, KFDA promulgated the DRL for pediatric chest x-ray in Korea [16], and DRLs for pediatric CT are under investigation. Research in 149 radiological facilities revealed some interesting facts, namely; (1) imaging facilities using age-specific radiologic units showed significantly lower exposure; and (2) imaging facilities not using automatic exposure control (AEC) were associated with twice as much exposure (Table 22.7).

Justification includes the elimination of unnecessary procedures and substitution of CT with other suitable alternatives, such as MRI or ultrasonography. Radiologic procedures should only be used when the potential benefit from an indicated procedure is recognized and documented, and is far greater than the potential radiation risk. Optimization includes avoiding redundant procedures and employing CT protocols with minimal radiation dose while maintaining diagnostic accuracy. Most of the time, a single phase is sufficient for a diagnosis in pediatric CT and in

Table 22.7 Factors affecting pediatric radiation dose, based on a 2010 Korean survey

Factors	ESD (µGy)
With pediatric exam. Room	104.7
Without pediatric exam. Room	148.0
Use automatic exposure control (AEC)	78.5
Do not use automatic exposure control (AEC)	168.1
DRL	100

ESD entrance surface dose, *DRL* diagnostic reference level many cases the pre-contrast phase is unnecessary. As shown in Table 22.7, the use of AEC reduces dose by a significant amount. From time to time, a repeat view or phase may be required as the images may not be diagnostic due to patient anxiety or movement. Improving communication and providing a more relaxed atmosphere can minimize additional radiation exposure as a result of fear. The reduction of radiation dose in pediatric imaging calls for a closer co-operation between the pediatric health care professionals and the patients, thus minimizing unnecessary exposure.

4   Conclusion

In Korea, the KFDA is leading and continuing its enthusiastic efforts to reduce population exposure by establishing DRLs for CT and conventional x-rays and investigating current practice of radiology. Unfortunately, not all medical personnel are aware of radiation exposures from different imaging modalities or are well informed of radiation protection and radiation safety. Promoting awareness and informing the general population of possible radiation hazard, through the use of accurate radiation-dose estimation and radiation dose card, can lead to reduction of radiation dose. Inter-organization networks such as KARSM are expected to play a central role in reducing unwanted radiologic exposure through education of radiation risk and promotion of preventive measures. The most pressing matters in the reduction of exposure are by raising awareness of possible radiation-related hazards among the referrers and educating the radiologists for a better understanding of radiation equipment operation and more effective use of radiation protection measures.

References

1. Balter S, Hopewell JW, Miller DL et al (2010) Fluoroscopically guided interventional procedures: a review of radiation effects on patients' skin and hair. Radiology 254(2):326–341
2. Borgen L, Stranden E, Espeland A (2010) Clinicians' justification of imaging: do radiation issues play a role? Insight Imaging 1(3):193–200
3. Curtis JR (2010) Computed tomography shielding methods: a literature review. Radiol Technol 81(5):428–436
4. International Atomic Energy Agency (IAEAa) Radiation Protection of Patients. 10 Pearls: radiation protection of patients in fluoroscopy (2013). Available from: https://rpop.iaea.org/RPOP/RPoP/Content/Documents/Whitepapers/poster-patient-radiation-protection.pdf.   Accessed 1 October 2013
5. International Atomic Energy Agency (IAEA) (2013) Radiation Protection of Patients. 10 Pearls: radiation protection of staff in fluoroscopy. Available from: https://rpop.iaea.org/RPOP/RPoP/Content/Documents/Whitepapers/poster-staff-radiation-protection.pdf. Accessed 15 July 2012
6. Jeong WK (2011) Radiation exposure and its reduction in the fluoroscopic examination and fluoroscopy-guided interventional radiology. J Korean Med Assoc 54(12):1269–1276

7. Jung AY (2011) Medical radiation exposure in children and dose reduction. J Korean Med Assoc 54(12):1277–1283
8. Korea Food and Drug Administration (2007) Evaluation of patient dose in interventional radiology. Korea Food and Drug Administration, Seoul
9. Korea Food and KFDA Drug Administration (2009) Guideline for diagnostic reference level in CT examination in Korea: radiation safety series no. 19. Korea Food and Drug Administration, Seoul
10. Korea Food and KFDA Drug Administration (2009) Guideline for diagnostic reference level in chest PA in Korea: radiation safety series no. 19. Korea Food and Drug Administration, Seoul
11. Korea Food and Drug Administration KFDA (2010) Technical report for pediatric radiography: radiation safety series no. 22. Korea Food and Drug Administration, Seoul
12. Korea Food and Drug Administration KFDA (2011) Korea Food and Drug Administration 2010 Report. Occupational radiation exposure in diagnostic radiology: radiation safety management series no. 27. Korea Food and Drug Administration, Seoul
13. Krille L, Hammer GP, Merzenich H et al (2010) Systematic review on physician's knowledge about radiation doses and radiation risks of computed tomography. Eur J Radiol 76(1):36–41
14. Lee CI, Haims AH, Monico EP et al (2004) Diagnostic CT scans: assessment of patient, physician, and radiologist awareness of radiation dose and possible risks. Radiology 231(2):393–398
15. National Health Insurance Corporation NHIC, Health Insurance Review and Assessment Service (2006) 2005 National health insurance statistical yearbook. National Health Insurance Corporation, Seoul
16. National Institute of Food and Drug Safety Evaluation NIFDS (2010) Technical standard for the performance of pediatric radiography. National Institute of Food and Drug Safety Evaluation, Seoul
17. Ng KH, Lean ML (2012) The Fukushima nuclear crisis reemphasizes the need for improved risk communication and better use of social media. Health Phys 103(3):307–310
18. Ng KH, Rehani MM (2006) X ray imaging goes digital. BMJ 333(7572):765–766
19. O'Sullivan J, O'Connor OJ, O'Regan K et al (2010) An assessment of medical students' awareness of radiation exposures associated with diagnostic imaging investigations. Insights Imaging 1(2):86–92
20. Park MY, Jung SE (2011) CT radiation dose and radiation reduction strategies. J Korean Med Assoc 54(12):1262–1268
21. Ron E (2002) Ionizing radiation and cancer risk: evidence from epidemiology. Pediatr Radiol 32(4):232–237
22. Sung DW (2011) Investigation of patient dose for diagnostic reference levels (DRL) in radiographic examination: national survey in Korea. Final report. Korea Food and Drug Administration, Seoul
23. United Nations Scientific Committee on the Effects of Atomic Radiation UNSCEAR (2010) Sources and effects of ionizing radiation. UNSCEAR 2008 report to the General Assembly with scientific annexes. United Nations, New York
24. Valentin J (2000) Avoidance of radiation injuries from medical interventional procedures. Ann ICRP 30(2):7–67
25. Valentin J (ed) (2007) ICRP Publication 105. Radiological protection in medicine. Ann ICRP 37 (6). Elsevier, Oxford
26. You JJ, Levinson W, Laupacis A (2009) Attitudes of family physicians, specialists and radiologists about the use of computed tomography and magnetic resonance imaging in Ontario. Health Policy 5(1):54–65

Chapter 23
Euratom Initiatives and Perspective on Advancing Radiation Protection in Radiology

Georgi Simeonov, Remigiusz Baranczyk, and Augustin Janssens

Abstract The European Union has a long history of successful dealings with radiation protection of patients in radiological imaging. This includes the adoption in the 1980s of specific Euratom legislation and its subsequent updates, the publication of the European guidance, the support of research and the facilitation of information exchange and stakeholder involvement. Today the rapidly developing medical imaging technology together with an increasing number of patients undergoing radiological procedures have led to a global increase in population doses from medical exposure. There is growing concern from stakeholders regarding the justification and optimization of some radiological procedures, the training and education of medical radiation practitioners and the scarcity of reliable information. In response to these challenges, the European Commission's Directorate-General for Energy is undertaking a series of initiatives and actions to address these concerns.

Keywords Euratom legislation • European Commission Directorate-General for Energy • Policy implementation • Radiation protection

1 Introduction

Together with the rapid advances in technology and the increase in new applications for medical imaging, there is a significant escalation in the use of ionizing radiation in medicine and the associated population exposure in recent years. There is a growing stakeholders concern regarding the appropriate use of

G. Simeonov (✉) • R. Baranczyk • A. Janssens
Radiation Protection Unit, Directorate-General for Energy, European Commission,
Luxembourg, Luxembourg
e-mail: Georgi.Simeonov@ec.europa.eu

L. Lau and K.-H. Ng (eds.), *Radiological Safety and Quality: Paradigms in Leadership and Innovation*, DOI 10.1007/978-94-007-7256-4_23,
© Springer Science+Business Media Dordrecht 2014

radiological procedures, the lack of timely and trustworthy information and the training of radiation medicine practitioners. The European Union (EU) has a long history of leadership, commitment and success in implementing actions in the radiation protection of patients in radiological imaging. The European Commission's Directorate-General for Energy is undertaking a range of initiatives and actions to strengthen its on-going efforts in advancing quality care and radiation safety.

2 Legal Basis

2.1 The European Framework

The EU is an economic and political partnership between 28 European countries, i.e. "Member States" (MS) with around 500 million inhabitants that together cover much of the continent. The EU traces its origins from the European Coal and Steel Community, the European Economic Community and the European Atomic Energy Community (Euratom), i.e. "the European Communities", which were formed in the 1950s. The Maastricht Treaty established the Union under its current name in 1993. The latest amendment to the constitution, the Treaty of Lisbon, came into force in 2009.

The EU operates through a system of supranational independent institutions, e.g. the European Parliament, the European Council, the Council of the European Union (often simply referred to as the Council), the European Commission, the Court of Justice of the European Union, the European Central Bank and the Court of Auditors.

The European Commission is the EU's executive body and represents the interests of Europe as a whole. It is responsible for proposing legislation, implementing decisions, upholding EU treaties and the general day-to-day running of the Union. The term "Commission" refers to both the College of Commissioners and the institution, which has its main offices in Brussels (Belgium) and Luxembourg.

2.2 Euratom Treaty

The treaty that underpinned the European Atomic Energy Community [4], commonly referred to as the Euratom Treaty, is *binding primary law* for the 28 MS. Following its entry into force in 1958 the Euratom Treaty provides the framework on which the European institutions and MS share their competencies and discharge their respective responsibilities [19]. The main tasks under the Euratom Treaty are defined in Article 2, one of which is to protect the health of the workers and the public against the dangers arising from ionising radiation.

Chapter III of the Euratom Treaty entitled "Health and Safety" offers the legal framework for the establishment of the Euratom Basic Safety Standards (BSS) for the protection of the health of workers and general public. The first Euratom BSS dated back to 1959 and the latest version, Council Directive 96/29/Euratom [1], was published as Euratom secondary law in 1996.

Additional binding instruments supplement the Euratom BSS. The relevant legislation with respect to medical exposure is Council Directive 97/43/Euratom: "Medical Exposures Directive" (MED) [2]. The EC also issues other documents of non-binding nature, which have different status in the hierarchy of EU-instruments. These include EC recommendations and communications.

2.3 Legislation Development and Implementation

In accordance with Article 31 of the Euratom Treaty, the EC drafts radiation protection legislations after obtaining advice from a group of independent radiation protection experts from EU MS. The proposal is presented to the Council of the European Union for its consideration and adoption before becoming a binding Euratom legal act, e.g. as a Council Directive. The Council may amend the proposal if required, as part of the adoption process (Fig. 23.1).

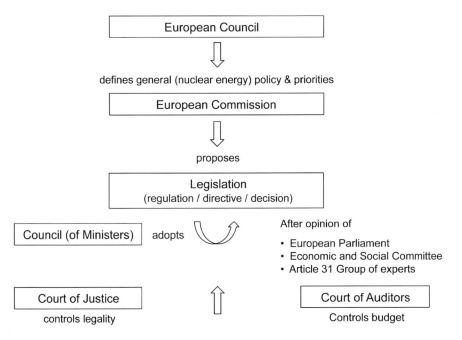

Fig. 23.1 Euratom legislation development process [19]. This chart outlines the steps from proposal to the adoption of Euratom legislation, including the radiation protection Directives. Please refer to the text for a more detailed description

The Council Directives are binding for the EU MS. For their national implementation, each MS has to adopt transposition measures, i.e. legislation, regulations and administrative arrangements. The EC has certain obligations and powers over national transposition and implementation.

Further, the EC publishes radiation protection guidelines to assist MS to implement Euratom legislations in a consistent way and to facilitate the harmonization in the respective areas. Radiation protection publications are published by the EC after approval by the Article 31 Group of Experts (GoE). These guidelines do not have any legal status, i.e. they are non-binding and they are not formally adopted by the EC. In fact, there is a disclaimer in these radiation protection publications that they "do not necessarily represent the opinion of the Commission". Therefore, on the one hand, the guidance often follows the requirements outlined in the Euratom legislation but, on the other hand, has no official standing in the hierarchy of the EU/Euratom instruments.

3 EC Policy Initiatives in 2010

3.1 Adoption of COM (2010) 423

In August 2010 the EC adopted a Communication COM [11] 423 "on medical applications of ionizing radiation and security of supply of radioisotopes for nuclear medicine" [11] to the EU Council and the European Parliament. The Communication is focused on two problem areas, the radiation protection of patients and staff and the supply of radioisotopes for nuclear medicine. This document provides an overview of the main challenges and the Community actions undertaken so far and aims at stimulating discussions in the EU on the need for further actions and allocation of resources and responsibilities. The scope and coverage of COM [11] 423 is not restricted to the Euratom Treaty but expands to other EU policies, e.g. the European legal framework on medical devices [3] and European policies for research and innovation [6].

3.2 Proposed Way Forward

COM [11] 423 proposes a way forward for the EC and for the EU MS to address the identified issues in the radiation protection of patients and staff. The proposed actions in the different areas are discussed below.

Strengthening the Existing Regulatory Framework

Although timely and relatively advanced at the time, the EC emphasises that the revision of the Euratom legislation for radiation protection of patients alone will not guarantee an efficient regulatory framework in the EU. Therefore, further actions are proposed to strengthen the regulatory supervision of medical applications, to provide tools to facilitate the practical implementation of legislations, to enhance the role of the Medical Physics Experts (MPE) and to continuously monitor the trends in medical exposures in the European Union.

Raising Awareness and Developing Safety Culture

The EC points out that the medical professionals should not only receive adequate and up-to-date training and education, but should also be aware of their responsibilities and are committed to a radiation safety culture to ensure good patient care and adequate radiation protection. Further awareness raising is needed among national healthcare policymakers, patients and the general population.

Fostering Radiation Protection in Medicine through Research

There is a need for better understanding of the health risks from exposure to low doses of ionizing radiation, such as individual sensitivity to radiation exposure, thereby leading to optimised health protection. Future actions under the Euratom framework should be undertaken within the Multidisciplinary European Low-Dose Initiative [21]. The more general Health theme under the EU Framework Programme should integrate appropriate protection of patients and staff with considerations of future developments of medical radiation technology.

Integration of Policies

The EC advocates the importance of integration of policies in public health, research, trade and industry for the benefit of patients. An area of special interest is the regulation of medical devices under the EU legislation [3] where a standing platform to examine radiation protection features of such devices is proposed.

International Co-operation

The EC is committed to co-ordinate efforts with the International Atomic Energy Agency (IAEA), the World Health Organization (WHO) and other players and to support other international programs to improve radiation protection of patients.

4   Revision of the Euratom BSS

4.1   Introduction of the Revised BSS

Soon after the publication of the 2007 Recommendations of the International Commission on Radiological Protection [18], the EC launched a revision of the Euratom BSS. This included a simplification of the Euratom legislation on radiation protection by integrating five current Directives, in particular the Medical Exposure Directive (MED) included, into a single revised Euratom BSS Directive [22]. The EC adopted a final proposal for revised Euratom BSS in May 2012 [12].

4.2   Proposed Changes to Medical Exposure

Even though the MED has been widely recognized as one of the most advanced pieces of legislation on radiation protection in medical exposure, some changes to the current requirements were proposed in the revised Euratom BSS. The reasons for these changes include: the objective to develop a coherent text in the revised Euratom BSS, the experience gained and the lessons learnt in the last 10 years from the implementation of the previous legislation and the unforeseeable advances in medicine and technology. The changes, as described in the following sub-sections, are mostly found in Chapter VII of the revised Euratom BSS.

Definition of Medical Exposure

Medical exposure is now defined as exposure to patients resulting from treatment, diagnosis or health screening and exposure to biomedical research volunteers, carers or comforters. Current "medico-legal procedures" are treated separately as part of the newly defined "non-medical imaging exposure". The new definition proposed by the Commission incorporates the notion of *intended benefit* to the exposed individual, which refers not only to the health but also the *well-being* of the individual in an attempt to cover cases of sport and recreational medicine, diagnosis of child abuse, etc.

Justification

The Commission proposed that *staff* exposure shall also be taken into account in justifying a type of medical procedure. New requirements were proposed with respect to the medical exposure of *asymptomatic individuals* for early detection of disease. They shall only be exposed under an approved health-screening programme or, if this is not the case, the justification shall be specifically documented (for that individual) by the radiology practitioner in consultation with the referrer. In the latter case, appropriate guidelines from the professional bodies or competent authorities shall be followed and special attention shall be paid to the provision of information. The radiology practitioner shall ensure that the *patients are provided with adequate information* relating to the benefits and risks associated with the medical exposure to enable informed consent.

Optimization

The requirements in this area remain substantially unchanged except that the diagnostic reference levels (DRLs) were *expanded to include interventional radiology*, when appropriate.

Dose Recording and Reporting

The Commission proposed that *any system* used for interventional radiology and computed tomography (CT) shall have a device or a feature informing the radiology practitioner of the quantity of radiation produced by the equipment during the procedure. Any other medical radiodiagnostic equipment shall have such a device or feature or equivalent means of determining the quantity of radiation produced. The radiation dose shall form part of the report of the examination.

Medical Physics Expert

The proposed new definition and detailed description of the MPE's responsibilities aim to provide a link between their required competences and the assigned responsibilities in relation to medical exposure. The requirements for the involvement of the MPE in medical exposure procedures have been changed to strengthen her/his presence in high-dose radiological imaging examinations.

Information, Education and Training

The requirements for the education and training of medical professionals were strengthened and a course on radiation protection in the basic curriculum of medical and dental schools was proposed as a *mandatory* requirement. New legal provision requires the MS to ensure that mechanisms are in place for a timely dissemination of appropriate information relevant to radiation protection in medical exposure on lessons learned from significant events.

Accidental and Unintended Exposures

The revised Euratom BSS Directive defines several new requirements on accidental and unintended medical exposures. The radiotherapy quality assurance programs will include a *study of risks* of accidental or unintended exposures. The operators of diagnostic and therapeutic radiological equipment shall implement a *registration and analysis* system of events involving or potentially involving accidental or unintended medical exposures. The operators shall *declare to the competent authorities* the occurrence of significant events, as defined by the authorities, including the results of their investigation and the corrective measures to avoid such events in the future. The referring and the radiological medical practitioners as well as the patient shall be informed about unintended or accidental exposures.

5   Other EC Activities

In addition to its legislative initiatives, the EC carries out other activities to support the implementation of the Euratom legislation and, as a final objective, improve radiation protection of patients and staff. These include the promotion of exchange of experience between MS, publication of guidelines and information material, and co-operation with international organizations, professional bodies and stakeholders, etc.

The EC funds several projects each year in different areas relating to the implementation of the MED. In the recent years, the EC published guidelines on estimating population doses from medical procedures [7], clinical audit [9], acceptability criteria for radiological equipment [13] and cone-beam computed tomography [14] as well as the proceedings of scientific seminars on radiation induced circulatory diseases [10] and medical overexposure [8]. Further material on the education and training of medical professionals in radiation protection [20], qualification framework and curriculum for MPE [16] and a summary report on European population doses from radiodiagnostic procedures [5] are planned for 2013. The programme for the following years will cover the issues of accidental exposure in radiotherapy, referral guidelines for medical imaging and DRLs for paediatric and young patients. Between 2010 and 2012 the EC also provided support for the European Medical ALARA Network [15], which is now a self-sustainable network with the ambition to become a single source of information on radiation protection for the health professionals in Europe.

6   Regulatory Cooperation in Europe

A unique network of European radiation protection authorities operates in Europe since 2007 under the name of HERCA (Heads of Radiation protection Competent Authorities). HERCA is a voluntary association in which *the heads* of radiation protection authorities of 31 European countries, including the 28 EU Member States, work together to identify the common issues and to propose practical solutions for these issues. According to the statement on the HERCA website [17], "HERCA is working on topics generally covered by provisions of the Euratom Treaty".

HERCA has currently an active Working Group (WG) on Medical Applications, which is working on several topics including the involvement in a dialogue with CT manufacturers, justification of radiological procedures and regulatory inspection of optimization. The EC is a permanent observer to the HERCA WG on Medical Applications and participates actively in the debate on the role of regulators in this area.

7 Conclusion

To narrow the gap between the availability of information, the implementation of policies and their use in everyday practice, a range of actions is needed. The EC achieves this by regular reviews and updates of legislation, the publication of studies and guidance to facilitate practical implementation of legislation and engagement in dialogue with stakeholders and national regulators.

In the EU the radiation safety standards are proclaimed in the Euratom Treaty and the secondary legislation, e.g. the Euratom Directives, issued under it. MS transpose these Directives into national regulations, thus ensuring consistency. In addition to the preparation of these binding legislations, the EC publishes guidelines, information and implementation guidance tools for a more effective and consistent national implementation of radiation protection policies and actions.

The EC advocates and facilitates: research into the benefits and risks of radiation use in medicine; awareness raising; development of a radiation safety culture; exchange of experience between MS, and collaboration and teamwork between organizations and professional bodies. For example, MELODI, EMAN and HERCA are effective platforms for stakeholder engagement and co-operation.

Good policies and actions are based on reliable and up-to-date information, e.g. on population exposure trends and on the health effects of medical radiation. Hence, a system to monitor the trends in medical exposures in the EU shall be established and maintained [5]. Other actions to facilitate the recording and reporting of individual patient exposures shall be undertaken.

The public should be provided with evidence-based information about the benefits and risks of medical radiation. In addition to generic web-based information and publications from reputable professional organizations and agencies, it is desirable for patients to receive procedure-specific information, which is documented in concise and plain language outlining the issues, benefits and risks including radiation exposure. While such fact sheets would normally cover the most common questions, the facility's staff should be able to answer other questions and provide explanations prior to any procedure.

The involvement with different professional organizations and agencies in Europe is crucial not just to define common education and training strategies but also to ensure broader collaborations towards better radiation protection of patients. Both MEDRAPET and EMAN initiatives are important and the EC looks forward to their future as self-sustainable collaborative networks.

This chapter outlined the steps, which the EC Directorate-General for Energy has taken and is implementing to promote radiation protection in the EU. As a result of its leadership and the range of innovative actions, the EC is confident that the sum of its efforts and those from other stakeholders will further strengthen and improve the appropriate use of medical radiation and radiation safety within the EU and beyond.

References

1. Council of the European Union (1996) Council Directive 96/29/Euratom of 13 May 1996 laying down basic safety standards for the health protection of the general public and workers against the dangers of ionizing radiation. OJEC 1996/L 159/1
2. Council of the European Union (1997) Council Directive 97/43/Euratom of 30 June 1997 on health protection of individuals against the dangers of ionizing radiation in relation to medical exposure, and repealing Directive 84/466/EURATOM. OJEC 1997/L 180/22
3. Council of the European Union (2007) Directive 2007/47/EC of the European Parliament and of the Council of 5 September 2007 amending Council Directive 90/385/EEC on the approximation of the laws of the Member States relating to active implantable medical devices, Council Directive 93/42/EEC concerning medical devices and Directive 98/8/EC concerning the placing of biocidal products on the market. OJEU 2007/L 247/21
4. Council of the European Union (2010) The High Contracting Parties of EURATOM (1957), Consolidated version of the Treaty establishing the European Atomic Energy Community. Official Journal of the European Union (OJEU) 2010/C 84/01
5. Dose Datamed 2 (2013) http://ddmed.eu/. Accessed 26 Mar 2013
6. EUROPA (2013) Policy area of research and innovation. http://europa.eu/pol/rd/index_en. htm. Accessed 26 Mar 2013
7. European Commission (2008) Radiation Protection 154. European Guidance on Estimating Population Doses from Medical X-Ray Procedures. EU Publications Office. http://ddmed.eu/ _media/backg round_of_ddm1:rp154.pdf. Accessed 26 May 2012
8. European Commission (2008) Radiation Protection 149. EU Scientific Seminar 2003 "Medical overexposures." EU Publications Office. http://ec.europa.eu/energy/nuclear/radiation_-protection/doc/publication/149.pdf. Accessed 22 Dec 2011
9. European Commission (2009) Radiation Protection 159. European guidelines on clinical audit for medical radiological practices (diagnostic radiology, nuclear medicine and radiotherapy). EU Publications Office. http://ec.europa.eu/energy/nuclear/radiation_protection/doc/publication/159.pdf. Accessed 22 Dec 2011
10. European Commission (2009) Radiation Protection 158. EU Scientific Seminar on Emerging evidence for radiation induced circulatory diseases. EU Publications Office. http://ec.europa. eu/energy/nuclear/radiation_protection/doc/publication/158.pdf Accessed 22 Dec 2011
11. European Commission (2010) COM/2010/0423 Communication from the Commission to the European Parliament and the Council on medical applications of ionizing radiation and security of supply of radioisotopes for nuclear medicine. eur-lex.europa.eu/LexUriServ/ LexUriServ.do?uri=CELEX:52010DC0423:EN:HTML. Accessed 26 May 2012

12. European Commission (2012) Proposal for a Council Directive laying down basic safety standards for protection against the dangers arising from exposure to ionising radiation. http://ec.europa.eu/energy/nuclear/radiation_protection/doc/2012_com_242.pdf.   Accessed 12 June 2012
13. European Commission (2012) Radiation Protection 162. Radiation criteria for acceptability of medical radiological equipment used in diagnostic radiology, nuclear medicine and radiotherapy.   http://ec.europa.eu/energy/nuclear/radiation_protection/doc/publication/162. pdf. Accessed 26 Mar 2013
14. European Commission (2012) Radiation Protection 172. Cone beam CT for dental and maxillofacial radiology. Evidence Based Guidelines. http://ec.europa.eu/energy/nuclear/ radiation_protection/doc/publication/172.pdf. Accessed 26 Mar 2013
15. European Medical ALARA Network (EMAN) (2013) http://www.eman-network.eu/. Accessed 26 Mar 2013
16. Guidelines Medical Physics Expert (MPE) (2013) http://portal.ucm.es/web/medical-physics-expert-project. Accessed 26 Mar 2013
17. Heads of European Radiological protection Competent Authorities (HERCA) (2013) http:// www.herca.org/. Accessed 26 Mar 2013
18. International Commission on Radiological Protection (2007) ICRP Publication 103. The 2007 recommendations of the International Commission on Radiological Protection. Elsevier, Oxford
19. Kilb W (2010) The European Atomic Energy Community and its primary and secondary law. In: International nuclear law: history, evolution and outlook. Nuclear Energy Agency, Organization for Economic co-operation and Development, Paris
20. Medical Radiation Protection Education and Training (MEDRAPET) (2013) http://www. medrapet.eu/. Accessed 26 Mar 2013
21. Multidisciplinary European Low Dose Initiative (MELODI) (2013) http://www.melodi-online.eu/. Accessed 26 Mar 2013
22. Mundigl S (2011) Revision of the euratom basic safety standards directive – current status. Radiat Prot Dosimetry 144(1–4):12–16

Chapter 24
Bridging the Radiological Healthcare Divide with Social Entrepreneurship

Evelyn Lai-Ming Ho

"The test of our progress is not whether we add more to the abundance of those who have much; it is whether we provide enough for those who have too little."

Franklin D. Roosevelt, 20 Jan 1937

Abstract Radiology is an important pillar of good healthcare. Yet, it is estimated that two-thirds of the world's population has no or inadequate access to basic X-ray examinations and/or ultrasound. Most of these occur in low to middle income or developing countries. The main reason would be the cost of radiology services, being equipment intensive and resource hungry. Inroads in meeting the acute needs of these countries have been made by some non-profit organizations. One possible innovation to meet this challenge and improve public access to quality radiology and radiation protection is by applying social entrepreneurship. Social entrepreneurship is the application of an unusual or innovative approach to solve a social or environmental problem. The emphasis is on positive social or environmental impact. In practical terms for bridging the radiological divide, it will be through leadership, collaboration and participation. Examples of social radiology include the Physicians Ultrasound in Rwanda Education Initiative, RAD-AID International and the Malaysian College of Radiology's Value Added Mammogram Program. However, infrastructure, finance, education, policy, equipment maintenance, safety and radiation protection issues must also be addressed to reduce waste and inefficiency. Social entrepreneurship offers an innovative solution to meet societal needs, which is sustainable, pervasive and positive. Financial sustainability may be achieved through grants, donations, a viable business model or a combined approach. Microfinance together with training, impact investments or social impact

E.L.-M. Ho (✉)
Imaging Department, Sime Darby Medical Centre ParkCity,
2, Jalan Intisari Perdana, Desa ParkCity, Kuala Lumpur, 52200 Malaysia
e-mail: evelynlmho@gmail.com; webeditor@radiologymalaysia.org

L. Lau and K.-H. Ng (eds.), *Radiological Safety and Quality: Paradigms in Leadership and Innovation*, DOI 10.1007/978-94-007-7256-4_24,
© Springer Science+Business Media Dordrecht 2014

bonds could help the social entrepreneur to achieve the goals faster. The focus is on pro-active and positive social impact, not just financial goals. It produces results faster, provides an alternative, or assists governments and public sectors to achieve more equitable coverage of quality radiological healthcare including those at the base of the pyramid.

Keywords Base of pyramid • Developing countries • Impact investments • Low income • Medical imaging • Microfinance • Non-profit organizations • Quality radiology • Radiation protection • Radiological healthcare divide • Radiology services • Social entrepreneur • Social enterprise • Social impact bonds

1 Introduction

Radiology is an exciting field of medicine that is constantly changing by technological breakthroughs, assisted by the boom in computing power. One can "fly through" the bowel in virtual 3-D reconstruction of datasets from a computed tomography (CT) scan even sans bowel preparation. CT or Magnetic Resonance Imaging (MRI) enables virtual autopsies without mutilating the body, rendering it acceptable to all cultures and religions. Image guided intervention, image guided treatment and imaging monitoring of treatment response have revolutionised and provided more management options for various diseases. In cardiac imaging, there is improvement in the detection and risk stratification of disease.

Radiology plays a critical role in the surveillance, prevention and diagnosis of disease; and the monitoring of treatment including theranostics and image guided minimally invasive therapy. Without doubt, radiology is an important pillar of modern healthcare.

2 The Radiological Healthcare Services Disparity

According to the World Health Organization (WHO), 80–90 % of diagnostic problems can be easily solved with basic X-ray examinations and/or ultrasound. It estimates that two-thirds of the world's population has no or inadequate access to these services [29]. Where equipment is available, equipment maintenance, practitioner training, mismatched acquisition with the equipment not used and left to collect dust, questionable radiation protection, safety and imaging quality issues may exist [15].

There is great disparity in radiological healthcare services between high, middle and low income countries. The WHO 2010 medical devices survey [30,31] found the density of mammography units per million population (pmp) for low income countries was 0.27, whilst that for high income countries was 22.93. Between upper middle and high income countries, the CT scan density pmp was 7.16 versus 44.31

[27]. We should bear in mind that only slightly more than a quarter of the world's countries are considered high income countries.

Intra-country differences exist between rural and urban areas, different states and various ethnic groups. For example, the density of MRI was 8.8 pmp in Kuala Lumpur, compared to the Malaysian national density of 2.9 and Organization for European Economic Cooperation and Development (OECD) average of 11.0 [35]. For CT it was 12.9 pmp in Kuala Lumpur, 4.22 in Perlis, West Malaysia and 1.98 in Pahang and Terengganu.

Equipment availability is only one indicator for radiology services. The radiologist workforce in Malaysia was about 0.13 per 10,000 in 2008–2009 [2]. This varied from as low as 0.03 per 10,000 in Sabah and Labuan to 0.55 per 10,000 in the Federal Territory Kuala Lumpur. The distribution of radiology practitioners was very uneven between the various states, partly due to the larger size of some or the geographical terrains.

3 Factors Influencing the Provision of Quality Radiology and Radiation Protection

Radiology services are equipment intensive and resource hungry - financial, human and infrastructure. Assessing the radiological needs for the developing, low income countries or the rural, service-deficient areas within countries is the initial step to finding solutions. Each country or area has its own needs or gaps that may encompass all or some of the factors below:

1. Infrastructure, e.g. roads, electricity and water;
2. Laws and Health Policies, i.e. guidance for good radiology practice, radiation safety and radiation protection should be integrated into the healthcare system;
3. Affordability, i.e. financial resources;
4. Adequately trained radiology specialists, radiology practitioners, and medical physicists and their continuing professional development;
5. Expertise in equipment selection to best meet its needs, which is appropriate for the infrastructure and local expertise to avoid a mismatch;
6. Expertise in equipment maintenance and repair, to ensure the equipment is in optimal working condition; and training in quality control, quality assurance and quality improvement processes;
7. Radiation safety and radiation protection expertise;
8. Information and Communications Technology (ICT) expertise and infrastructure, digital readiness;
9. Management of radiology services; and
10. Access to radiology services.

A social solution for radiology services may target one, some, or perhaps all of these factors. Integration of the services is important. For example, a breast cancer screening program would not work without adequate surgical and oncological support or a tuberculosis screening program would be ineffective if anti-tuberculous drugs are not available. WHO has estimated that in general, 20–40 % of a country's resources spent on health are lost through waste and inefficiency [33]. Efficiency would mean doing the things right and doing the right things.

"Technological edge" advertising by equipment vendors influences purchases, because the public and even philanthropic bodies or individuals are attracted to the marketing spin. A decision based on the "wow factor" or prestige might result in a costly purchase that could fast become a white elephant. Donors and purchasers may be unaware that there is no expertise available to use or maintain the equipment, resulting in frequent breakdowns or worse, decommissioning. In addition, CT scanners with higher number of slices or higher field strength MRI scanners do not guarantee better images, and are often only appropriate for academic or research centres.

Overuse and overdose issues exist in developed and developing countries which may compromise patient safety [6]. Education and training are crucial in ensuring proper installation, maintenance, quality assurance, radiation protection and safe use of equipment [15]. The digitalization of radiology brings advantages and challenges [16]. It has enabled teleradiology and eliminated hazardous processing chemicals but introduced medicolegal, dose creep, i.e. an imperceptible increase in exposure over time owing to auto-optimisation associated with digital technique, workflow and image interpretation issues.

4 Examples of Social Radiology

Many developing countries already have healthcare plans in place, while some are restructuring their systems. Bodies such as the WHO are ready to help countries achieve a safe level of health services including essential diagnostic imaging through its programmes and projects [32]. The International Atomic Energy Agency (IAEA) has in place programmes for improving radiation safety and protection. The International Organization for Medical Physics (IOMP) collaborates with professional bodies such as the International Radiation Protection Association (IRPA), International Commission on Radiological Protection (ICRP) and others to promote the development of medical physics and safe use of radiation and radiological equipment.

Yet poor and developing countries have acute needs. For these communities, more urgent actions are required to improve access to basic radiology services. Public sector restructuring to improve quality and safety, and to overcome financial and other infrastructure challenges will take some time. Fortunately, inroads have been made in the provision of quality radiological services and meeting the needs in various settings, either within a country or across international borders in social radiology [7].

4.1 Physicians Ultrasound in Rwanda Education Initiative

Physicians Ultrasound in Rwanda Education Initiative (PURE) is provided by an international group of physicians who are dedicated to the training of Rwandan physicians in the use "point of care" ultrasound (at the bedside). They are helping the Rwandans to develop the necessary tools to teach ultrasound to other Rwandan healthcare practitioners, i.e. "train the trainers" scheme. PURE materialised in October 2010. In Rwandan hospitals, many ultrasound units are not in use because of lack of expertise. PURE has mainly emergency medicine physicians in their volunteer pool and the members come from nine countries. The training curriculum is tailored to the local needs. By 2011, PURE had completed two assessment and planning visits to Rwanda, and developed an action plan in collaboration with the Rwandan Ministry of Health. Part of the assessment of effectiveness will be an analysis of PURE's training approach, which will contribute to the global experience for ultrasound education [19].

4.2 RAD-AID International: Radiology Serving the World

RAD-AID International (RAD-AID [20]) is a non-profit organization that aims to promote and improve global health care by increasing the access to radiology services in developing nations. RAD-AID is the brain child of Daniel J. Mollura who is the Founder, President and Chief Executive Officer of the organization. His varied and broad background in government and international relations, finance and radiology has provided him with the necessary skills and experience to advance his vision and ambition for RAD-AID [3].

The leadership, organizational structure and multipronged approach by RAD-AID reflect the organization's determination of not just being a social service provider, where once the expertise or service leaves, the original state returns in the underserved country. RAD-AID activities encompass education, training and support in clinical radiology, business administration, public health, the use and implementation of health care technologies and international policy. Radiology-Readiness™ was developed by RAD-AID to avoid or reduce waste and inefficiency. Other healthcare facilities and infrastructure must be in place before implementing radiology services for this to be effective and be part of healthcare delivery. RAD-AID will help the country to find and develop the resources they need before implementing the targeted radiology service.

RAD-AID has advisors from leading academic institutions, financial institutions and consulting firms to provide guidance to best serve their goals. RAD-AID functions with the aid of volunteers, donations and collaborations with specific partners and organizations, such as Imaging the World (ITW), Engineering World Health, Jefferson University Research and Education Institute, American College of Radiology (ACR) and the World Health Imaging Telemedicine and Informatics

Alliance (WHITIA). They report their activities on their Blog, Facebook and Twitter in addition to their official website. Since 2009, their annual conference on International Radiology for Developing Countries, hosted by Johns Hopkins, provides the platform for various stakeholders from healthcare, academic, business, humanitarian and radiological organizations to meet, discuss and develop ways to improve medical imaging in the developing world.

Collaboration is important to maximise and optimise the use of resources. RAD-AID, Project HOPE (Health Opportunities for People Everywhere) and Philips Healthcare joined forces to assess the ability of the communities in western China and northern India in the use of CTs, MRIs and other imaging equipment to improve healthcare [5]. RAD-AID works with Project Hope to study how radiology could be part of the clinics operating throughout Project HOPE's healthcare delivery system. Project HOPE has been working since 1958 to bring long term sustainable healthcare, through health education and assistance programmes to many nations [18]. Their network and experience provided RAD-AID with a good partnership. RAD-AID and Columbia Business School worked with Project HOPE to study the expansion of imaging services in North India in 2011. RAD-AID also has projects in Africa, Latin America and the Caribbean.

4.3   Malaysian College of Radiology Value Added Mammogram Program

Malaysia does not have a mammogram screening program although breast cancer is the top cancer amongst women. A woman in Malaysia has a 1 in 20 chance of getting breast cancer in her lifetime. Amongst the Chinese population in Malaysia, the cumulative life time risk is higher, at 1 in 16 women [11, 14]. There is unequal distribution of mammography services, being concentrated in the major cities on the west coast of West Malaysia. The east coast of West Malaysia and East Malaysia is generally underserved.

In 2001, the Malaysian College of Radiology (CoR) together with the Malaysian Medical Association Foundation Radiology Fund (MMARF) launched a nation-wide breast cancer awareness campaign and its inaugural subsidized mammogram program. The campaign included breast health education, and harnessed the assistance of breast surgeons and breast cancer support groups for its media thrust [21]. A sub web (on its homepage), the Breast Health Information Centre in both English and Bahasa Malaysia was set up [22]. The Bahasa Malaysia site has proven very popular from the feedback, queries and comments. Bahasa Malaysia is the national language of Malaysia.

Poor quality mammography would be ineffective. Therefore, voluntary basic accreditation of mammography providers by the CoR was introduced to educate the providers and elevate the quality of the mammography service. The CoR organizes annual multidisciplinary breast seminars/workshops to provide the opportunity for radiologic technologists, radiologists and medical physicists to share and update their skills. Public reports and updates of this program are available from the CoR's homepage.

The program does not just provide an access to mammography. If an abnormality is detected, immediate extra mammogram views and/or adjunct ultrasound are done. Where necessary, the program also funds image-guided biopsy, hence the term "value added" mammogram program. When breast cancer is diagnosed, the patient is referred to the nearest public facility or a hospital of their choice and the local breast cancer support team is on hand to look into the psychosocial needs. For the program participants from Segamat and the nearby villages in northern Johor, the program funds transportation to and from the nearest accredited breast imaging center, the International Islamic University Breast Center in Kuantan, Pahang.

The program makes use of the "slack" in mammography utilization in private and academic centers primarily; although in the first year a public hospital (after hours service) was included. With experience gleaned over the years, the program has been tweaked so that the underserved population was better targeted. By 2011, the program has been running for a decade and currently is concentrated in the east coast of West Malaysia. The program is funded through grants and donations. The CoR is in a fortunate position that it does not have to focus on fund raising. The program would not be possible without the aid of the breast cancer support groups, the volunteer mammography providers and the local breast surgical and oncology services. Electronic communications are used to coordinate the program. This minimises the "waste" of donor funds on administration and maximises the use of the funds for early cancer detection.

The program has been successful even though it is not available in all of the underserved areas. All aspects of breast cancer management, e.g. screening/diagnostic services, treatment and psychosocial support must be available for the program to work effectively. The CoR's Value Added Mammogram Program has served as a model for other programs such as the National Family and Population Development Board's subsidized mammogram program.

5 Social Entrepreneurship in the Provision of Quality Radiology and Radiation Protection: Heading in the Right Direction

5.1 What Is Social Entrepreneurship?

Social entrepreneurship in radiology could possibly be defined as collaborative public efforts to improve sustainable access to radiology services and radiation protection by individual and organizational leadership and end-user participation through innovative delivery models.

Ashoka "Innovators for the Public", an organization founded in 1980 by Bill Drayton in Washington DC, United States of America is committed to investing in new solutions for the world's toughest problems. Ashoka believes social entrepreneurs will create innovative and successful solutions rather than waiting for the government or business sectors to meet society's needs. Social entrepreneurship is here to stay and organizations such as Ashoka [1], Schwab [23] and Skoll Foundations [24] promote, advance and foster social entrepreneurship to maximise impact.

Others echo the sentiment that social entrepreneurship could solve the world's woes. In the opinion of Ian MacMillan, director of Wharton's Sol C. Snider Entrepreneurial Research Center, social entrepreneurship may in situations be "an alternative to governments undertaking the task of solving societal problems." [10]. This neatly sums up why social entrepreneurship is that bridge for the haves and have-nots in radiology.

Ever since Muhammad Yunus and the Grameen Bank won the Nobel Peace Prize in 2006 [26], social entrepreneurship has catapulted into the limelight. Yunus was exemplary of a change agent that has impacted many in Bangladesh. He lifted many Bangladeshi out of poverty through an innovative solution by providing microcredit and teaching basic financial principles to enable people to help themselves. Yet, social entrepreneurship is not a new concept, having been around for 30 years or so. Although there are various definitions of social entrepreneurship, they all have a few key features that identify them as such. A case for having a stricter definition of social entrepreneurship has been debated to enable resources to better focus on this field [12].

Social entrepreneurship is the application of an unusual or innovative approach to solve a social or environmental problem. It creates positive social or environmental benefit. The aim is not just to make money or to minimise negative impact, such as by using organic materials or reducing the carbon footprint in manufacturing. It is more than the documentation of corporate and social responsibilities in the portfolios of almost every company. It does emphasise sustainability, including financial sustainability.

A social entrepreneur is a society's agent for change. It usually requires collaboration with various individuals and organizations including governments to bring about the positive change and may require years to achieve. It would be obvious that such a change agent must feature amongst others, strong leadership, ambition, persistence, creativity and innovative thinking.

A social enterprise may mix social and commercial goals. However, the emphasis is on the social impact, and the profits are either rechanneled into the organization or used for other social goals. Social enterprises can vary, from not-for-profit to for-profit. Financial sustainability may be based on donations/grants or a self-sustaining viable business model. Often, combinations are used. There are some innovative funding models for the social entrepreneurs working in radiology. Microfinancing coupled with education and mentorship could be applied, e.g. the Grameen Bank model, impact investments or social impact bonds.

5.2 Impact Investments

Impact investors support social enterprise by investing money in a manner where they may make some money or at least, will not lose money. It allows the social entrepreneur to achieve the goals faster. In the 2010 J.P. Morgan Global Research report, the authors have ventured that impact investments are emerging as an alternative asset class, which is an "appropriate and economically effective way to complement government and philanthropy efforts to solve certain social and environmental problems" [17].

Developments and progress have resulted in the commercialization of medicine including radiology. Making massive profits in medicine has not led to an improvement in healthcare services in general. It has resulted in the rich being able to access good quality care and the poor 'neglected' and unable to access even basic care. Research and development tends to follow where the money is, addressing the health issues of the richer strata of society – the lifestyle diseases.

It would be ideal if the equipment vendors and manufacturers could direct some focus on the development of robust and low cost equipment that would help to address the needs of the two-thirds of the world without basic radiology services. This is where impact investments may start to make a difference, as impact investments generally target the base of the economic pyramid, to improve lives of the poor and provide environmental benefits. The World Resources Institute defines the base of the pyramid as people earning less than USD3,260 a year [34].

A common language for the reporting of social and environmental performance, i.e. Impact Reporting and Investment Standards, has been drawn up by the Global Impact Investing Network (GIIN) [4]. GIIN is a not-for-profit organization dedicated to expand the use and improve the effectiveness of impact investing. Measuring returns and intended impact are vital to ensure continuing investment, attract new investors and review the outcome of the social program [9].

5.3 Social Impact Bonds

Social Finance, a United Kingdom based organization which conceptualized Social Impact Bonds (SIB) defines such bonds as: "A form of outcomes-based contract in which the public sector commissioners are committed to pay the financier a pre-defined sum for significant improvement in the social outcomes for a defined population". Social Finance launched the first SIB in September 2010 in the criminal justice sector [25].

The SIB agency will work through service providers on the targeted population towards pre-agreed outcomes, which would address a specified social need (such as the prevention and early intervention services) and save government expenditure. When the program has achieved the desired outcome, the social investors who have provided the funds would get their returns from the government, ideally with interest in addition to the initial capital. It is therefore a win-win-win (W^3) situation for all the parties involved, together with improved social outcome [8].

In America, they are known as "Pay for Success" bonds. This is another way to lower the "risk" for government spending on public programs by using a model that focuses on positive outcome. This may attract more investors and philanthropic organizations as the desired social outcome could be achieved more quickly. If a program fails; it will not receive continuing funding. On the corollary, positive outcomes ensure continuing or increase funding to further expand the social good (the virtuous cycle). The social service provider is thus incentivized to find more ways to achieve and improve the original outcome without needing to focus on fund raising.

6 Conclusion

"The journey of a thousand miles begins with a single step." – Lao Tzu, Chinese philosopher.

Radiology is expensive and resource-intensive yet an essential component of our healthcare system. The challenges are great but not insurmountable in narrowing the radiological healthcare divide. Dependence on donations or volunteering from radiologically advanced countries is at best temporary. Eventually, there should be technology and knowledge transfer whilst still maintaining the links for continuing professional development in all facets of setting up, provision and delivery of radiology services.

Sustainable solutions for improving radiology services in the developing countries should entail the following components: finance, public health, clinical radiology services, practitioner education and sustainable technology [28]. Business leadership training for radiology entrepreneurs is vital as radiology equipment is expensive to acquire and maintain. RAD-AID is exemplary in this respect with its concerted and coordinated efforts to produce local radiology entrepreneurs and, to develop low cost business models including assistance to finance sourcing. Loans to radiology entrepreneurs should be packaged with radiologic training, equipment donation and business administration mentorship [13].

In areas where radiology services are deficient, an acute need is often met with the provision of a service. However, the impact is limited to the local area and population. The key is to plan for scalability and the multiplier effect, which could result in a nationwide adoption of the solution. Throw in advocacy and social activism to influence the government, or the use of social impact bonds; the programme would fit into the definition of social entrepreneurship. In reality, most programs would be a combination of these facets.

What seems paradoxical is the application of this concept to medicine, in this case radiology or more broadly, imaging services. The origins of medicine could be based on "social entrepreneurships", e.g. Florence Nightingale set up the first nursing school and developed modern nursing practices. Perhaps, she can be considered as the "founder" of social enterprise in healthcare. Perhaps, it is time to return to our roots.

Social entrepreneurship for radiology is at the nascent stage. Time will tell if the current and new efforts turn out to be successful social entrepreneurships, i.e. the social impact is positive, pervasive and sustained. Existing efforts have evolved over the years, and become more focussed on sustainable strategies, which serve the local needs. Social radiology is about to take on a new dimension as social entrepreneurship with all the tools currently available and being developed may meet societies' needs faster and assist governments to achieve more equitable access to quality radiological healthcare.

References

1. ASHOKA (2011) www.ashoka.org. Accessed 26 Dec 2011
2. Clinical Research Centre (2011) National healthcare establishments and workforce statistics (hospital) 2008–2009. https://www.macr.org.my/nhsi/document/Hospitals_Report.pdf. Accessed 7 Jan 2012
3. Edelson M (2009) Intense vision. Johns Hopkins Dome Magazine. http://www.rad-aid.org/pdf/DOM-RADAID-Article.pdf. Accessed 2 Jan 2012
4. Global Impact Investing Network (GIIN) (2011) www.thegiin.org. Accessed 26 Dec 2011
5. Greenemeier L (2010) PET project: radiologists push imaging technologies in developing countries. Scientific American. http://www.scientificamerican.com/article.cfm?id=radiology-developing-countries. Accessed 2 Jan 2012
6. Ho ELM (2010) Overuse, overdose, overdiagnosis ... overreaction? Biomed Imaging Interv J 6(3):e8. http://www.biij.org/2010/3/e8/. Accessed 5 Jan 2012
7. Ho ELM (2012) Social radiology: where to now? Biomed Imaging Interv J 8(1):e9 (in press)
8. Hollmann D (2011) NexThought Monday: social impact bonds – an innovative mechanism for social change. Next Billion. http://www.nextbillion.net/blog/social-impact-bonds-innovative-financing. Accessed 3 Jan 2012
9. IRIS: Impact Reporting & Investment Standards (2011) http://iris.thegiin.org. Accessed 27 July 2013
10. Knowledge@Wharton (2003) Social entrepreneurs: playing the role of change agents in society. http://knowledge.wharton.upenn.edu/article.cfm?articleid=766. Accessed 16 Dec 2011
11. Lim GCC, Rampal S, Halimah Y (eds) (2008) Cancer incidence in Peninsular Malaysia 2003–2005. National Cancer Registry, Malaysia
12. Martin RL, Osberg S (2007) Social entrepreneurship: the case for definition. Stanford Social Innovation Review. http://www.ssireview.org/articles/entry/social_entrepreneurship_the_case_for_definition. Accessed 23 Dec 2011
13. Mollura DJ, Azene EM, Starikovsky A et al (2010) White paper report of the RAD-AID conference on international radiology for developing countries: identifying challenges, opportunities, and strategies for imaging services in the developing world. J Am Coll Radiol 7(7):495–500
14. National Cancer Registry (2012) http://www.radiologymalaysia.org/Archive/NCR/NCR2003-2005Bk.pdf. Accessed 27 July 2013
15. Ng KH, Mclean ID (2011) Diagnostic radiology in the tropics: technical considerations. Semin Musculoskelet Radiolo 15:441–445
16. Ng KH, Rehani MM (2006) X ray imaging goes digital. Br Med J 333:765–766

17. O'Donohoe N, Leijonhufvud C, Saltuk Y et al (2010) Impact investments: an emerging asset class. JP Morgan Global Research. http://www.jpmorgan.com/cm/BlobServer/impact_invest-ments_nov2010.pdf?blobkey=id&blobwhere=1158611333228&blobheader=application%2Fpdf&blobcol=urldata&blobtable=MungoBlobs. Accessed 26 Dec 2011

18. Project HOPE (2012) http://www.projecthope.org/what-we-do/. Accessed 27 July 2013

19. PURE: Physicians Ultrasound in Rwanda Education Initiative (2011) http://www.ultrasound-rwanda.org. Accessed 27 July 2013

20. RAD-AID International (2011) www.rad-aid.org. Accessed 2 Jan 2011

21. Radiology Malaysia (2011) College of Radiology Academy of Medicine of Malaysia Mammogram Program (2001–2011). http://www.radiologymalaysia.org/breasthealth/smp. Accessed 5 Jan 2012

22. Radiology Malaysia (2012) Breast Health Information Centre. http://www.radiologymalaysia.org/breasthealth. Accessed 6 Feb 2012

23. Schwab Foundation for Social Entrepreneurship (2011) www.schwabfound.org. Accessed 27 July 2013

24. Skoll Foundation (2011) www.skollfoundation.org. Accessed 27 July 2013

25. Social Finance UK (2012) http://www.socialfinance.org.uk/. Accessed 27 July 2013

26. The Nobel Peace Prize (2006) http://www.nobelprize.org/nobel_prizes/peace/laureates/2006/. Accessed 27 July 2013

27. Veláquez-Beruman A (2010) WHO perspective on health technologies: medical imaging. Presentation at ESR 2011, Vienna. http://www.who.int/diagnostic_imaging/imaging_modalities/WHOPerspectiveonHealthTech_MedicalImaging.pdf. Accessed 12 Dec 2011

28. Welling RD, Azene EM, Kalia V et al (2011) White paper report of the 2010 RAD-AID conference on international radiology for developing countries: identifying sustainable strategies for imaging services in the developing world. J Am Coll Radiol 8:556–562

29. World Health Organization (2011) Essential diagnostic imaging. http://www.who.int/eht/en/DiagnosticImaging.pdf. Accessed 12 Dec 2011

30. World Health Organization (2011) Baseline country survey on medical devices 2010. http://www.who.int/medical_devices/en/. Accessed 12 Dec 2011

31. World Health Organization (2011) Medical devices: analysis of results of the baseline country survey 2010. http://www.who.int/gho/health_technologies/medical_devices/medical_equipment/en/. Accessed 12 Dec 2011

32. World Health Organisation (2013) Diagnostic imaging. http://www.who.int/diagnostic_imaging/en/

33. World Health Report (2010) Health systems financing: the path to universal coverage. http://www.who.int/whr/2010/en/index.html. Accessed 16 Jan 2012

34. World Resources Institute (2006) Market of the majority: the BOP Opportunity Map of Latin America and the Caribbean. http://pdf.wri.org/market_of_majority_english.pdf. Accessed 4 Jan 2012

35. Zaharah M, Fatimah O, Sivasampu S (2011) Radiology Devices in Malaysian Hospitals. In: National Medical Device Statistics 2009. https://www.macr.org.my/nhsi/document/NMDS_Report.pdf. Accessed 7 Jan 2012

Chapter 25
From Volume to Outcomes: The Evolution of Pay for Performance in Medical Imaging

Richard Duszak Jr. and Ezequiel Silva III

Abstract Although long ago established as the basis for physician remuneration in the United States, fee for service reimbursement systems are increasingly challenged by payers and consumers alike. As various stakeholders seek avenues to promote quality, enhance safety and improve outcomes, value based health care purchasing models receive ever-increasing attention. Trends, challenges and possible innovations in pay for performance programs, particularly as they relate to the physician component of medical imaging, are discussed.

Keywords Accreditation • Fee for service • Maintenance of Certification • Outcomes • Pay for performance • Physician payment • Physician Quality Reporting System • Quality • Resource Based Relative Value System • Safety • Value based purchasing

Abbreviations

ABR	American Board of Radiology
ACO	Accountable Care Organization
ACR	American College of Radiology
CMS	Centers for Medicare & Medicaid Services
CPT	Current Procedural Terminology
FFS	Fee for Service

R. Duszak Jr. (✉)
Harvey L. Neiman Health Policy Institute, 1891 Preston White Drive, Reston, VA 20191, USA
e-mail: rduszak@neimanhpi.org

E. Silva III
South Texas Radiology Group, University of Texas Health Science Center,
San Antonio, TX, USA

L. Lau and K.-H. Ng (eds.), *Radiological Safety and Quality: Paradigms
in Leadership and Innovation*, DOI 10.1007/978-94-007-7256-4_25,
© Springer Science+Business Media Dordrecht 2014

GDP   Gross Domestic Product
IAC   Intersocietal Accreditation Commission
MIPPA   Medicare Improvements for Patients and Providers Act
MOC   Maintenance of Certification
MPPR   Multiple Procedure Payment Reduction
MQSA   Mammography Quality Standards Act
P4P   Pay for Performance
PPACA   Patient Protection and Accountable Care Act
PQRI   Physician Quality Reporting Initiative
PQRS   Physician Quality Reporting System
RBRVS   Resource Based Relative Value System
RUC   RVS Update Committee
RVU   Relative Value Unit
SGR   Sustainable Growth Rate
TJC   The Joint Commission
UCR   Usual Customary, and Reasonable
US   United States

1   Introduction

Although individual health insurance entities establish their own fee schedules for physician and facility payments, Medicare's Resource Based Relative Value System (RBRVS) derived fee schedule currently serves as the benchmark for most healthcare payers throughout the United States (US). As a result of this *de facto* national payment system, free market principles – wherein supply, demand, service, and outcomes would normally drive price – play relatively little role in healthcare purchasing. Compounding this disconnect between price and quality is the insulation of patients – the end consumers of healthcare services – from individual service payment through insurer and employer intermediaries.

The current prevailing fee for service (FFS) payment system in the US was designed to uniformly compensate physicians for unit services based on the time and intensity of typical services and associated costs [27]. The downside of such uniformity, however, is that the model inadequately considers service variation at the individual patient and provider level as well as ultimate overall patient outcome.

Increasingly, stakeholders are actively seeking value-based models for healthcare payment as an alternative to current transaction-based fee schedules. Under these arrangements, compensation is tied to achievement in pre-defined goals such as appropriateness, service, satisfaction, quality, and outcomes, rather than sheer volume. Such a transformation, however, will necessitate overcoming numerous operational, regulatory, political, and cultural obstacles, which create considerable inertia in maintaining the *status quo* of the US healthcare payment system. This chapter focuses on US models, but many of the concepts discussed will apply to payment systems in place or in development in other countries as well.

In this chapter, we will explore the ongoing evolution of physician payment systems as they pertain to pay for performance (P4P). Although many of the concepts discussed will apply to healthcare in general, our focus will be on professional services in medical imaging.

2 Fee for Service Payment: Past and Present

Physician payment systems have evolved dramatically over the last several decades. Prior to the widespread availability of health insurance in the US, payment for healthcare services was in many ways similar to that for other professional services. Price was largely agreed upon by the provider and recipient, and often rendered contingent on one's ability to pay. In many cases in the country's history, physicians accepted non-traditional compensation (such as bartered farm goods) in return for their professional services [30].

The national landscape for healthcare payment changed dramatically in 1965 with amendments to the Social Security Act, resulting in the creation of Medicare and Medicaid – the country's first nationalized healthcare insurance programs. Although these programs were primarily designed to extend healthcare access to senior citizens and the poor, respectively, they have in many ways established policy precedents for private payers. Physician payments under these programs were initially based on historical usual, customary and reasonable (UCR) rates. Although somewhat arbitrarily derived, they became codified by carriers in fee schedules, marking the first step toward establishing a national fee schedule [1].

Some two decades later, the Resource Based Relative Value System (RBRVS) was introduced, moving the US even closer toward national fee schedule uniformity. The system was designed to value physician services using relativity metrics which would, at least in theory, address differences in specialty, training, and a variety of other factors [25, 26]. A detailed review of RBRVS methodology is beyond the scope of this chapter, but nicely outlined in other publications [46, 50]. In short, the system specifically considers physician time, intensity, malpractice expenses, and selected practice expenses when assigning Relative Value Units (RVUs) to individual services. Using a national Congressionally determined conversion factor, along with regional geographic practice cost indices, RVUs are converted into fee schedule payment amounts [50]. Although private payers are not bound by RBRVS methodology, most use it in some form, with individually negotiated conversion factors, as the basis of their fee schedules [45].

Fee schedules for imaging services typically recognize the fact that radiologists frequently provide interpretative services for facilities or institutions in which they have no ownership. As such, the professional component of service payment is made to a radiologist for his interpretive or procedural services. A separate technical or facility payment is made to an imaging center or hospital for a variety of non-interpretive services involved in the provision of medical imaging (e.g., equipment, supplies, and administrative and clinical staff). When both professional and technical services are provided by the same entity (such as a radiologist owned imaging center), these components are bundled into a global payment [1].

Although RBRVS methodology supporting FFS was intended to promote stable and uniform payments, the political milieu in which these systems have evolved has morphed the Medicare Physician Fee Schedule through a hobbled patchwork of ever changing components. The sustainable growth rate (SGR), which serves as the basis of the annual conversion factor for translation of RVUs into dollars, has been challenged as fundamentally flawed [16]. Each year, Congressional action has been required to avert drastic conversion factor-related physician fee cuts, which could compromise patient access to medical imaging services in which Medicare and Medicaid were intended to address [31]. Medical imaging has in particular been the target of a disproportionate number of rules and laws, all created outside of original RBRVS design, to reduce both technical and professional payments. The most recent of these has been Medicare's Multiple Procedure Payment Reduction (MPPR) policy [10] which assumes efficiencies in the delivery of radiology services, is not supported by the peer reviewed literature [3], and is never considered as part of RBRVS.

Despite ongoing challenges, the FFS payment system has withstood the test of time as a practical basis for physician compensation. Based on the nationally recognized Current Procedural Terminology (CPT) code system for reporting services by physicians and other healthcare providers, FFS-based payment systems are broadly understood by most parties involved, ranging from physicians to claim-processing staff to large insurers. The current system permits considerable physician input into ultimate fees given relativity negotiations between specialties through the American Medical Association's RVS Update Committee (RUC). Nonetheless, FFS fundamentally incentivizes volume – rather than quality – and it is that shortcoming which is largely responsible for increased societal interest in value-based payment systems.

3   The Case for Value-Based Purchasing

By financially rewarding volume, rather than quality or outcomes, the current US FFS system is believed by many to have contributed to the recent escalation of healthcare expenditures. Currently, growth in healthcare spending is rising at a rate much faster than that of the gross domestic product (GDP) and is tracking on an unsustainable trajectory [12]. Spending on medical imaging, in particular, has risen at a rate even faster than other physician services [29], creating particular interest in radiology payment reform.

This impending healthcare spending crisis looms at a time of increased public scrutiny on safety and outcomes, which is further compounded by budget restrains due to a slow recovery after the recent global financial crisis. Although a recent Institute of Medicine report describing thousands of preventable medical error related deaths each year [32] has been disputed by many [36], it has nonetheless fueled public concern about receiving – and paying for – services which may be suboptimal at best or dangerous at worst. More recently, the lay press has focused attention on perceived unnecessary over-imaging [4], and in doing so has converged discussions about quality, safety and cost squarely in radiologists' domain.

In its purest form, FFS ignores quality and outcomes, basing physician compensation solely on volume. FFS paradoxically permits physicians to align their perceived ethical responsibilities, by delivering more care to their patients, with their own economic self-interests, since increased services translate to increased earnings. This concept is particularly pertinent to medical imaging, as it may explain much of the reason for perceived overutilization spurred by self-referral [34]. Additionally, pure FFS payment systems may actually unintentionally but perversely reward suboptimal care, since such care can create a demand for future payable physician and facility services. Examples include ignoring appropriate preventive care or inadequate attention to error minimization.

When *de facto* fee schedules are imposed on private payers through government influence, traditional market forces are disrupted. Price is no longer dependent on the supply of or demand for healthcare services, nor their real or perceived quality. Additionally, bureaucratically derived pricing may create rank order anomalies in physician payment, preferentially driving services perceived as lucrative, while restricting access to those perceived as unrewarding.

Under FFS, healthcare spending appears more dependent on the supply of providers in a region, rather than actual patient needs [21]. Additionally, payments to physicians and other healthcare providers under FFS paradoxically do not necessarily translate into better outcomes [21]. In fact, inappropriate over-utilization may actually result in worse outcomes [23].

The perceived failure of traditional FFS to appropriately incentivize quality and outcomes has created broad-based increased interest in the pursuit of value-based purchasing models. Although robustly defining value in healthcare is complicated, it is conceptually illustrated with a very simple equation [40]:

$$\text{Value} = \frac{Quality}{Cost}$$

At any given cost, if quality increases, so too does value. And, even if quality remains constant, value increases when costs are reduced. This intimate relationship between cost and quality is what fundamentally distinguishes value-based purchasing from traditional FFS.

Pay for performance (P4P) is a mechanism to financially incentivize physician behavior through either bonuses or penalties. It represents a specific form of value-based purchasing, and will likely become a larger component of future healthcare payment systems. Under P4P, physician and facility payment is not based on the delivery of services *per se*. Instead, payment is dependent on specific characteristics of those services, such as the types of services performed, their indications, how they were performed, or even the outcomes that resulted. P4P can be binary (i.e., payment contingent on specific criteria) or graded based on a variety of performance metrics [19].

The greatest challenge for P4P development will be defining quality in a robust, measurable, reliable, and widely accepted manner that can serve as the basis for such systems. Such a goal has not yet been achieved, and may remain elusive for many years to come. Current early US P4P systems and other models, which could impact medical imaging in the future are discussed below.

4 Current US Pay for Performance Programs

Although P4P has garnered considerable interest as an alternative to FFS, few mature models are currently operational within the US. The programs presently impacting – or most likely to soon impact – radiologists and radiology departments are discussed below.

4.1 Accreditation

Accreditation refers to rigorous third party scrutiny resulting in certification of quality systems. When applied to radiology facilities, accreditation indicates adherence to numerous requirements and standards for adequacy, competency, safety, and overall quality pertaining to equipment, staff, and institutional processes. The accrediting organizations most familiar to US radiologists include the American College of Radiology (ACR), the Intersocietal Accreditation Commission (IAC), and The Joint Commission (TJC). Although these organizations have similar requirements for certification, their specific criteria vary, and are beyond the scope of this chapter.

Third party accreditation is increasingly being applied as a precondition for payment for the office or technical (i.e., not physician professional) component of medical imaging services [13]. For many years, the Mammography Quality Standards Act (MQSA) has required site accreditation as a requirement for billing Medicare for mammographic examinations [22]. Other accreditation requirements have gradually promulgated amongst individual Medicare carriers, particularly for vascular ultrasound. Recently, the Medicare Improvements for Patients and Providers Act (MIPPA) has made facility payment for advanced imaging studies contingent on imaging center accreditation by a federally deemed organization [8]. That recent legislation, however, curiously exempts hospitals from this accreditation requirement.

With regard to service payment, accreditation is a binary precondition: a center is either accredited (and thus eligible for payment) or not. No incremental financial incentives currently exist for facilities exceeding minimal standards. Current accreditation payment policies additionally apply only to the technical component of medical imaging services. Office accreditation, however, usually includes requirements for interpreting physicians (such as training, continuing education and experience, and ongoing peer review). Payment to those physicians, though, is not directly contingent on accreditation – an incentive mismatch, which could potentially create conflicts between independent radiologists and the facilities they staff.

Current US accreditation programs affecting Medicare are modality specific. This creates considerable administrative burdens for facilities since separate accreditation is required for each imaging modality. An entity must also first possess a machine since image review is necessary. This requirement may create operational and financial challenges for facilities given the lag time between the purchase of a machine and the ability to submit images necessary for accreditation and, hence, payment [43]. Additionally, the accreditation process itself is rather costly; facilities incur considerable expenses when both preparing for and applying for accreditation. Nonetheless, facility-based accreditation P4P programs are currently well established for Medicare and many private payers, and are likely to be maintained.

4.2   Never Events

"Never events" are defined as entirely preventable patient safety occurrences, which pose serious harm to patients [38]. In a recent consensus document, the National Quality Forum created a list of over two dozen such acts or omissions. These include surgical mishaps (e.g. wrong site surgery), care management failures (e.g. medication errors resulting in harm or death), patient protection deficiencies (e.g. infant discharged with wrong parents), and environmental injuries (e.g. device-related patient electrocution). Based on the presumption that these are absolutely preventable occurrences, Medicare now disallows hospitalization payment when certain such incidents occur [9]. As a corollary to accreditation-based P4P wherein specific criteria define eligibility for payment, never event-based P4P program criteria establish specific criteria for which payment is denied. This is not insignificant, since the program not only withholds payment for the event resulting in the adverse outcome – payment for the entire admission is denied.

Although financially rewarding safety is a laudable goal, the practicality of such programs remains uncertain. Even at centers of excellence, such events do in fact occasionally occur [49]. Their real – albeit uncommon – incidence in such best practice environments challenges the payment model's presumption that such events are truly and absolutely preventable in complex healthcare environments. Early studies further indicate that the financial impact of such programs on centers of excellence may be substantial [49]. This portends unfavorably on how smaller community hospitals will fare as such programs are expanded, and raises concern that such payment models may create unintended patient access hurdles if it renders some such facilities financially insolvent.

Currently, this never event-based P4P program applies only to hospital inpatients. Such a model, however, could easily provide the framework for expanded payment denial programs for hospital outpatient facilities, imaging centers, or even physician services (e.g. alleged misinterpretations).

4.3   Physician Quality Reporting System

Medicare's Physician Quality Reporting System (PQRS) represents formal codification of its Physician Quality Reporting Initiative (PQRI), the earliest program by a large US payer to transition professional compensation from FFS toward P4P. Under PQRS, participating physicians are paid under Medicare's fee schedule, but receive bonuses for compliance with specific reporting quality metrics.

Current reporting requirements most likely to impact radiologists include documentation of (1) specific findings for cross sectional neuroimaging, (2) reference standards when reporting carotid stenosis, (3) sterile barrier technique when placing central venous catheters, (4) radiation time or dose for fluoroscopic guided procedures, (5) appropriate screening mammography classifications, and (6) bone scan correlation with previous studies [17].

Program participation details are complicated and still somewhat ambiguous. In short, when specific metric thresholds are achieved for billable services, and so reported to the Centers for Medicare and Medicaid Services (CMS), a physician becomes bonus eligible. Although the list of eligible metrics appropriate for radiologists is short, the bonus implications are much larger. Bonuses are applied to all of the Medicare services rendered by a physician – not just those for which metrics exist.

Existing PQRS metrics vary considerably in validity as surrogates of quality. Some, such as the documentation of the presence or absence of hemorrhage, mass, or infarct on brain CT and MRI studies, are based on well-established reporting guidelines. Another, which requires reporting fluoroscopy dose or time, may at first glance seem appropriate. However, the metric actually results in a bonus for merely reporting fluoroscopy time – even if excessive – rather than legitimate efforts to reduce radiation exposure. This illustrates the challenges of using payment claims surrogates of quality, which are easy to measure, but of clinically dubious value.

Despite such shortcomings, Medicare's PQRS program will soon have even greater influence on physician payment as the system evolves from bonuses to penalties. By 2016, physicians failing to meet selected PQRS metrics will be penalized 2 % on all Medicare services – not just those for which metrics apply [42]. One could easily envision this evolving into a system in which compliant reporting becomes requisite for participation in the Medicare Program, not dissimilar to the never event exclusion program.

PQRS will soon receive considerable attention in the public domain. The CMS a "Physician Compare" website will post information on physician conformance with PQRS measures. Whether patients will be able to differentiate such claims reporting metrics from true measures of quality remains to be determined.

4.4   Maintenance of Certification

Physician board certification is nearly universally recognized throughout the US as an important affirmation of competency. Given obvious parallels to accreditation for facilities, it should be of no surprise that Maintenance of Certification (MOC) has recently emerged as Medicare's newest P4P incentive program.

The new MOC bonus system reflects an expansion of PQRS, and eligibility actually requires successful PQRS participation. The program will reward physicians with bonuses for participation in their medical specialty board's formal MOC program. Currently, bonuses are promised at 0.5 % of total Medicare revenue, and require participation at higher levels than required for board certification in areas pertaining to professional standing, lifelong learning, cognitive experience, and professional quality improvement [44].

No experience has yet been reported with this brand new program, and at present it is unclear how certificants will communicate requirement fulfillment to their specialty boards and to CMS for payment. It is anticipated that the American Board of Medical Specialties will soon coordinate a registry for such reporting and communication, which should facilitate physician participation, but at present, no such national mechanism exists.

For time-unlimited "grandfathered" American Board of Radiology (ABR) certificants, bonuses will be contingent on participation in MOC. For more recent time-limited certificants, bonuses will be contingent on continued MOC participation, but at a higher level than otherwise required by the ABR.

MOC participation will likely have greater relevance in the future since Medicare has indicated that MOC practice assessment may be incorporated in its "composite of measures of quality" for future value-based payment modifiers [6]. Although MOC-based P4P is currently bonus-based, like PQRS, this could easily be transformed into a penalty system, or could conceivability evolve into a precondition for physician payment eligibility.

5   Future Value-Based Payment Models for Radiologists

Robustly linking payment to quality of care, patient satisfaction, and overall outcomes will likely remain an elusive goal for many years to come. The environments in which healthcare services are delivered are incredibly complex, and each patient and each encounter differ greatly. Accordingly, valuing the contribution of individual providers and individual healthcare entities in any episode of care will be challenging.

As radiologists begin to identify their role in future value-based payment systems, present models may be useful in predicting the future – at least for the short term. Several concepts and programs are currently in various stages of development for the US healthcare enterprise as a whole, and these may provide insight into likely future directions for medical imaging in the US and other countries. These include expansion of existing facility-based P4P programs to professional services, and implementation of more significant non-conformance penalties. They additionally include integration of radiologists' services into accountable care organizations (ACOs) and bundled episodic payment models. Such concepts will be introduced below.

Given the dynamic political milieu in which healthcare policy is crafted and enacted, the time frame for implementation of such programs – or whether they will be implemented at all – remains uncertain, but will likely hinge at least in part on leadership and advocacy by individual radiologists, radiology practices, and professional societies. Ideally, however, durable and robust value-based payment systems should be applicable to physicians and institutions alike, and should align financial incentives and outcome goals for all healthcare delivery stakeholders.

5.1 Evolution of Existing P4P Programs

Current US P4P programs largely fall into two categories: (1) modified FFS, with fee schedule payments adjusted through either bonuses or penalties (e.g., PQRS), and (2) conditional FFS, in which payment is contingent on satisfactory fulfillment of expectations (e.g., accreditation or never events). In the short term, these two groups will likely serve as the basis for ongoing P4P initiatives.

Present programs apply to either professional or facility payment, but not both. Accordingly, one could easily envision crossover. For example, PQRS-like adjustments to the technical component of imaging payments, either in a bonus or penalty fashion, could soon be indexed to particular quality metrics (e.g., equipment, staffing, patient satisfaction). Similarly, conditional facility-based "play to pay" models, such as accreditation, could soon be applied to professional reimbursement. For example, PQRS or MOC could be established as criteria for Medicare physician payment eligibility.

One related concept, which has been considered, but at present not gained substantial political traction, is tying the annual legislative SGR "fix" to quality metrics. Each year, the US Congress adjusts the Medicare conversion factor to address inadequacies in SGR methodology. Legislatively linking different conversion factors to specific performance metrics could essentially create various national quality-based fee schedules. Such an approach would not be dissimilar to the value-based payment modifier system which has recently been proposed by Medicare to increase or decrease provider payments using composite measures of quality.

5.2   *Shared Savings Through Accountable Care*

An Accountable Care Organization (ACO) is a group of providers, facilities, and other healthcare stakeholders which work together with aligned financial interests to deliver coordinated high quality patient care. Shared savings programs encourage physicians, hospitals and certain other types of providers to form ACOs with the goal of providing cost-effective and coordinated care. The potential role of radiologists in such organizations has been previously discussed [5]. The implications for medical imaging professionals are not insignificant [2]. An ACO would assume clinical and financial responsibility for the care of its participants, and would share in savings, when achieved, when particular quality thresholds are met.

Quality and performance standards measures are an integral component of Medicare's vision of an acceptable ACO [7]. The Agency's requirements include dozens of metrics, many related to patient satisfaction, which must be achieved to receive shared savings payments. Well-respected integrated healthcare systems, such as the Mayo Clinic, expressed concerns for the ACO model described in Medicare's proposed ACO rule [7]. In response, in its Final Rule, CMS eased the quality requirements from 65 measures to assess ACO quality in 5 "domains" to 33 measures in 4 domains. This change will make successful participation in the shared savings program more attainable. Time will tell whether the ACO concept will, after bureaucratic intervention, continue to be perceived as so burdensome and impractical, however, to prevent widespread – or even limited – implementation.

A recent authoritative review of the role of physicians in ACOs focuses largely on primary care, with little discussion of specialties, and no mention at all of radiology [35]. Accordingly, the ultimate role of radiologists in these models remains unclear, in part related to the difficulty in measuring and valuing radiologist contributions to overall outcomes and quality. The relevancy and success of radiologists in ACOs will require their proactive participation, and will likely be contingent on them developing and assuming ongoing responsibility for the organization's imaging utilization and quality management activities [2].

5.3   *Bundled Episodic Payment*

Under a bundled episodic payment model, an institution or organization is paid a predetermined amount for the entirety of professional and facility services rendered to a patient over the course of an illness or episode of care. The goal, of course, would be to incentivize the most cost-efficient delivery of care by capping total spending. The distribution of payments to various providers would then be determined by that organization – rather than a national fee schedule – as it determines appropriate. Such local financial control puts the organization in a position, at least in theory, to better coordinate and direct efficient care.

Several bundled payment models are now being tested nationally; of these, Prometheus has attracted the most attention and interest. These models package payments around a comprehensive episode of care that covers all services related to a single illness or condition [14]. As such, they are conceptually easiest to apply to hospital acute and post-acute care [47]. Since these models address only a single episode of care without assuming long term clinical and financial care responsibility, they are potentially operationally and organizationally less complex to implement than ACOs. Nonetheless, early pilot projects have found their logistics more complicated than expected [28], with some of the largest challenges related to unambiguously defining an episode and addressing the considerable variability in complexity between specific patients and encounters [37].

As with ACOs, the ultimate role of radiologists and radiology departments in bundled episodic payment models is yet unknown. In CMS demonstration projects to date, however, there has been no radiologist involvement at all. In fact, in a majority of these, radiology practice leaders at participating institutions were not even aware of their facility's involvement [48]. Given the likelihood of such payment systems moving forward, initiatives by radiology professional societies will be necessary to spur the economic modeling necessary to quantify the value of medical imaging as part of bundled payments.

6   Challenges for Pay for Performance Development and Implementation

In concept, tying payment to outcomes is both laudable and straightforward, but the widespread implementation of value-based payment models is proving problematic for a variety of reasons. In totality, the challenges to providers, policy makers, and other healthcare delivery system stakeholders are daunting. Key obstacles to radiology P4P policy promulgation are outlined below. The relative immaturity of P4P systems makes it impossible to predict the future, but awareness of such challenges will hopefully spur the radiology community at large to identify and test innovations which may serve as key components of durable future payment systems.

6.1   Identification of Quality Metrics

The impact of medical imaging on individual patients and individual encounters is extremely difficult to quantify, given the enormity of downstream care variables (e.g., how referring physicians use results). Even noncontroversial and achievable metrics such as radiation dose reduction may not have demonstrable outcomes for many years. Identifying and accurately measuring the long-term impact of various imaging modalities and strategies will thus remain an ongoing challenge – one augmented by the frequently transient relationships between patients and their providers and insurers. Nonetheless, considerable interest exists in identifying such metrics and various initiatives by radiology organizations will likely serve as the source of those ultimately adopted by government and private payers [39].

Under current claims-based processing systems, only immediately quantifiable – and thus usually perfunctory – metrics can be applied as the basis for payment. These easily measurable quality metrics, at least with regard to PQRS, have not been perceived by physicians in general [20] or radiologists in particular [18] to be meaningful with regard to patient care or clinical outcomes. Unless physicians perceive P4P programs as credible, and not simply bureaucratic exercises to garner payment, widespread clinical acceptance will be doubtful.

6.2 Funding Necessary Research and Policy

Expanded health services and comparative effectiveness research will be imperative to establish a solid foundation for successfully implementing relevant and durable value-based purchasing models. At present, even fundamental questions remain incompletely answered. How exactly does one define quality? What exactly is an encounter or episode of care? Although elementary, questions such as these must be rigorously answered before such parameters can be robustly incorporated into payment policy.

Research is both time consuming and expensive, and historically relatively little funding has been directed toward payment policy endeavors. Public and private insurers will be the key beneficiaries of such research. These entities, however, have traditionally directed their funds almost exclusively toward beneficiary care expenditures. Monetary allocations for future policy and payment systems, in contrast, have been sparse.

Recent initiatives offer some promise for necessary financial support. CMS just announced $1 billion in grants to investigators with compelling ideas to improve care and lower cost [11]. High-level interest as well has been expressed in directing at least some of the shared savings from evolving value-based models toward funding future quality initiatives [35]. Private practices and radiology organizations might serve as sources for individual project funding as well.

6.3 Education and Cultural Changes

Traditionally, US medical training has focused largely on acute care, much to the neglect of long-term outcomes and the coordination of chronic care. Physicians have been trained to serve as zealous advocates for their patients, and in doing so, often ignore costs in their decision-making. Similarly, the country's extremely litigious environment encourages ever increasing testing, often of marginally incremental value. Even if fundamental changes could be rapidly enacted to refocus priorities in undergraduate and graduate medical education, generations of "old school" physicians, with long ingrained behavior, continue to practice. Successful P4P programs will need to be sufficiently credible and robust to overcome these considerable hurdles. Medical schools and residency programs can take a leadership role in catalyzing such cultural change, and innovative tort reform initiatives for automobile insurance, such as "no fault" coverage or "limited tort" discounts, may help overcome some of these obstacles [33].

6.4 Political and Legislative Inertia

Through decades of legislation and regulation, the current Medicare FFS system has gained considerable regulatory and bureaucratic inertia. Additionally, since many private payers use Medicare, at least in part, as the basis of their systems, implementation of value-based payment systems will require considerable coordinated change by numerous independent corporate entities as well. The recently enacted Patient Protection and Accountable Care Act (PPACA) has been – and remains – politically polarizing. Even the constitutionality of one of its core requirements – the individual funding mandate – remains uncertain [41], and that alone could conceivable nullify this massive legislative initiative. Without legal, political and cultural consensus – which currently seem to be lacking – durable healthcare payment reform may be doomed.

6.5 Costs Exceed Rewards

Current provider and payer operational systems are designed to process claims under traditional FFS. Even early P4P initiatives have proven costly, in terms of money, time, and effort. Preliminary observations suggest that such programs may result in net financial losses – rather than gains – for providers. For primary care practices, administrative costs alone exceed potential financial rewards [24]. And, for radiology practices, physician time investments far exceed the relative return [18]

As more and more payers introduce P4P programs, each presumably with unique rules and requirements, the administrative burden to physician practices will only increase. Payer non-uniformity alone could make widespread P4P implementation an impractical and unwieldy proposition.

6.6 Moral Hazard

Moral hazard arises when financial incentives within the healthcare system lead to either inappropriate demands for care by patients or inappropriate supply of care by providers [15]. Under bundled payment models, physicians, facilities and other healthcare entities may be unintentionally incentivized financially to withhold appropriate care. To that end, the development of appropriate quality and outcome safeguards will be imperative. As value-based purchasing models take hold, ensuring physician and institutional duty to patients, free of economic bias, will remain an ongoing challenge.

7   Summary

FFS – the longstanding basis for US physician payment – will almost surely remain the prevailing healthcare payment model, to some degree, for the foreseeable future. Value-based purchasing, however, continues to gain rapid inroads, and is increasingly becoming incorporated in a blended fashion into existing payment systems. How rapidly this transformation will occur and how robustly P4P will be reflected in future payment models remain to be seen. Patients are almost sure to benefit, however, from increased alignment of healthcare payment with quality and outcomes.

References

1. Allen B Jr (2007) Valuing the professional work of diagnostic radiologic services. J Am Coll Radiol 4(2):106–114
2. Allen B, Levin D, Brant-Zawadski M, Lexa F, Duszak R (2011) ACR white paper: strategies for radiologists in the era of health care reform and accountable care organizations. A report from the ACR Future Trends Committee. J Am Coll Radiol 8:309–317
3. Allen B Jr, Donovan WD, McGinty G, Barr RM, Silva E III, Duszak R Jr, Kim AJ, Kassing P (2011) Professional component payment reductions for diagnostic imaging examinations when more than one service is rendered by the same provider in the same session: an analysis of relevant payment policy. J Am Coll Radiol 8(9):610–616, Epub 2011 Jun 29
4. Bogdanich W, McGinty JC (2011) Medicare Claims Show Overuse for CT Scanning. The New York Times, 17 June 2011. Available at http://www.nytimes.com/2011/06/18/health/18radiation.html?pagewanted=all. Accessed 21 Oct 2011
5. Breslau J, Lexa FJ (2011) A radiologist's primer on accountable care organizations. J Am Coll Radiol 8(3):164–168
6. Centers for Medicare & Medicaid Services (2010) Medicare program; payment policies under the physician fee schedule and other revisions to Part B for CY 2011: final rule. Fed Regist 75(228):73169–73860
7. Centers for Medicare & Medicaid Services (2011) Medicare program; Medicare Shared Savings Program: Accountable Care Organizations. Final rule. Fed Regist 76(212): 67802–67990
8. Centers for Medicare and Medicaid Services (2011) Advanced diagnostic imaging accreditation. https://www.cms.gov/MedicareProviderSupEnroll/03_AdvancedDiagnosticImagingAccreditation.asp. Accessed 21 Oct 2011
9. Centers for Medicare and Medicaid Services (2011) Eliminating serious, preventable, and costly medical errors – never events. https://www.cms.gov/apps/media/press/release.asp?Counter=1863. Accessed 21 Oct 2011

10. Centers for Medicare & Medicaid Services (2011) CY 2012 physician fee schedule final rule with comment period. Available at https://www.cms.gov/physicianfeesched/. Accessed 4 Dec 2011
11. Centers for Medicare and Medicare Services (2012) Health care innovation challenge. Available at http://innovations.cms.gov/initiatives/innovation-challenge/index.html. Accessed 2 Dec 2012
12. Congressional Budget Office (2011) The long-term outlook for health care spending. Available at http://www.cbo.gov/ftpdocs/87xx/doc8758/maintext.3.1.shtml. Accessed 19 Oct 2011
13. Conway PH, Berwick DM (2011) Improving the rules for hospital participation in Medicare and Medicaid. JAMA 306(20):2256–2257, Epub 2011 Oct 18
14. de Brantes F, Rosenthal MB, Painter M (2009) Building a bridge from fragmentation to accountability–the Prometheus Payment model. N Engl J Med 361(11):1033–1036, Epub 2009 Aug 19
15. Donaldson C, Gerard K (1989) Countering moral hazard in public and private health care systems: a review of recent evidence. J Soc Policy 18(2):235–251
16. Dorman T (2006) Unsustainable growth rate: physician perspective. Crit Care Med 34(3 Suppl):S78–S81
17. Duszak R (2009) P4P: pragmatic for practice. J Am Coll Radiol 6:477–478
18. Duszak R, Saunders WM (2010) Medicare's Physician Quality Reporting Initiative (PQRI): incentives, physician work, and perceived impact on patient care. J Am Coll Radiol 7:419–424
19. Emmert M, Eijkenaar F, Kemter H, Esslinger AS, Schöffski O (2011) Economic evaluation of pay-for-performance in health care: a systematic review. Eur J Health Econ (2012) 13(6):755–767. doi:10.1007/s10198-011-0329-8. Epub 10 Jun 2011
20. Federman AD, Keyhani S (2011) Physicians' participation in the Physicians' Quality Reporting Initiative and their perceptions of its impact on quality of care. Health Policy 102 (2-3):229–234, Epub 2011 May 31
21. Fisher ES, Wennberg DE, Stukel TA, Gottlieb DJ, Lucas FL, Pinder EL (2003) The implications of regional variations in Medicare spending. Part 2: health outcomes and satisfaction with care. Ann Intern Med 138(4):288–298
22. Food and Drug Administration (2011) Frequently asked questions about MQSA. Available at http://www.fda.gov/Radiation-EmittingProducts/MammographyQualityStandardsActandProgram/ConsumerInformation/ucm113968.htm. Accessed 1 Dec 2011
23. Gawande A (2009) The cost conundrum. The New Yorker, 1 June 2009. Available at http://www.newyorker.com/reporting/2009/06/01/090601fa_fact_gawande. Accessed 23 Oct 2011
24. Halladay JR, Stearns SC, Wroth T, Spragens L, Hofstetter S, Zimmerman S, Sloane PD (2009) Costs to primary care practices of responding to payer requests for quality and performance data. Ann Fam Med 7:495–503
25. Hsiao WC, Braun P, Becker ER, Thomas SR (1987) The Resource-Based Relative Value Scale. Toward the development of an alternative physician payment system. JAMA 258(6):799–802
26. Hsiao WC, Braun P, Kelly NL, Becker ER (1988) Results, potential effects, and implementation issues of the Resource-Based Relative Value Scale. JAMA 260(16):2429–2438
27. Hsiao WC, Levy JM (1990) A national study of resource-based relative value scales for physician services: phase II final report. Harvard School of Public Health, Boston
28. Hussey PS, Ridgely MS, Rosenthal MB (2011) The PROMETHEUS bundled payment experiment: slow start shows problems in implementing new payment models. Health Aff (Millwood) 30(11):2116–2124
29. Iglehart JK (2009) Health insurers and medical-imaging policy–a work in progress. N Engl J Med 360(10):1030–1037
30. Jackson & Coker Research Associates (2011) Special report: bartering for medical services. Available at http://www.jacksoncoker.com/physician-career-resources/newsletters/monthlymain/des/BarteringforMedicalServices.aspx. Accessed 28 Nov 2011

31. Jennifer L (2010) The fix is back. . .but jury-rigged SGR solution still frustrates docs. Mod Healthcare 40(10):10
32. Kohn LT, Corrigan JM, Donaldson MS (eds) (2000) To err is human: building a safer health system. National Academy Press, Washington, DC
33. Lascher EL, Powers MR (eds) (2001) The economics and politics of choice no-fault insurance. Kluwer, Boston
34. Levin DC (2006) The 2005 Robert D. Moreton lecture: the inappropriate utilization of imaging through self-referral. J Am Coll Radiol 3(2):90–95
35. McClellan M, McKethan AN, Lewis JL, Roski J, Fisher ES (2010) A national strategy to put accountable care into practice. Health Aff (Millwood) 29(5):982–990
36. McDonald CJ, Weiner M, Hui SL (2000) Deaths due to medical errors are exaggerated in Institute of Medicine report. JAMA 284(1):93–95
37. Mechanic RE (2011) Opportunities challenges for episode-based payment. N Engl J Med 365 (9):777–779, Epub 2011 Aug 24
38. National Quality Forum (2007) Serious reportable events in health care 2006 update: a consensus report. National Quality Forum, Washington, DC
39. Parmenter D (2007) Key performance indicators: developing, implementing, and using winning KPIs. Wiley, Hoboken
40. Porter ME (2010) What is value in health care? N Engl J Med 363(26):2477–2481
41. Rosenbaum S, Gruber J (2010) Buying health care, the individual mandate, and the Constitution. N Engl J Med 363(5):401–403, Epub 2010 Jun 23
42. Silva E III (2010) PQRI: from bonus to penalty. J Am Coll Radiol 7(11):835–836
43. Silva E III (2011) Accreditation and you. J Am Coll Radiol 8(9):600–601
44. Silva E III (2011) MOC for Dollars. J Am Coll Radiol 8(11):746–748
45. Sigsbee B (1997) Medicare's resource-based relative value scale, a de facto national fee schedule: its implications and uses for neurologists. Neurology 49(2):315–320
46. Smith SL (2010) Medicare RBRVS 2010: the physicians' guide. American Medical Association, Chicago
47. Sood N, Huckfeldt PJ, Escarce JJ, Grabowski DC, Newhouse JP (2011) Medicare's bundled payment pilot for acute and postacute care: analysis and recommendations on where to begin. Health Aff (Millwood) 30(9):1708–1717
48. Steele JR, Reilly JD (2010) Bundled payments: bundled risk or bundled reward? J Am Coll Radiol 7(1):43–49
49. Teufack SG, Campbell P, Jabbour P, Maltenfort M, Evans J, Ratliff JK (2010) Potential financial impact of restriction in "never event" and periprocedural hospital-acquired condition reimbursement at a tertiary neurosurgical center: a single-institution prospective study. J Neurosurg 112(2):249–256
50. Woody IO (2005) The fundamentals of the US Medicare physician reimbursement process. J Am Coll Radiol 2(2):139–150

Chapter 26
Global System-Based Quality Improvement in Radiation Medicine and Medical Imaging

Lawrence Lau

Abstract Radiation medicine (RM) covers the fields of diagnostic radiology, interventional radiology, nuclear medicine and radiotherapy. Medical imaging (MI) comprises all imaging techniques with or without using ionizing radiation, i.e. ultrasound and magnetic resonance imaging are included. Radiation medicine and medical imaging (RMMI) are indispensible in modern medicine, improve patient-centered care, and their use has expanded worldwide. While advances in techniques and technologies have led to major improvements in RMMI and in the diagnosis and treatment of many conditions, inappropriate or unskilled use of radiation can lead to unnecessary or unintended exposures and potential health hazards to patients and staff. From a public health perspective, it is important to maximize the benefits and minimize the risks when radiation is used in medicine.

The stakeholders have successfully conducted many actions to improve quality in RMMI and radiation safety in different settings. However, examples of prospectively designed and system-based initiatives covering a comprehensive range of integrated actions are less common. A global example is the World Health Organization's Global Initiative on Radiation Safety in Health Care Settings and a national example is the Australian Quality Use of Diagnostic Imaging Program. The Global Initiative provides an inclusive global platform to engage and mobilize the stakeholders towards a safer and more effective use of radiation in health through the implementation of radiation safety actions. This chapter describes the key features of these initiatives. The challenges and solutions of a system-based quality improvement initiative in RMMI are discussed.

Keywords Quality Use of Diagnostic Imaging Program • Radiation protection • Radiation safety • System-based initiative • System-based quality improvement

L. Lau (✉)
International Radiology Quality Network, 1891 Preston White Drive,
Reston, VA 20191-4326, USA
e-mail: lslau@bigpond.net.au

L. Lau and K.-H. Ng (eds.), *Radiological Safety and Quality: Paradigms in Leadership and Innovation*, DOI 10.1007/978-94-007-7256-4_26,
© Springer Science+Business Media Dordrecht 2014

• World Health Organization • WHO Global Initiative on Radiation Safety in
Health Care Settings

1 Global and System-Based Perspective

1.1 System-Based Improvement

Quality in RMMI can be characterized in different ways [1, 9]. Patient safety,
including radiation safety, and appropriate use are key quality elements. In this
chapter, a system-based initiative is defined as: (1) a body leading and conducting
actions(s) which are relevant and applicable to a healthcare system rather than to an
individual facility or a group of facilities; and (2) an orderly assembly of related
actions which are structured under a coordinated scheme to tackle a range of issues
and to improve quality in a comprehensive way. Most system-based processes and
improvement actions are generic and are applicable to both RMMI [24].

1.2 Emerging Trends

The access to and resources for RMMI vary markedly between different regions:
being close to none in some and over-utilized in others. There has been a significant
increase in the use of radiation in medicine in recent years [20] and it has become
the main contributor to general population exposure from artificial sources in many
countries [16]. The escalation in healthcare expenditure is threatening system
sustainability in some countries. The causes, issues and possible solutions for
increased use were reported.

Although an individual's long-term cancer risk from the use of radiation in
RMMI is in general very low, population exposure is becoming a public health
concern due to its widespread use. The safe and effective use of medical radiation is
one of the cornerstones for good medical practice. Despite existing efforts, more
research is needed to improve the understanding of radiation health risks and data
collection on population exposure and procedure utilization. Improved knowledge
in these areas underpins the development and implementation of evidence-based
recommendations and policies.

1.3 World Health Reform

The World Health Report documented values and principles to guide the development of healthcare systems [22]. Four sets of reforms are recommended to improve
primary healthcare:

• Universal coverage reforms to improve equity;
• Service delivery reforms to make health systems more people-centered;

- Leadership reforms to make authorities more reliable; and
- Public policies reforms to promote and protect the health of communities.

Applying these values and principles to RMMI practice will improve quality of care, radiation safety and appropriate use of procedures. System-based quality improvements in RMMI support these reforms by strengthening the healthcare infrastructure.

2 WHO Global Initiative on Radiation Safety in Health Care Settings

2.1 Radiation Safety Initiative

The World Health Organization (WHO) is the coordinating authority for health within the United Nations (UN) and its core functions are summarized in Table 26.1. In line with these core functions, it undertakes medical radiation and health related actions by contributing to the development of radiation safety norms and standards and facilitating their adoption and application; evaluating global trends in RMMI procedures; facilitating workforce education and training; providing advice for the incorporation of appropriate technologies; and publishing, co-sponsoring and disseminating guidance tools and technical documents.

The WHO is responsible for the development of evidence-based public health policy recommendations, and provides technical support to and facilitates capacity building of its Member States (MS). Under the Radiation and Environmental Health program, the WHO launched a Global Initiative on Radiation Safety in Health Care Settings (Global Initiative) in 2008 [23].

2.2 Objectives

The Global Initiative's objective are to improve the implementation of international radiation safety standards in patient care and promote safe and more effective use of

Table 26.1 Core functions of the WHO

Providing leadership on matters critical to health and engaging in partnerships where joint action is needed;

Shaping the research agenda and stimulating the generation, translation and dissemination of valuable knowledge;

Setting norms and standards, and promoting and monitoring their implementation;

Articulating ethical and evidence-based policy options;

Providing technical support, catalyzing change, and building sustainable institutional capacity; and

Monitoring the health situation and assessing health trends.

radiation in medicine by identifying the needs of its Member States and improving capacity, identifying priorities, and defining how best to complement other international, regional and national actions to improve radiation safety in healthcare settings. It aims to complement the activities under the International Action Plan for the Radiological Protection of Patients and the International Action Plan on Occupational Radiation Protection developed by the International Atomic Energy Agency (IAEA).

The objectives will be achieved by developing and facilitating the implementation of evidence-based policies and recommendations covering diagnostic radiology, interventional radiology, nuclear medicine and radiotherapy; focusing on the public health aspects and considering the risks and benefits of the use of radiation in healthcare. It aims to improve countries' capacity to: assess health risks; develop and implement policies that take into account of the potential health impacts, costs and benefits; monitor and evaluate the effectiveness of policies and interventions; and engage and communicate with stakeholders.

2.3 Platform

The Global Initiative provides a common platform for the engagement of and collaboration with the stakeholders from health and relevant non-health sectors. This inclusive approach fosters positive relationships and facilitates teamwork to coordinate and conduct projects. The participants include individual experts and representatives from UN agencies, international organizations, professional organizations, scientific societies, academic institutions, research institutions, consumer organizations, regulatory authorities and health ministries.

2.4 Framework

The Global Initiative's framework is:

- Patient-focused;
- Multi-tiered, i.e. projects are appropriate, relevant, and prioritized and address specific issues;
- Multi-dimensional, i.e. projects catering for the needs of specific stakeholder groups are included;
- Integrated and coordinated, i.e. projects are inter-related and add value to other projects, the initiative or other relevant health or cross-sector initiatives; and
- Cost-effective and sustainable, i.e. projects are conducted without wasting resources and with capacity to support future projects through savings from the resultant actions.

Table 26.2 The areas of work and prioritized projects of the Global Initiative [23]

Areas of work	Prioritized projects
Risk assessment	Develop tools for population medical exposure estimation and facilitate national surveys
	Shape and promote a research agenda on the effects of medical radiation exposure
Risk communication	Raise awareness on radiation benefits, risks and safe use of radiation in healthcare, focusing in children
	Develop communication strategy including advocacy tools
	Develop guidance on communication of risk to patients
	Develop risk communication tools for patients, family and public
Risk management	Promote the use of evidence-based appropriateness criteria and referral guidelines for justification of radiological procedures

2.5   Areas of Work and Prioritized Projects

The prioritized projects are listed under three areas of work (Table 26.2) [23]:

- Risk assessment: assess risks and potential impacts;
- Risk communication: engage and communicate with the stakeholders; and
- Risk management: implement policies and health interventions.

The projects are globally oriented but end-user targeted. A work plan detailing the projects, deliverables, time schedule, partners and end-users guides the development and implementation of an individual project and the initiative. The projects are advanced through conferences, consultations, and workshops.

Risk Assessment

- *Radiation risk research agenda*

The aim is to shape and promote global research on the effects of medical radiation exposure. The actions include: raising awareness and strengthening advocacy; adopting a research strategy guided by quality, impact and inclusiveness; fostering a global research agenda with the initiative as an inclusive platform for stakeholder engagement and collaboration; building national capacity; and facilitating education and training [17].

- *Population medical exposure survey*

The United Nations Scientific Committee on the Effects of Atomic Radiation (UNSCEAR) conducts regular population-based surveys on the use of medical radiological procedures and levels of exposure. However, national participation is still limited and data quality could be improved. The Global Initiative aims to facilitate survey participation and data estimation by developing tools in collaboration with UNSCEAR.

Risk Communication

• *Awareness raising and advocacy*

The aim is to raise awareness with a positive message; improve communication on the benefits and risks of RM; and strengthen advocacy towards a safer and more appropriate use of radiation in healthcare and the prevention of risks by using guidance tools.

• *Radiation protection in paediatric medical imaging*

The radiation protection of children is one of the priorities under the radiation risk communication strategy. Risk communication is an exchange of views between those responsible for assessing, minimizing and controlling radiation risks and those who may be affected. The deliverable is a communication package consisting of advocacy messages and a radiation risk communication toolkit for healthcare practitioners, patients, family and the public. These resources provide the stakeholders with evidence-based information and trustworthy advice to improve their understanding of and dialogue in radiation risks versus benefits in paediatric MI.

Risk Management

• *Procedure justification*

The Global Initiative aims to improve MI practice, radiation protection and radiation safety among radiological and other medical practitioners through procedure justification by the use of evidence-based appropriateness criteria and referral guidelines. The project partners with imaging guidelines developers from the International Radiology Quality Network (IRQN) and over thirty international organizations and agencies [7, 25]. The project plans to develop, pilot, implement and evaluate the use of referral guidelines. The imaging experts drafted a set of consensus guidelines and the registration for the pilot program has begun.

2.6 Staged Implementation

The Global Initiative is delivered in stages. Time is required to finalize priorities; assess national capacities; identify realistic interventions and targets; prepare work plans and timelines; allocate work, roles and responsibilities; mobilize resources; and develop indicators to monitor progress and evaluate impact. One area that must be strengthened is bridging the gap between evidence and practice. The Global Initiative advocates policy recommendations to the health ministries and decision makers on the one hand and provides evidence-based practical guidance tools to the practitioners on the other to facilitate their implementation in practice.

2.7   Evaluation

Impact assessment and project evaluation form part of the Global Initiative to determine its effectiveness. Adjustments may be required based on feedback from the pilot programs. While the Global Initiative could only be evaluated at a much later stage, indicators will be developed early to monitor progress and to improve the design of future projects.

2.8   Resources Mobilization

Resources mobilization includes fund raising, cooperation and contribution from MS when undertaking specific tasks. Secondment could be an innovative means of providing resources. The WHO works with national and international partners to raise funds from donors and identify collaboration opportunities with others. The collaborators will mobilize resources at country level and focus on health authorities. For an individual project, a MS may be interested in the entire proposal or a particular deliverable. The private sector could possibly contribute, but interaction should be transparent.

3   Quality Use of Diagnostic Imaging Program

An example of a national system-based quality improvement initiative is the Australian Quality Use of Diagnostic Imaging (QUDI) Program [10]. Under this program, a well-structured and comprehensive framework ensures an effective use of resources and a timely completion of a range of projects used to improve quality and safety in MI.

3.1   Objectives

The program was launched in 2004 and contributes to the national efforts to promote the quality of healthcare in Australia. Its objective is to champion quality, safety, and appropriate and sustainable use of MI.

3.2   Framework and Projects

An integrated and coordinated framework supports a range of projects focusing on four streams of work covering consumer, referrer, provider and economic sustainability issues. The projects aim to strengthen research, awareness, education, infrastructure and standards. For example, the research projects include a national pediatric CT exposure survey and a fellowship program to translate research into practice. Awareness on appropriate use is promoted through the development and hosting of the 'InsideRadiology' website to provide consumers and referrers with information on MI [18]. In addition to 'InsideRadiology', public information is available from The Royal Australian and New Zealand College of Radiologists [19] website.

Education and training include an optimization project on CT exposure and the development and provision of evidence-based medicine and critical appraisal skills educational material. System infrastructure is strengthened by a review and development of better referral forms and standardized reports to improve communication between referrers and providers; and the use of the Radiology Events Register database to collate and analyze adverse events and near misses. Standards and policies are updated by documenting recommendations on teleradiology, practice accreditation, professional supervision, role evolution and new technology assessment etc.

3.3   Funding

The QUDI program is funded by the Australian Government and managed by the Royal Australian and New Zealand College of Radiologists (RANZCR).

3.4   Impact

The program conducts projects, publishes reports and informs stakeholders. For example, the RANZCR has prepared responses to the findings and informed its Council and the profession about the critical issues, especially when these impact on the training curriculum and the RANZCR Standards of Practice.

4   Discussion

RMMI are indispensible in modern healthcare. Quality and safety improvement is an integral part of good medical practice. The stakeholders, needs, issues, settings, resources, and priorities vary, so are the solutions and actions at global, national or local levels. Despite these variations, the elements underpinning success are common. A system-based approach to safety and quality in RMMI is more integrated and has the potential to achieve wider improvements across the healthcare system. The following discussions cover the features, issues and solutions for such system-based improvement initiatives.

4.1 System-Based Quality Improvement

There are more stakeholders involved and more issues covered in system-based initiatives, resulting in wider impact. However, the lead-time to achieve tangible outcome is thus longer and will test the commitment of the stakeholders and the patience of the funding sponsor(s). The challenge is to convince the skeptics and lead the converted via an inclusive platform to achieve common goals. Leadership from the initiator and participating stakeholder groups and innovative actions designed under an effective framework are essential.

Quality improvement is a proactive process that covers the analysis, development, and implementation of on-going improvement actions for each and every step in the ways a procedure is done. There is no end point for quality improvement. It is a continuum. While a solution may work well for a certain setting, the parameters affecting performance and the way things are done will change with time. Therefore, on-going review and updates are required to improve the solutions or initiative processes so these elements stay relevant.

4.2 Stakeholders and Objectives

The stakeholders in RMMI are listed in Table 26.3 [9]. Despite varying needs, they share common objectives: to improve the quality, safe, and appropriate and sustainable use of procedures. In practice, this means to do the RIGHT procedure by justification and to do the procedure RIGHT by optimization and error minimization.

4.3 Inclusive Platform and Teamwork

Actions to improve quality in RMMI and radiation safety are system-based and require the involvement of all the stakeholders and good teamwork. Each stakeholder plays a different and valuable role, e.g. collaborator, partner or end-user. For example, individual experts prepare evidence-based recommendations and guidance tools, organizations and agencies advocate their adoption by regulatory authorities and use by the practitioners (Fig. 26.1). An inclusive multi-disciplinary and inter-sectorial platform encourages stakeholder engagement and collaboration; provides synergy and minimizes duplication; enables knowledge sharing and collaborative resources mobilization. A team approach overcomes discontinuity due to inevitable changes of leadership and team members. Better communication between initiatives and other actions will add value to each other and improve outcome.

Table 26.3 Stakeholders in radiation medicine and medical imaging

Consumers: patients and general public;

Referrers: general practitioners, specialists and other eligible providers;

Providers: radiologists, radiation oncologists, nuclear medicine specialists, radiographers, nuclear medicine technologists, medical physicists, and other eligible providers;

Payers: public authorities, private insurers, social services, individuals and others;

Regulators: governments, health ministries, competent authorities, other related sectors, policy and decision makers;

Research and academic institutions;

International organizations and UN agencies;

Professional, academic and scientific organizations;

Medical defence organizations, malpractice insurers; and

Equipment manufacturers and vendors.

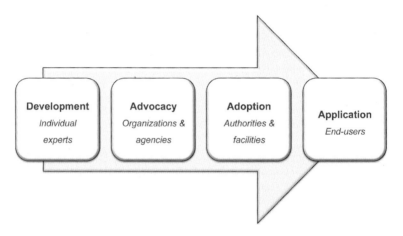

Fig. 26.1 Teamwork in system-based quality improvement. Each stakeholder group plays a unique role and adds value along the stages of a project or action, cumulating in policy implementation and change in practice. Professional experts prepare evidence-based recommendations and guidance tools, organizations and agencies advocate their adoption by regulatory authorities and use by the practitioners

4.4   Integrated Framework

Many actions are used to improve quality in RMMI [12]. Most past projects, proactive or reactive, are stand-alone actions and rarely grouped under a coordinated framework. An integrated, comprehensive, multi-dimensional and system-based framework enables the selection, prioritization and development of patient-focused actions to meet different stakeholder needs. The framework used by the Global Initiative is shown in Table 26.4 and a generic framework is shown in

Table 26.4 The Global Initiative framework [23]. The projects for a stakeholder group are listed under a column. When projects are applicable to more than one group, they are shown across columns. The projects belonging to the same area of work are listed in sequence in the same section

Areas	Key Stakeholder Groups			
	Consumers	Referrers	Providers	Regulators & Payers
Risk communication	Promote actions for stakeholders engagement			
	Raise awareness on radiation benefits, risks and safe use radiation in health care			
	Develop communication strategy including advocacy tools			
	Develop guidance on informed consent and communication of risk to patients			
	Develop risk communication tools for patients, family and public			
Risk assessment				Develop tools for population medical exposure estimation and facilitate national surveys
				Shape and promote a research agenda on the effects of medical radiation exposure
Risk management	Develop information tools on the benefits and risks of new technologies and the use of justification criteria	Promote the use of evidence-based appropriateness criteria and referral guidelines for justification of radiological procedures		Support Member States in the implementation of policies to reduce unnecessary medical exposures
			Develop guidance tools on optimization including the use of DRL	Support Member States in the implementation of international safety standards and national regulations, foster cooperation between health authorities and other competent regulatory authorities
			Develop guidance on good practice standards on pediatric CT (justification, optimization)	
			Promote clinical audits by developing and disseminating guidance and conducting training activities	Support Member States in the implementation of clinical audit and quality improvement programs
			Harmonize the criteria for error reporting including taxonomy, analyze the risk profiles and disseminate the lessons learnt	Support Member States in the prevention of unintended medical exposures by capacity building through education and training
			Review occupational health risk profiles; develop a toolkit on radiation risk management at the workplace; and disseminate guidance for the implementation of policies to protect workers	
			Develop a strategy to address the needs of radiation medicine workers, including workforce, workload, role extension etc.	
		Promote safety culture in health care settings and develop multi-disciplinary training, knowledge transfer, and training models; disseminate guidance on radiation protection including translation of existing materials		
		Advocate the inclusion of radiation protection topics in medical and public health curricula and foster co-operation between health and nuclear / radiological authorities towards medical education		

Table 26.5. By mapping the stakeholders, end-users, areas of work, and actions in a grid provide the stakeholders with a clearer overview and facilitate gap analysis. In an integrated framework, an individual action improves knowledge in and practice of a certain area. By informing each other, an action adds value to related actions. Collectively these actions will improve many different issues across the healthcare system.

L. Lau

Table 26.5 A generic system-based quality improvement framework. An integrated, comprehensive and multi-dimensional framework facilitates the design, prioritization and development of patient-focused actions to meet different needs. By showing the stakeholders and end-users, e.g. consumers, referrers, providers, regulators and payers etc.; areas of work, e.g. justification, optimization, error minimization, practice improvement etc.; and individual projects clearly in a grid provides the collaborators an overview and facilitates gap analysis. For example, Projects A1, B1 and C1 are of interest to all stakeholders, while Project B4 and Project C5 are of interest to the stakeholders in Group 2 and Group 3 respectively

Areas of work	Stakeholders			
	Stakeholder Group 1	Stakeholder Group 2	Stakeholder Group 3	Stakeholder Group 4
Area A	Project A1			
		Project A2		
	Project A3	Project A5	Project A6	Project A8
	Project A4		Project A7	
Area B	Project B1			
		Project B2		
	Project B3	Project B4	Project B5	Project B8
			Project B6	
			Project B7	
Area C	Project C1			
	Project C2		Project C3	
		Project C4	Project C5	Project C6
				Project C7
				Project C8

4.5 Health System Issues

Some of the issues, which affect the safe and appropriate use of RMMI include: lack of awareness; inadequate education and training; poor system infrastructure; scarce population exposure data; weak policy framework; insufficient funding; fragmented care; workforce shortage; workload volume and complexity; unjustified or inappropriate procedures; inappropriate technique optimization; human error and unintended exposure; occupational radiation protection; and special needs of the vulnerable groups etc. [11]

From a public health perspective, the use of RMMI should be rational and not be different to any other valuable health resources. Inappropriate use will lead to unnecessary or unintended exposure and cost. Therefore, a procedure should only be used if and when clinically indicated and the minimum exposure is used to achieve a diagnosis, i.e. by applying the 'As Low as Reasonably Achievable' (ALARA) principle [6], which is described in medicine as 'managing the radiation dose to commensurate with the medical purpose'.

There is increasing concern about the fragmentation and over-specialization of healthcare, therefore efforts are needed to maintain and strengthen an informational, relational and managerial continuity of care. Unregulated commercialization of health could be inefficient and expensive, exacerbate inequality, compromise the quality of care and contribute to the perception that health authorities are less capable in serving the public.

Patients, especially those with chronic illnesses, consult an ever-expanding range of practitioners. The practitioners who are not aware of the previous records may refer patients to repeat procedures. A controlled availability of confidential medical records across the facilities will reduce unnecessary procedures and exposures.

Health and non-health authorities, i.e. economic, educational, environmental and industrial, have to tackle overlapping issues and develop policies with potential health, economic or social implications, which could be beyond their competence. There should be better communication and collaboration between sectors to improve awareness, coordinate action and achieve better outcome. Informing each other about the possible impact of their proposed plans in other sectors and taking appropriate steps when developing policies could prevent un-intended consequences. A multi-disciplinary and multi-sectorial approach by engaging stakeholders from other non-health sectors is particularly relevant when addressing radiation safety issues in health.

Countries should implement and monitor the use of radiation safety regulations. A closer working relationship between health authorities, other national competent regulatory authorities and professional organizations is desirable. There must be a system of on-going assessment of the effectiveness of radiation protection actions. The application of new and rapidly evolving technologies has raised new issues that require solutions. For example, the implementation of quality assurance programs is essential to improve clinical outcome and ensure radiation safety. Measures to address the shortage of qualified practitioners, particularly of radiologists, radiographers and medical physicists are needed in many countries. Referrers, providers and trainees of RMMI procedures should be properly and regularly trained in radiation protection.

4.6   Evidence-Based Advocacy

A better understanding of radiation risks, procedure utilization and population exposure improves evidence-based advocacy, recommendations, policies and improvement actions. There are issues when attributing health effects to medical radiation exposure such as uncertainties associated with risk assessment at low doses, and insufficient statistical power in epidemiological studies [8, 15, 21].

A global research agenda encourages collaborations in radiation risk assessment and facilitates larger epidemiological studies. One of the challenges in global health planning is trend estimation based on quality data. The Global Initiative in collaboration with UNSCEAR and IAEA aims to improve population-based survey participation and data analysis.

Data from many countries, particularly those with lower resources, are not forthcoming. An innovative solution to facilitate data modeling and trend estimation is by placing countries with similar profiles into a common category. When estimating data and extrapolating trend for a country where data is unavailable, the average value from other countries within the same category could be used. There are two ways to group countries into categories: based on a 'physician to population ratio' which is used by UNSCEAR [14, 20] or 'gross national income (GNI) per capita' which is used by the World Bank and WHO. The GNI model might better reflect a country's economy and its healthcare expenditure and explain the variations in procedure utilization.

4.7 Closing the Loop

A series of steps are required to convert an idea into policy and to change practice (Fig. 26.2). Closing the project loop requires leadership and commitment from authorities, collaboration with stakeholders and participation of end-users. A generic process is applicable to all projects or actions:

1. Define objectives;
2. Identify stakeholders;
3. Develop framework;
4. Design actions to match stakeholder needs;
5. Collaborate and develop actions;
6. Develop tools and recommendations;
7. Trial and evaluate;
8. Refine tools and recommendations;
9. Advocate policy and change in practice; and
10. Implement policy.

Many projects or actions have completed a set of recommendations or guidance tools, but due to various reasons, uptake was slow and change to practice limited. The time and efforts invested are wasted. Referral guidelines and the International Basic Safety Standards for Protection against Ionizing Radiation and for the Safety of Radiation Sources (BSS) [4] are examples where uptake could be improved. The uptake of BSS by the health sector is slow and the engagement of health authorities in its applications is inadequate and insufficient. Countries vary greatly in the regulatory infrastructure for the medical use of radiation and workforce training in radiation protection. A radiation safety agency, nuclear safety agency or health authority may regulate the medical use of radiation, but some countries do not have

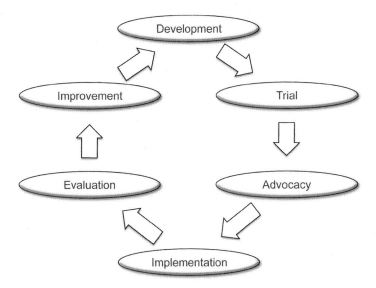

Fig. 26.2 Closing the project loop. A series of steps are involved to develop a concept into recommendations and guidance tools, to trial and advocate for their adoption as policy before finally use in practice. Closing this loop bridges the gap between evidence and practice

any authority or regulation. To improve the implementation of the BSS is a challenge that requires cooperation between the competent national authorities and concerted regional and global efforts.

Further, it is quite possible after some time, the same or another group would re-start the cycle for the same action. Poor communication could result in conducting similar projects by different groups, thus duplicating effort. The teams should be conscious of these steps at an early stage, and allocate resources and strategies to close the loop. The eventual indicator of success is change and improvement in practice. This is an area where leadership and innovative planning will make a difference.

4.8   Global Leadership and Advocacy

To achieve the desired outcome for a system-based initiative, it is important to inform the stakeholders and end-users by the most effective way. The most effective way to increase the profile of RMMI, to mobilize the health sector and to improve radiation safety is to engage the health ministries and radiation regulatory authorities. The IAEA is one of the key partners for the WHO Global Initiative and complementary actions and synergies are expected. The strength of the WHO and IAEA is based on their national counterparts. The IAEA partners are mainly the radiation protection regulatory authorities while the WHO counterparts are the ministries of health, which in many countries are also responsible for radiation

488u

LL. Lau

8448

protection matters. Cooperation between the ministries of health and radiation safety authorities has to be fostered.

By interacting with the ministries of health, the Global Initiative provides an important link between the regulatory authorities and the health ministries to jointly address radiation safety issues in healthcare settings. It is essential to explain to the health ministries the public health implications of the medical use of ionizing radiation: medical exposures are increasing rapidly, some are unjustified or unnecessary, and the implementation of BSS in healthcare is inadequate. A system-based initiative can facilitate the implementation of the BSS by delivering the radiation protection message to the local health community. One important supporting action is the translation of the existing material into local languages. There are other excellent existing documents and training materials, which need to be distributed to the medical community. The WHO and IAEA could be a clearinghouse to facilitate global co-operation. The Global Initiative will facilitate this link and evaluate those materials that should be harmonized and/or adapted.

In addition to local ministries and regulatory authorities, the UN agencies cooperate with a large network of professional organizations, research bodies and academic institutions, which are potential participants and collaborators. Radiation safety should be integrated into health programs to ensure patient protection and in occupational health programs for the protection of practitioners. The advocacy for the strengthening of MI and inclusion of radiation protection topics into medical school curricula is important. The ministries of education could be approached for consideration of the provision of key messages at school level. The education and training of health professionals in radiation protection is a topic of collaboration between WHO Global Initiative and the European Commission [2].

There are many crosscutting issues in a system-based initiative, e.g. cancer prevention, accident prevention, patient safety, primary healthcare, universal access to health technology, and cost-effectiveness of healthcare etc., which require a multi-sectorial approach. There are limitations in resources, experience, time and expertise for an individual organization.

An example of a recent global action is the 'International Conference on Radiation Protection in Medicine – Setting the Scene for the Next Decade', organized by the IAEA, co-sponsored by the WHO and hosted by the Government of Germany. This is a follow-up to the 2001 IAEA Malaga Conference, whose action plan continues to guide actions to improve radiation protection worldwide. The delegates supported the conference communiqué 'Bonn Call-for-Action' [5], which will guide improvements in radiation protection in medicine worldwide in the coming years.

4.9 Resources Mobilization

Funding is required to support the breadth and depth of actions in a system-based initiative and to ensure the project loop for each action is closed. However, the funding available to improve quality and safety in RMMI is low when compared to investments in other enterprises or industries with similar or lower budget. This could be due to poor awareness of the issues and appreciation of the benefits of proactive initiatives compared to the costs of reactive problem fixing. The funding for most projects is traditionally ad hoc. Projects were usually reactive and designed to fix a particular problem.

System-based quality improvement actions in RMMI compete with other non-health projects for public funding which is generally limited. Within health, there are other projects with perceived higher profile and impact. For the same reasons, funding from private philanthropic sources is low. However, awareness of the role of RMMI in modern healthcare and the importance of radiation safety is improving.

Donors expect cost-effectiveness, value, tangible results and deliverables, more so after the Global Financial Crisis. To improve their interest in and commitment to funding, the stakeholders should demonstrate:

1. Solidarity and advocacy towards a common goal;
2. Value through a well-designed and integrated framework, which is supported by all stakeholders;
3. Action are prioritized to meet the needs of the stakeholders;
4. Tangible deliverables;
5. Measurable indicators and outcomes;
6. Cost effectiveness; and
7. Advantages of a prospective and system-based approach.

Resources can be financial and non-financial and the stakeholders could contribute in different ways. Experience in technical know-how and non-technical elements are equally valuable. For example, knowing what will and will not work or how best to advocate and implement actions saves time and resources.

Resources mobilization is more effective if advocated by known entities with good track records or undertaken jointly rather than by individuals or smaller organizations. A collaborative approach implies a common mandate rather than vested interests. UN agencies are accountable and have established processes to handle funds ear-marked for such purposes. This is an advantage and provides assurance to public or private donors.

4.10 Other Influencing Factors

In a system-based initiative, in addition to the contributions from knowledgeable experts, there are other factors which could significantly influence its success or otherwise. The challenges threatening performance and outcome vary pending on an individual action. These include a range of societal, legal and organizational factors [13] and technical and non-technical skills for individuals [3].

Some examples of societal and legal factors are: timing; awareness; community values, needs, norms, expectations and demands; political priorities; economic

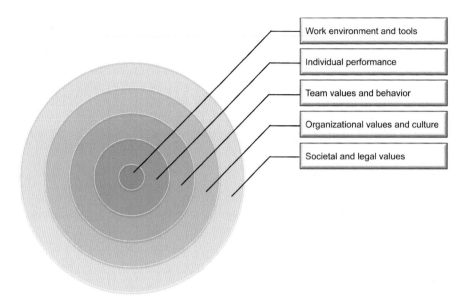

Fig. 26.3 Factors influencing system-based quality improvement. A range of societal, legal, organizational, team, individual and work factors affect the performance and outcome of an individual action or the initiative [13]

pressures; rules, laws, regulations, and policy implementation; and partnership barriers (Fig. 26.3). The organizational factors include: management values, goal, and style; authority hierarchy; safety policy and culture; and commitment to and involvement in quality and safety improvement actions. The performance of teams depends on group value; goodwill; communication; decision-making; coordination; cooperation; collaboration; bias; cohesiveness; personnel turnover; and perception of responsibility. The factors affecting individual performance are: training; knowledge; skill; competency; perception; awareness; interest; and participation. The physical environment at the facility; setting and layout; and the available equipment and tools are other local factors. In addition, strong leadership and innovative thinking are paramount and are applicable at all levels. Their influence on performance and outcome in a system-based quality improvement initiative should not be underestimated.

5 Conclusion

Concerted efforts are required to improve the quality and appropriate use of RMMI, radiation safety, and sustainability of the healthcare system. Quality improvement in healthcare systems includes actions in radiation safety. These actions are indispensable and support professionalism and risk minimization.

As awareness increases, communication improves, goals converge, resources become finite, rationalization makes sense, and collaboration strengthens, an innovative system-based framework offers great value. While the precise form may vary, a collaborative approach will become more common. Strong leadership supports this emerging trend. Synergy is achieved by jointly designing actions under an inclusive platform and an integrated framework. The stakeholders can undertake an individual project, and will inform each other and share outcome. The generic findings will benefit all projects while the specific recommendations will be useful to a specific setting.

From a public health perspective, the impact of a system-based initiative will be seen in the longer term. Time is needed to promote awareness; to assemble teams; to develop framework, systems, templates, and processes; and to trial, implement and refine recommendations and guidance tools. A system-based initiative such as the WHO Global Initiative on Radiation Safety in Health Care Settings is a unique opportunity to build global partnerships to improve radiation safety culture and change practice. Regular review and evaluation are integral components because the needs vary between settings and safety and quality improvement is a continuum.

Closer collaboration between the stakeholders to improve quality, safety and cost-effectiveness in healthcare are needed, now more than ever. Ministries, competent authorities, policy and decision makers should lead and collaborate by identifying and engaging the key stakeholders. Individual leaders representing organizations and agencies should facilitate and encourage collaboration between these initiatives and other related projects.

A collaborative, coordinated, comprehensive and system-based approach to and end-user participations in quality improvement will lead to an increased awareness of the benefits and risks of the medical use of radiation; a better understanding of the risks and evolving trends to underpin evidence-based policy and decision making; a reduction of unnecessary medical radiation exposures; an enhancement of knowledge, skills and safety attitude of practitioners; and better prevention of unintended exposures. These changes in practice will result in better outcomes in the quality of care, radiation safety and the use of RMMI in different settings worldwide.

Acknowledgement The author is most grateful to the World Health Organization for an invitation to participate in the WHO Global Initiative on Radiation Safety in Health Care Settings and Dr. María del Rosario Perez MD from the Department of Public Health and Environment World Health Organization for providing valuable advice and feedback to this manuscript.

References

1. Blackmore CC (2007) Defining quality in radiology. J Am Coll Radiol 4(4):217–223
2. European Commission (2013) Medical radiation protection education and training (MEDRAPET). http://www.medrapet.eu/about.html. Assessed 20 Feb 2013

3. Flin R, O'Connor P, Crichton M (2008) Safety at the sharp end: a guide to non-technical skills. Ashgate Publishing, Farnham

4. International Atomic Energy Agency (2011) Radiation protection and safety of radiation sources: international basic safety standards – interim edition, IAEA Safety standards series GSR Part 3 (Interim). IAEA, Vienna

5. International Atomic Energy Agency (2012) International conference on radiation protection in medicine – setting the scene for the next decade. http://www.iaea.org/newscenter/news/2012/workerprotection.html. Accessed 31 Dec 2012

6. International Commission on Radiological Protection (2007) ICRP Publication 105. Radiation protection in medicine. Ann ICRP 37(6). Elsevier, Oxford

7. International Radiology Quality Network (2010) Referral guidelines project. http://www.irqn.org/work/referral-guidelines.htm. Accessed 23 June 2012

8. Jacob P (2012) Uncertainties of cancer risks estimates for applications of ionizing radiation in medicine. In: International conference on radiation protection in medicine – setting the scene for the next decade. IAEA, Bonn

9. Lau LSW (2006) A continuum of quality improvement in radiology. J Am Coll Radiol 3 (4):233–239

10. Lau LSW (2007) The Australian national quality program in diagnostic imaging and interventional radiology. J Am Coll Radiol 4(11):849–855

11. Lau LSW, Perez MR, Applegate KE et al (2011) Global quality imaging: emerging issues. J Am Coll Radiol 8(7):508–512

12. Lau LSW, Perez MR, Applegate KE et al (2011) Global quality imaging: improvement actions. J Am Coll Radiol 8(5):330–334

13. Moray N (2000) Culture, politics and ergonomics. Ergonomics 43(7):858–868

14. Mettler FA Jr, Davis M, Kelsey CA et al (1987) Analytical modeling of worldwide medical radiation use. Health Phys 52(2):133–141

15. Müller W (2012) Can we attribute health effects to medical radiation exposure? In: International conference on radiation protection in medicine – setting the scene for the next decade. IAEA, Bonn

16. National Council on Radiation Protection and Measurements (2009) Report no.160 Ionizing radiation exposure of the population of the United States. NCRP, Bethesda

17. Perez MR, Lau LSW (2010) WHO Workshop on radiation risk assessment in paediatric health care. J Radiol Prot 30(1):105–110

18. The Royal Australian and New Zealand College of Radiologists (2013) InsideRadiology. http://www.insideradiology.com.au/. Accessed 1 Jan 2013

19. The Royal Australian and New Zealand College of Radiologists (2013) Quality use of diagnostic imaging program. http://www.ranzcr.edu.au/quality-a-safety/qudi. Accessed 1 Jan 2013

20. United Nations Scientific Committee on the Effects of Atomic Radiation (2010) Sources and effects of ionizing radiation. UNSCEAR 2008 report to the General Assembly with scientific annexes. United Nations, New York

21. United Nations Scientific Committee on the Effects of Atomic Radiation (2013) UNSCEAR 2012 report to the General Assembly, with annexes. United Nations, New York (in press)

22. World Health Organization (2008) The world health report 2008 – primary health care. WHO, Geneva

23. World Health Organization (2008) WHO global initiative on radiation safety in health care settings technical meeting report. WHO, Geneva

24. World Health Organization (2009) Systems thinking for health systems strengthening. Alliance for health policy and systems research. WHO, Geneva

25. World Health Organization (2010) Medical imaging specialists call for global referral guidelines. http://www.who.int/ionizing_radiation/medical_exposure/referral_guidelines.pdf. Assessed 23 June 2012

About the Editors

Lawrence Lau, FACR (Hon), FRANZCR, FRCR, FAMS (Hon), DDR, DDU, is a radiologist from Melbourne and the founding Chairman of the International Radiology Quality Network, an organization focusing on collaborative actions to improve the quality, safety and appropriate use of radiology for practitioners, facilities, and healthcare systems.

Kwan-Hoong Ng, Ph.D., MIPEM, DABMP, is a Professor of Biomedical Imaging and Intervention in the Department of Biomedical Imaging, University of Malaya Research Imaging Centre, University of Malaya, in Kuala Lumpur.

L. Lau and K.-H. Ng (eds.), *Radiological Safety and Quality: Paradigms in Leadership and Innovation*, DOI 10.1007/978-94-007-7256-4,
© Springer Science+Business Media Dordrecht 2014

Biography

Dr. Lawrence Lau MBBS, FACR (Hon), FRANZCR, FRCR, FAMS (Hon), DDR, DDU, is the founding Chair of the International Radiology Quality Network. At Monash University in Melbourne, he was Clinical Associate Professor and Chair of the Professional Advisory Committee, Medical Imaging and Radiography Course; and served as Director and Chairman of the Department of Diagnostic Imaging for Southern Health. Within the Royal Australian and New Zealand College of Radiologists (RANZCR), he served as President, founding Chair of the Accreditation Guidelines and Quality Committee, founding Director and Chief Medical Advisor of the Australian Quality Use of Diagnostic Imaging Program and founding Editor of the RANZCR "Imaging Guidelines". He was the founding Chairman of the Advisory Committee for the Australian Medical Imaging Accreditation Program of the RANZCR and National Association of Testing Authorities. In the last 30 years, Dr. Lau collaborated with local, national, regional, and international stakeholders in actions towards better quality, safety, and more appropriate use of diagnostic imaging.

Dr. Kwan-Hoong Ng PhD, DABMP, FInstP (UK), FIPM (Mal), CSci (UK), AMM, is Professor in the Department of Biomedical Imaging and a Senior Consultant at the University of Malaya Medical Center in Kuala Lumpur, Malaysia. He received his M.Sc. in Medical Physics from the University of Aberdeen and his Ph.D. in Medical Physics from the University of Malaya. He is certified by the American Board of Medical Physicists. Dr. Ng has authored or coauthored more than 200 papers in peer-reviewed journals and 20 book chapters. He has presented over 450 scientific papers, more than 200 of which were invited lectures. He has directed several workshops on radiology quality assurance, digital imaging, and scientific writing. He is the co-founder and co-Editor in Chief of the Biomedical Imaging and Intervention Journal. His main research contribution has been in breast imaging, radiological protection, and radiation dosimetry. Dr. Ng has been serving as an International Atomic Energy Agency expert, a member of International Advisory Committee of the World Health Organization and a consulting expert for the International Commission on Non-Ionizing Radiation Protection. Dr. Ng is the immediate Past President of the Asia-Oceania Federation of Organizations for Medical Physics.

L. Lau and K.-H. Ng (eds.), *Radiological Safety and Quality: Paradigms in Leadership and Innovation*, DOI 10.1007/978-94-007-7256-4, © Springer Science+Business Media Dordrecht 2014

Remigiusz Baranczyk MSc, is Policy Officer in the Radiation Protection Unit of the European Commission's Directorate-General for Energy. He is in charge of the Unit's activities on radiation protection in the medical applications of ionizing radiation.

Theocharis Berris MSc, is a Medical Physicist and works at the International Atomic Energy Agency. His main work focuses on the development of new training and information material for the Radiation Protection of Patients Unit website and the management of its social media campaign.

Dr. C. Craig Blackmore MD, MPH, is a Radiologist and Director of the Center for Health Services Research at the Virginia Mason Medical Center. His research interest is in understanding the appropriate and efficient use of imaging and other medical technologies, and in determining the effectiveness of Lean methodology in healthcare delivery. He chairs the Washington State Health Technology Clinical Committee.

Dr. Caridad Borrás DSc, DABR, DABMP, FACR, FAAPM, has worked as a Medical Physicist in Spain, the USA, and Brazil. For 15 years, she was responsible for the radiological health program of the Pan American Health Organization. She is an Adjunct Assistant Professor of Radiology at the George Washington University School of Medicine and Health Sciences and a consultant. She chairs the Health Technology Task Group of the International Union for Physical and Engineering Sciences in Medicine.

Dr. Rethy Kieth Chhem MD, PhD (Edu), PhD (History), is Director of the Division of Human Health of the International Atomic Energy Agency and has worked to improve member state capacity and capabilities in the prevention, diagnosis, and treatment of diseases through nuclear techniques. He has published more than a 100 papers and chapters and edited several books on musculoskeletal ultrasound, radiology education, and paleoradiology.

Dr. Byung Ihn Choi MD, MS, PhD, is Professor and Past Chairman of the Department of Radiology, College of Medicine, at Seoul National University. He is immediate Past President of the Asian Oceanian Society of Radiology. In recognition of his outstanding leadership and achievements, he received honors from 14 international and regional societies including the American College of Radiology, European Society of Radiology, and Radiological Society of North America.

Ms. Kelly Classic MS, MA, DABMP, is Assistant Professor of Radiologic Physics at the Mayo Clinic, Rochester, Minnesota. She is Associate Editor for the Health Physics Journal and its supplement, Operational Radiation Safety. Ms. Classic serves as the public outreach liaison for the Health Physics Society.

Malcolm J. Crick MA, MSRP, CRadP, is Secretary of the United Nations Scientific Committee on the Effects of Atomic Radiation with over 30 years experience in radiation protection. He is a member of the editorial board of the Journal for Radiological Protection.

Dr. Adrian K. Dixon MD, FRCR, FRCS, FRCP, FMedSci, FRANZCR (Hon), FACR (Hon), FFRRCSI (Hon), is a Consultant Radiologist at Addenbrooke's Hospital in Cambridge. He is Emeritus Professor of Radiology, University of Cambridge, and Editor-in-Chief of *European Radiology*. Dr. Dixon is the Master of Peterhouse, the oldest of the constituent colleges in the University of Cambridge.

Dr. Keith J. Dreyer DO, PhD, is Vice Chair for Radiology Informatics at the Massachusetts General Hospital and has continuously led a substantial team of project managers, analysts, and programmers who design, build, and maintain the ROE-DS System.

Dr. Richard Duszak Jr. MD, FACR, FRBMA, is CEO and Senior Research Fellow at the Harvey L. Neiman Health Policy Institute and practices radiology in Memphis. He serves on the American Medical Association's Current Procedural Terminology Editorial Panel and as Associate Editor of the Journal of the American College of Radiology. His research interests focus on physician payment systems and quality.

Dr. Xiaoyuan Feng MD, PhD, is Professor and Chairman of the Department of Radiology at the Huashan Hospital, Fudan University. He is also the Vice President of the Fudan University and President of the Chinese Society of Radiology.

Dr. Donald P. Frush MD, FACR, FAAP, is Professor of Radiology and Pediatrics; Interim Chairman, Radiology; Member, Medical Physics Graduate Program; Division of Pediatric Radiology at the Duke Medical Center, Durham. His research interests are in pediatric body CT, including technology assessment; MDCT techniques; image quality assessment; dosimetry; and patient safety.

Robert George ARMIT, Dip Pract Man, FIR, FAAPM, was a Diagnostic Radiographer and Practice Manager of a large private radiology practice in Adelaide. He is Past President of the Australian Institute of Radiography and the International Society of Radiographers and Radiological Technologists. His interests are in the protection of patients and staff from radiation and clinical risk management.

Dr. Hassen A. Gharbi MD, is Professor of Radiology and Medical Biophysics in Tunis with an interest in pediatric radiology, tropical disease, and use of ultrasound in developing countries. He is President of the African Society of Radiology and the World Federation for Ultrasound in Medicine and Biology. He was Founding Director of the Tunisian National Centre of Radiation Protection and Founding President of the Mediterranean and African Society of Ultrasound.

Dr. Lucy Glenn MD, is the Chair, Department of Radiology, at the Virginia Mason Medical Center, an institution which has been applying the Lean Principles to re-engineer healthcare delivery. She is a certified Lean Leader since 2002 and has been actively involved in improving radiological quality by organizing and lecturing at conferences hosted by the American College of Radiology and Radiological Society of North America.

Dr. Daniel F. Gutierrez PhD, is Physicist and Research Scientist in the PET Instrumentation and Neuroimaging Laboratory at the Geneva University Hospital. He has authored over 15 peer-reviewed journal papers and 20 conference proceedings on observer models, Monte Carlo simulation, medical imaging system characterization and dose optimization particularly in clinical and preclinical hybrid PET-CT.

Dr. Azza Hammou MD, is Professor of Radiology in the Medical School of Tunis and Director of the Tunisian National Center of Radiation Protection. She is Past President of Mediterranean and African Society of Ultrasound and a member of the International Atomic Energy Agency and World Health Organization advisory committees. She has conducted many regional training programs and conferences, and participated in research and publications in ultrasound, pediatric imaging, and radiation protection.

Dr. Evelyn Lai-Ming Ho MBBS, MMed (Radiology), FAMM, FAMS (Hon), works as a full-time Clinical Radiologist, is active in advocacy programs for breast health and end-of-life care, and has led the College of Radiology, Academy of Medicine of Malaysia, Mammogram Programmes since 2001. She is immediate Past President of the College of Radiology and Editor of College's website.

Dr. Augustin Janssens PhD, joined the European Commission in 1985 and became Head of the Radiation Protection Unit in 2004. He is in charge of legislations on radiation protection, including public, occupational, and medical exposure as well as emergency preparedness, response, and food controls.

Hannu Järvinen MSc, is Principal Advisor on Radiation Practice Regulation to the Radiation and Nuclear Safety Authority in Finland (STUK). He provides guidance to issues on the medical use of radiation covering research and training, clinical audit, mammography screening, regulatory control, policy development and implementation, and international cooperation. He has published extensively and participated in many national and international committees.

Dr. Kenneth R. Kase PhD, FACR, FHPS, FAAPM, Dipl ABHP, is the immediate Past President of the International Radiation Protection Association, a former Faculty Member of Radiation Oncology at the Harvard Medical School and University of Massachusetts Medical School, and a former Associate Director of the Stanford Linear Accelerator Center. He is an Honorary Senior Vice President of the National Council on Radiation Protection and Measurements.

Dr. Michael G. Kawooya MBChB, MMed (Radiology), PhD, is Professor of Radiology, Director of Ernest Cook Ultrasound Research and Education Institute and Director of World Federation for Ultrasound in Medicine and Biology Center of Excellence in Kampala. He is General Secretary of the African Society of Radiology. In the last 20 years, he has been involved in the training and assessment of radiologists, sonographers, and radiographers in Uganda and other African countries.

Dr. Lizbeth M. Kenny MBBS, FRANZCR, FACR (Hon), FBIR (Hon), FRCR (Hon), is a Senior Radiation Oncologist and her main interests are Head and Neck and Breast Cancer. Within the Royal Australian and New Zealand College of Radiologists, she has served as the Dean of The Faculty of Radiation Oncology and the College President. She has received Honorary Memberships or Fellowships from five international societies.

Dr. Bernard Le Guen MD, PhD, is Radiation Protection and Industrial Safety Vice-President of the Electricité de France (EDF), Nuclear Power Plant Operations, and the Executive Officer of the International Radiation Protection Association. He advances radiation protection by lecturing at universities, advising the French nuclear regulatory authority and coordinating global actions including the radiation protection culture initiative of the International Radiation Protection Association.

John Le Heron BSc, FACPSEM, is a Medical Physicist certified in Diagnostic and Interventional Radiology. He worked for the New Zealand radiation protection regulatory body for over 25 years, before joining the International Atomic Energy Agency where he is involved in the development of international radiation protection standards and their implementation worldwide.

Dr. Osnat Luxenburg MD, MPH, MBA, is Director of the Medical Technology and Infrastructure Administration at the Israeli Ministry of Health. She is responsible for the regulation and licensing of all pharmaceuticals, medical devices and medical technologies, and the regulation of radiation protection in Israel. She is involved in national health policy decision-making.

Dr. Catherine Mandel MBBS, FRANZCR, is Consultant Radiologist at Peter MacCallum Cancer Centre, Melbourne, and a Councilor for The Royal Australian and New Zealand College of Radiologists. She trained in neuroradiology in the United Kingdom. Her research interests are in neuro-oncological radiology, human factors, and patient safety. She teaches patient safety to radiology registrars and helped to develop the patient safety module of the RANZCR radiology curriculum.

Dr. Michal Margalit PhD, works in the Medical Technology Administration at the Israeli Ministry of Health in the field of radiation protection of patients. Her research interests are in microbiology, human fertility, and radiation protection.

Dr. Arl Van Moore Jr. MD, FACR, FSIR, FAHA, was the President of Charlotte Radiology for 15 years and is currently the Chairman and CEO of Strategic Radiology, Vice President of Charlotte Radiology and Chief of Radiology at Carolinas Medical Center, Mercy. He is a Past President and Past Chairman of the Board of Chancellors of the American College of Radiology. He is an Assistant Clinical Professor in the Department of Radiology at Duke University Medical Center, Durham, North Carolina.

Dr. Richard L. Morin PhD, FACR, FAAPM, FSIIM, is the Brooks-Hollern Professor and Consultant in the Department of Radiology at the Mayo Clinic in Florida. His research interests include all aspects of electronic imaging in medicine and radiation dose in Medical Imaging. Dr. Morin is interested in the performance and applications of volume CT and mammography physics.

Dr. Thomas N.B. Pascual MD, MHPed, is a Nuclear Medicine Physician with an interest in pediatrics and education. He has received awards from the Asian Regional Cooperative Council for Nuclear Medicine for outstanding research. Since 2007, he has been the Executive Director of the Asian School of Nuclear Medicine and has recently joined the International Atomic Energy Agency.

Dr. Madan M. Rehani PhD, is Director of Radiation Protection for the European Society of Radiology; Secretary, International Commission on Radiological Protection Committee 3; and Secretary General, International Organization of Medical Physics. He was Radiation Safety Specialist, International Atomic Energy Agency; Professor of Medical Physics, All India Institute of Medical Sciences in New Delhi; and Head, WHO Collaborating Centre on Imaging Technology and Radiation Protection.

Dr. Soveacha Ros EdD, is a Consultant advising the Division of Human Health of the International Atomic Energy Agency on education principles and implementation strategies. He is Adjunct Professor of education at the Royal University of Phnom Penh and Pannasastra University of Cambodia. His interests include quality assurance in education, competency-based curricula framework, learning how to learn and green university policy.

Dr. Daniel I. Rosenthal MD, is Associate Radiologist in Chief at the Massachusetts General Hospital, since 1989. He is the Executive Clinical Leader of the ROE-DS System having served in that capacity since its inception.

Dr. William Runciman BSc (Med), MBBCh, PhD, FANZCA, FJFICM, FHKCA, FRCA, is Professor of Patient Safety and Human Factors at the University of South Australia; Research Fellow at the Australian Institute of Health Innovation, University of New South Wales; and Clinical Professor at the Joanna Briggs Institute, University of Adelaide. He is also President of the Australian Patient Safety Foundation.

Dr. Ferid Shannoun Dipl-Ing, MSc, MPH, PhD, is Scientific Officer of the United Nations Scientific Committee on the Effects of Atomic Radiation. His main research interests focus on public health issues related to radiation protection and to population dose estimation. He is a member of the Scientific Committee of the International Organization of Medical Physics.

Dr. Esti Shelly MSc, is Senior Assistant to the Head of the Medical Technology Administration at the Israeli Ministry of Health. Among her duties are upgrading regulations and legislation of medical exposure to ionizing radiation and radiotherapy, and coordinating national and regional projects.

Dr. Paul C. Shrimpton MA, PhD, is a Senior Scientific Group Leader at Public Health England. He has been closely involved for over 30 years in the wider development and promotion of patient dosimetry and protection for x-ray examinations, including the performance of national (UK) and international patient dose surveys, and the application of reference doses to facilitate improvements in practice.

Dr. Ezequiel Silva III MD, RCC, is Director of Interventional Radiology for the South Texas Radiology Imaging Centers, and Adjunct Assistant Professor at the University of Texas Health Science Center at San Antonio. He writes on quality initiatives in his column "Reimbursement Rounds" for the Journal of the American College of Radiology.

Georgi Simeonov MSc, is Policy Officer in the Radiation Protection Unit of the European Commission's Directorate-General for Energy. He is in charge of the Unit's activities on radiation protection of patients in diagnostic imaging, including proposals for Euratom legislation; study and guidance development projects; and liaison with international organizations and other stakeholders.

Dr. Chris L. Sistrom MD, MPH, PhD, is a Radiologist from the Department of Radiology, University of Florida who spent a sabbatical year at and on the invitation of the Massachusetts General Hospital Radiology and the Physician's Organization during which he learnt about, analyzed and reported on the ROE-DS System. He has been a visiting research fellow at MGH since then.

Dr. Bin Song MD, is Professor and Director of the Department of Radiology and Director of Medical Imaging Center of the West China Hospital, Sichuan University. He is Vice Chairman of the Chinese Association of Radiologists and the Secretary General of the Chinese Society of Radiology.

Dr. Dong-Wook Sung MD, PhD, is Professor and Chairman of Thoracic Radiology in the Department of Radiology, Kyung Hee University in Seoul. He is Chairman of the Radiation Safety Committee of the Korean Radiology Society and General Secretary of the Korean Alliance for Radiation Safety and Culture in Medicine. He is involved in national surveys of radiation dose.

Dr. James H. Thrall MD, is Chairman Emeritus at the Massachusetts General Hospital and was the President of the American College of Radiology. His tireless efforts in supporting, advising, fostering and mentoring a multi-disciplinary team in the development and implementation of a comprehensive ROE-DS system at MGH must certainly be counted among his most outstanding professional achievements.

Dr. Eliseo Vano PhD, is Professor of Medical Physics at the Medical School of the Complutense University and Head of Medical Physics Service of the San Carlos University Hospital in Madrid. His research interests are in radiation protection in interventional radiology and cardiology. He is the Chair of Committee 3 (Protection in Medicine) of the International Commission on Radiological Protection.

Dr. Richard J. Vetter PhD, CHP, DABMP, FHPS, FAAPM, is Health Physics Consultant and Professor Emeritus at Mayo Clinic, Rochester, where he also served as Radiation Safety Officer for nearly 30 years. He is currently the Government Liaison for the United States Health Physics Society and a member of the United States National Academies Nuclear and Radiation Studies Board.

Dr. Jeffrey B. Weilburg MD, from the Department of Psychiatry, Massachusetts General Hospital, is Associate Medical Director of the Massachusetts General Physician's Organization. He is responsible for the fiscal and quality performance of several large services, including Radiology, and was instrumental in introducing and promoting the ROE-DS System to MGH.

Ms. Pamela A. Wilcox RN, MBA, joined the American College of Radiology in 1987 with the advent of the voluntary Mammography Accreditation Program. She is Assistant Executive Director of the Department of Quality and Safety and oversees all ACR quality initiatives, including accreditation for diagnostic imaging and radiation oncology, Appropriateness Criteria, and practice guidelines.

Dr. Wentao Wu MD, is Lecturer and Attending Doctor in the Department of Radiology at the West China Hospital, Sichuan University.

Dr. Habib Zaidi MSc, PhD, is Head of the PET Instrumentation and Neuroimaging Laboratory at the Geneva University Hospital and Professor at the University Medical Center of Groningen. He has authored over 300 publications, is a recipient of many awards and distinctions and has been an invited speaker of many keynote lectures at the international level.

Index

L. Lau and K.-H. Ng (eds.), *Radiological Safety and Quality: Paradigms
in Leadership and Innovation*, DOI 10.1007/978-94-007-7256-4,
© Springer Science+Business Media Dordrecht 2014